HISTORY OF ORIENTAL ASTRONOMY

ASTROPHYSICS AND SPACE SCIENCE LIBRARY

VOLUME 274

History of Oriental Astronomy

Proceedings of the Joint Discussion-17 at the 23rd General Assembly of the International Astronomical Union, organised by the Commission 41 (History of Astronomy), held in Kyoto, August 25–26, 1997

Edited by

S. M. Razaullah Ansari

President of IUHPS Commission for History of Ancient and Medieval Astronomy
Former President of IAU Commission for History of Astronomy
Former Professor of Physics, Aligarh Muslim University, Aligarh, India

KLUWER ACADEMIC PUBLISHERS
DORDRECHT / BOSTON / LONDON

Library of Congress Cataloging-in-Publication Data is available.

ISBN 1-4020-0657-8

Published by Kluwer Academic Publishers,
P.O. Box 17, 3300 AA Dordrecht, The Netherlands.

Sold and distributed in North, Central and South America
by Kluwer Academic Publishers,
101 Philip Drive, Norwell, MA 02061, U.S.A.

In all other countries, sold and distributed
by Kluwer Academic Publishers, Distribution centre
P.O. Box 322, 3300 AH Dordrecht, The Netherlands

Printed on acid-free paper

Cover picture
Details from an Indian celestial globe made by Ghulam Hussain Jaunpuri in 1816 AD
(Courtesy Prof. M. Mahdi Ansari, Aligarh)

Printed and bound in Great Britain by
Antony Rowe Limited

Dedicated to the Fond Memory of

Dr. Manali Kallat Vainu Bappu (1927–1982)

Founder Director, Indian Institute of Astrophysics, Bangalore
First Indian President of the International Astronomical Union (1979–82)

Contents

Preface

The History of Astronomy in the Orient has been vigorously researched in the last several decades. We may recall here the publications of Joseph Needham's monumental volumes on *Science and Civilisation in China*, one volume of which is devoted to Chinese Astronomy, S. Nakayama's *A History of Japanese Astronomy* (Tokyo, 1969), and the School of Edward Kennedy's writings on *Islamic Astronomy*,[1] which particularly culminated in the studies of the *Critique* of Ptolemaic Astronomy by the Islamic astronomers belonging to Naṣīruddīn Ṭūsī's School, established at Marāgha Observatory during the 13–14th centuries.[2]

In this backdrop of the emphasis on astronomy in the Orient, the *first* IAU Colloquium (No.91) on "*History of Oriental Astronomy*" was organised during the IAU General Assembly, held in New Delhi, Nov. 13–16, 1985. The Proceedings of the *Colloquium* were then published.[3] The *second* effort by this Commission was to organise another International Colloquium on *Interaction of European and Asian Astronomy*, held in Vienna in Sept. 1990. Unfortunately its Proceedings could not be published.[4]

Noteworthy is that the Far East or the East Asia did not lag behind in this endeavour. A group of historians of astronomy, Il Seong Nha (of Seoul) et al set up in 1992 an International Organising Committee to hold the *First International Conference of Oriental Astronomy* (ICOA), which was held in Seoul during Oct. 6–11, 1993. The main objective for instituting this series of conferences was the fact that "the study of the Astronomy in the Orient is essential in order to obtain a global view of the development of astronomy through the ages."[5] The *Second ICOA* was held in Yingtan (Jiangxi Province of China), during Oct. 15–21, 1995.[6] The *Third ICOA* was held in Fukuoka (Japan), Oct. 27–30, 1998,[7] and the *Fourth* in Nanyang (China), Aug. 19–25, 2001. Its *Proceedings* are under-preparation. We have gone here at length in order to record the efforts of historians of astronomy in studying the history of Oriental astronomy all the world over.

With all the aforementioned in mind, the undersigned, as the President of IAU Commission 41 (History of Astronomy) for the triennial 1994–97, submitted to the Programme Committee of 23rd IAU General Assembly (Kyoto, 1997) the proposal to organise a Joint Discussion (JD) "... befitting to Kyoto venue. The idea was to take stock of the research status of history of astronomy in Asia and the Far East, to understand the intellectual framework of the astronomers of traditional astronomy and the cross-cultural transmission of astronomical ideas among themselves, and to possibly stress the new directions and vistas opened for future in the field: History of Oriental Astronomy".[8] With these objectives in mind this Commission proposed *originally* the main title of the Joint Discussion as "Astronomy in Asia and the Far East". The Programme Committee approved the proposal with the stipulation that the title of the JD should be the "History of Oriental Astronomy". The Commission 41 accepted that suggestion naturally.

This Joint Discussion was held during Aug. 25–26, 1997 in Kyoto. It consisted of five sessions for oral presentations and a poster session. The talks were scheduled and distributed according to two broad themes: 1. *Oriental Astronomy during the Ancient and Medieval Period*, and 2. *Modern Astronomy in the Orient*. The former covered

Chinese, Indian, Islamic, Japanese and Korean traditional astronomy. The second theme was devoted to the transmission of European astronomy into the Non-European countries, and also to the modern astronomy as carried out presently in these countries. The detailed programme is given in these *Proceedings*, see p. 245. We may refer here also to the summary of the various contributions which were published in the *Highlights of Astronomy*.[9] However, since the published summaries were constrained only to a maximum of 3 pages, it was decided that the whole *Proceedings* should be published, comprising detailed contributions. The permission for this publication granted by J. Andersen, the General Secretary of the IAU, is hereby gratefully acknowledged.

These *Proceedings* comprise *nineteen* contributions. They cover the various cultural areas as follows: On the East Asian – Chinese, Japanese and Korean – indigenous astronomy *four* oral presentations, *four* additional contributions along with *two* talks on the transmission of European astronomy; on Ancient Indian astronomy *two* talks along with *two* on modern astronomy in India; *two* talks on transmission of Islamic astronomy in China, and *three* presentations on modern European astronomy. It may be mentioned that the additional contributions were supposed to be presented orally but for the respective scholars who could not participate due to some reasons.

It may be noted that whereas the East Asia and India is well represented, the contributions on astronomy in Islamic cultural area is rather nil, despite the fact that in the last half of the twentieth century enormous amount of researches in the Islamic astronomy have been carried out as mentioned above. In this respect our own personal efforts to rope in a couple of historians of Islamic astronomy did not bear any fruit either. The main reason appears to be, however, that the bulk of historians of astronomy of Islamic Middle Ages are not attached to the Commission 41. This is also true actually of many a historian of ancient and medieval astronomy. This fact has to be taken note by the Commission 41.

A few participants, who gave their talks, did not submit their detailed papers for these *Proceedings*. We are not including the abstracts of their talks, since the summaries of their talks have already been published in the *Highlights of Astronomy*. For the same reason we are not paraphrasing or enumerating also the essential points of all contributions published here, as is usual.

Finally, we may emphasise the importance of studies in the history of astronomy the world over by recalling the following. At the XXIst International Congress of History of Science (held in Mexico City, July 8–14, 2001), two specific sections were devoted to Physics and Astronomy. Further, the present author, in the capacity of the President of the Joint IUHPS and IAU Commission on History of Astronomy, organised a symposium: "*Astronomical Heritage of Non-European Cultural Areas*", which was well attended with 18 talks. In all, there were about 50 talks concerning history of astronomy. Further, the General Assembly of IUHPS approved the creation of a *new* Commission for the History of Ancient and Medieval Astronomy (President, S.M.R. Ansari). Here it will be appropriate to quote Edward Kennedy, one of the foremost living historians of Islamic Astronomy, "... if we delete (*which we should not*) the fundamental Hellenistic contribution, ... we can say that ancient and medieval astronomy was all Oriental."[10] Obviously, the aim of this new commission particularly is to include in its purview besides Oriental astronomy also Greek astronomy, astronomical history, astronomical iconography, archeoastronomy and even starlores, i.e., cultural ramification of astronomy in human society in general.

The editor submits his apology that the publication of these Proceedings became unusually late, due to his own unexpected preoccupations. It is neither the fault of the publisher nor that of authors. The co-operation of both is gratefully acknowledged here. Special thanks are due to Dr. H. Blom (of Kluwer) for his magnanimity. Despite the abovementioned drawback, we hope that these Proceedings will be very useful *per se*. The editor is also pleased to acknowledge the co-operation of Dr. Steve Dick (Washington), Prof. Michio Yano (Kyoto) for cordial assistance, Prof. S.R. Sarma (Aligarh) for the advice regarding the cover picture and Prof. M. Mahdi Ansari (Aligarh) for the permission to use the picture of Ghulam H. Jaunpuri's globe. Last but not the least my heartfelt appreciation of my son, Zia Nishat Ansari (Bochum) who assisted me in many ways to complete this Volume.

Aligarh (India) S.M. Razaullah Ansari
Feb. 6, 2002

Notes

1. Edward S. Kennedy, *Studies in the Islamic Exact Sciences*, (in collaboration with colleagues and former students), American University of Beirut, Beirut, 1983.
2. G. Saliba, *A History of Arabic Astronomy: Planetary Theories during the Golden Age of Islam*, New York University Press, New York, 1994.
3. G. Swarup, A.K. Bag, and K.S. Shukla (Eds.), *History of Oriental Astronomy*, Cambridge University Press, Cambridge, 1987. Noteworthy is that the Indian National Science Academy (New Delhi) published the "*History of Astronomy in India*" (1985) to commemorate that important meeting.
4. See the Report of the Colloquium by S. Débarbat in the *Transaction of the XXII nd IAU General Assembly*, held in The Hague, 1994.
5. Preface, p. v, in I.-S. Nha and F. R. Stephenson (Eds.), *Oriental Astronomy from Guo Shoujing to King Sejong* (Proceedings of the First International Conference on Oriental Astronomy, Oct. 6–11, 1993), Yonsei University Press, Seoul (Korea), 1997.
6. The Chairman of the Organising Committee was late Bo Shuren, to whom the *Proceedings* was to be dedicated. It might have been published by now.
7. Masanori Hirai (Ed.), *Proceedings of the Third International Conference on Oriental Astronomy*, Fukuoka University of Education, Munakata, Fukuoka (Japan), no date.
8. S.M.R. Ansari, Scientific Rationale of the Proposed Joint Discussion at the IAU/GA (Kyoto, 1997), document submitted to the Programme Committee of the 23rd IAU/GA.
9. J. Andersen (Ed.), *Highlights of Astronomy*, as presented at the XXIIIrd General Assembly of the IAU, 1997, Vol. 11B, Kluwer Academic Publisher, Dordrecht, 1998
10. E.S. Kennedy, Introductory Lecture, in G. Swarup *et al. loc. cit.*, p.3; emphasis ours.

Scientific Organising Committee

Supporting Commission: 41 (History of Astronomy): A Joint Commission of IAU and IUHPS
Co-Supporting Commission: 46 (Astronomy Education)

Scientific Organising Committee:

S.M. Razaullah Ansari (Chairman, India)

Susan Débarbat (France)

Steve J. Dick (Co-Chairman, USA)

B. Hidayat (Indonesia)

Owen Gingerich (USA)

Y. Maeyama (Germany)

Il-Seong Nha (Korea)

David Pingree (USA)

Y. Sobouti (Iran)

Xi Zezong (China)

Part 1

Oriental Astronomy during the Ancient and Medieval Period

1.1. The Two Supreme Stars, Thien-i and Thai-i, and the Foundation of the Purple Palace

Y. Maeyama
University of Frankfurt

Introduction

The two stars of Chinese astronomy which were the most worshipped and most frequently mentioned in Chinese literature were the small stars Thien-i[a] and Thai-i[b]. This phenomenon was evident for more than two millennia, apparently linked with mysticism rather than a direct result of the original significance of these two stars.

Their Chinese names, generally written in several different characters,[1] were rendered in Western languages by a wide variety of imposing expressions:

Thien-i: Coelum unum / Celestial Unique, Heavenly Unity / Unique du Ciel, Unité Céleste / Himmlische Monade.
Thai-i: Magnum unum / Great Unique, Grand Monad, August Unity, Supreme One / Archi-Première, Souverain des Cieux, Unité Suprême, Suprême Un, Unique Suprême, Grande Unité / das Erhabene Eine, Höchste Monade.

All these variations are intended to denote one and the same attribute, *supremacy and sovereignty in the heavens*. They also betray the fact that they are not purely linguistic translations of the original names, but also derive from the mystic backgrounds of what Ssuma Chhien (fl. 90 B.C.) and others tell us:

> "Of all Celestial Deities the Thai-i is the most venerable",
> "The Thien-i ... is the God of the Celestial Emperor. ... the Thai-i ... is also the God of the Celestial Emperor."[2]

Yet these lofty concepts relate to two tiny celestial objects, well known in the available star-maps from the 8th and later centuries (Figs. 1–3). In one of the star-maps (Fig. 1) two different sets apparently with the same names are depicted.

All this seems to have led us, since more than a thousand years ago, to the prevailing, tacitly accepted, mystic perplexity that the two stars, Thien-i and Thai-i, are of paramount importance in Chinese astronomy. However, the degree of that importance and the origin of the extraordinary significance of the two stars are not at all known.

Despite uncertainties, ambiguities and even apparent contradictions in the historical records we shall demonstrate below the following three theses, presenting what I believe to be an astronomically plausible solution to this enigma:

(1) Originally, the Thien-i was the Pole-Star.[3]
(2) Originally, the Thai-i was the unified celestial symbol of the Pole-Star and the terrestrial Emperor, designated to a star adjacent to the Thien-i.

S.M.R. Ansari (ed.), History of Oriental Astronomy, 3–18.

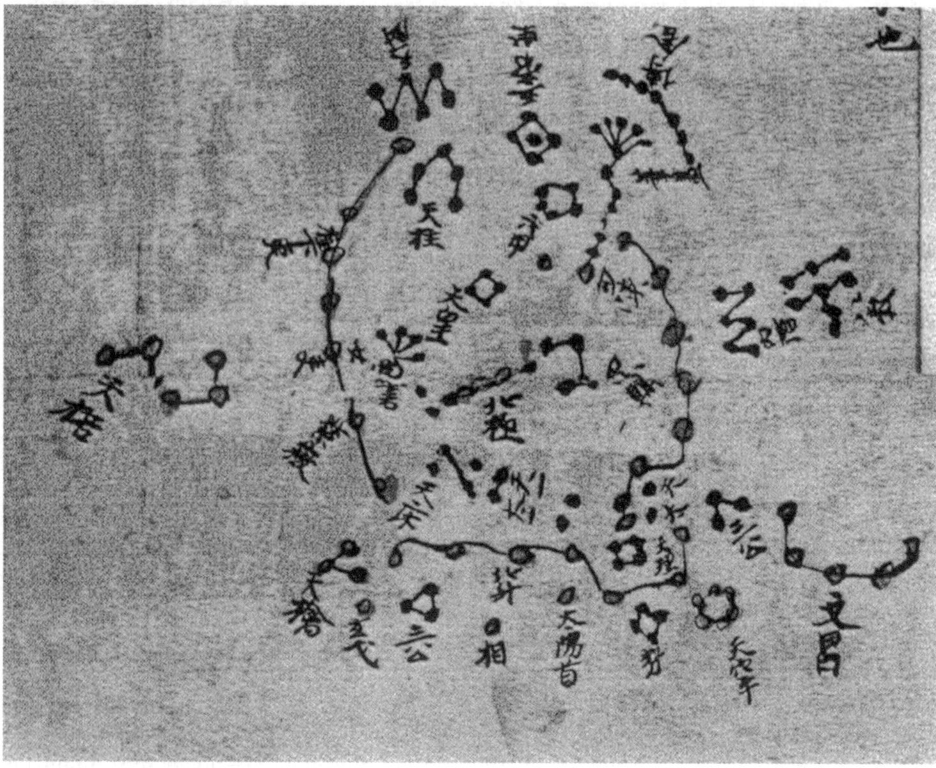

Figure 1 Tunhuang Star-Map A
(ca. A.D. 710; Brit. Mus. Stein no. 3326; copied from Xia Nai): Table 1, lines 2, 3

Note: The set of the two supreme stars is depicted twice, at the wide open gate of the Purple Palace (Thien-i, Thai-i) and outside the Western (Right) Boundary (Thien, Thai).

(3) The Purple-Palace[c] originated in connection with the "precession of the North-Pole". Its foundation is therefore inseparably connected with the two stars, Thien-i and Thai-i.

1. The Thien-i,[a] the Celestial Unique

In the bone-inscriptions (14th – 11th cent. B.C.) of the Yin[d]-dynasty we meet again and again the symbol of the Yin-people's highest Deity, *Ti*[e], which had been supposed to govern men's destinies.

In my paper (1992) I, inspired by Prof. H. von Dechend, claimed that the *Ti*, the later Emperor, must have been the Pole-Star.[4] As it turned out, the probable Pole-Star of that time, κ Dra, was exactly the star whose position had been measured in equatorial coordinates by one of the astronomical schools, the so-called Shih Shen or Shih Shih[f], under the name of Thien-i, the Celestial Unique, and according to my analysis, this was around 70 B.C.[5]

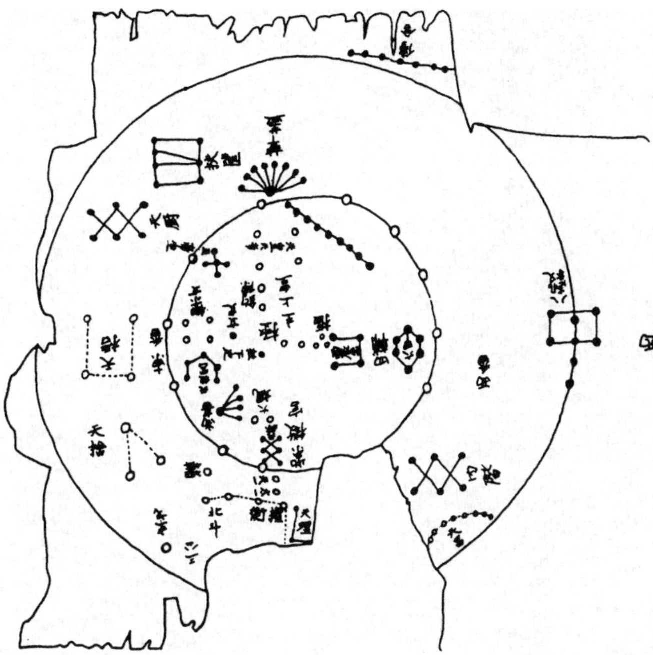

Figure 2 Tunhuang Star-Map B
(A.D. 10th cent.; Tunhuang Culture Center no. 076; copied from Deng Wenkuan p. 94 (modified)): Table, line 4

Note: The two boundaries, Eastern and Western, are connected together forming an enclosed Purple Palace, from which the two supreme stars, Thien-i and Thai-i, are excluded.

I have subsequently shown that, based on the Thien-i as the pivot of the celestial sphere and the Northern Dipper, the Chinese must have drastically improved their equatorial orientation by inventing the 28 lunar mansions (hsiu).[6]

In the same documents Shih Shen describes two other important stars as well as the Thien-i:

> No. 59 The Purple Tenuity Boundaries[g]
>
> Shih Shih says, the Purple Tenuity Boundaries consist of 15 stars, …, the Right Star[h] is north of the Pei Tou[i] …,
>
> No. 61 Thien-i, the Celestial Unique
>
> Shih Shih says, the star Thien-i is south of the Right Star, outside the gate of the Purple Palace, at the same degree (position in the same hsiu as the Right Star),
>
> No. 62 Thai-i, the Great Unique
>
> Shih Shih says, the star Thai-i is south of the Thien-i, they are near to one another (Khai-Yuan Chan Ching, ch. 67).

The records from some time later, though we scarcely know the epoch, tell us:

In the *old* times the Thien-i and Thai-i were near to one another in Chen (hsiu no. 28) but are *now* in I (hsiu no. 27). The star Thai-i is south of the Right Star at (outside of)

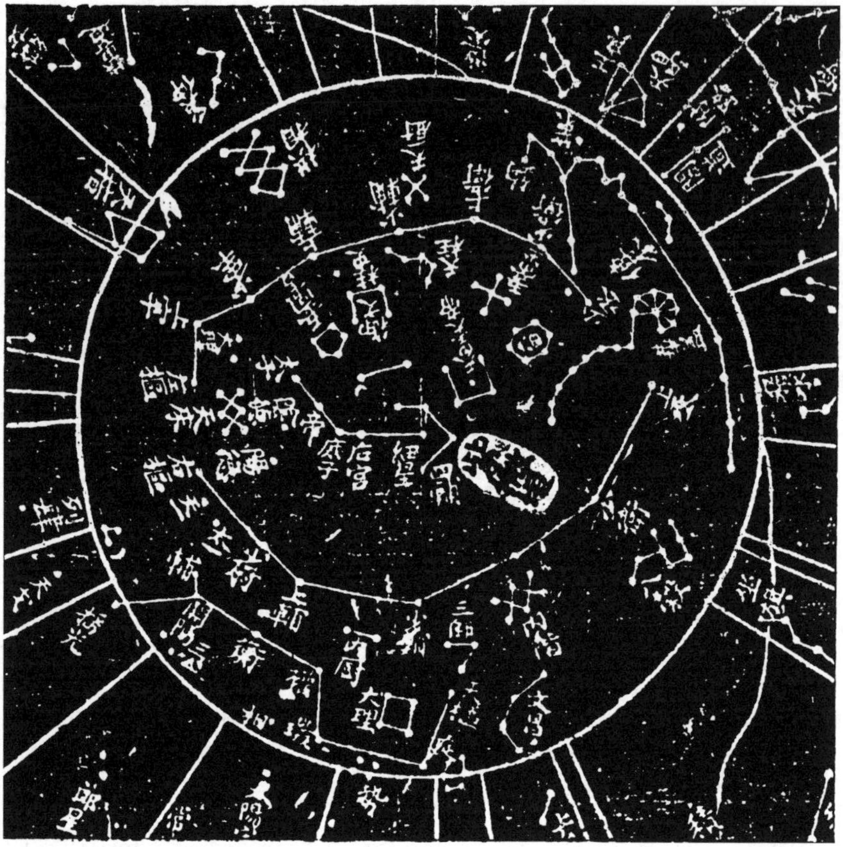

Figure 3 The Purple Palace of the Suchow planisphere, (A.D. 1193; copied from Saussure, p. 510): Table 1, line 5

Note: In relation to the Northern Dipper, the Western (Right) Boundary of the Purple Palace, in particular the range of its last three stars α (Right Pivot), 6 and λ Dra, is distorted. The two supreme stars, Thien-i and Thai-i, are depicted right between the Right Pivot (α Dra) and the early Right Star (6 Dra), now called Shao-wen, outside the Western (Right) Boundary.

the gate of the Purple Tenuity Palace. The star Thien-i is equally in the same mansion near the Thai-i (ibid., ch. 107). Based on the above descriptions and the numerically given star-positions, we saw only one possibility, and so came to our conclusive identification of the three stars of great importance as follows (Fig. 4):[7]

the Right Star (as opposed to the Left Star) at the gate of the Purple Palace = 6 Dra

the Thien-i (Celestial Unique), south of the Right Star = κ Dra

the Thai-i (Great Unique), south of the Thien-i = 4 Dra

Now the remarkable 1987 excavation of the "tomb no.45" from the Yangshao cultural stratum has shown us that even in those remote times of ca. 3000 B.C. the constellation

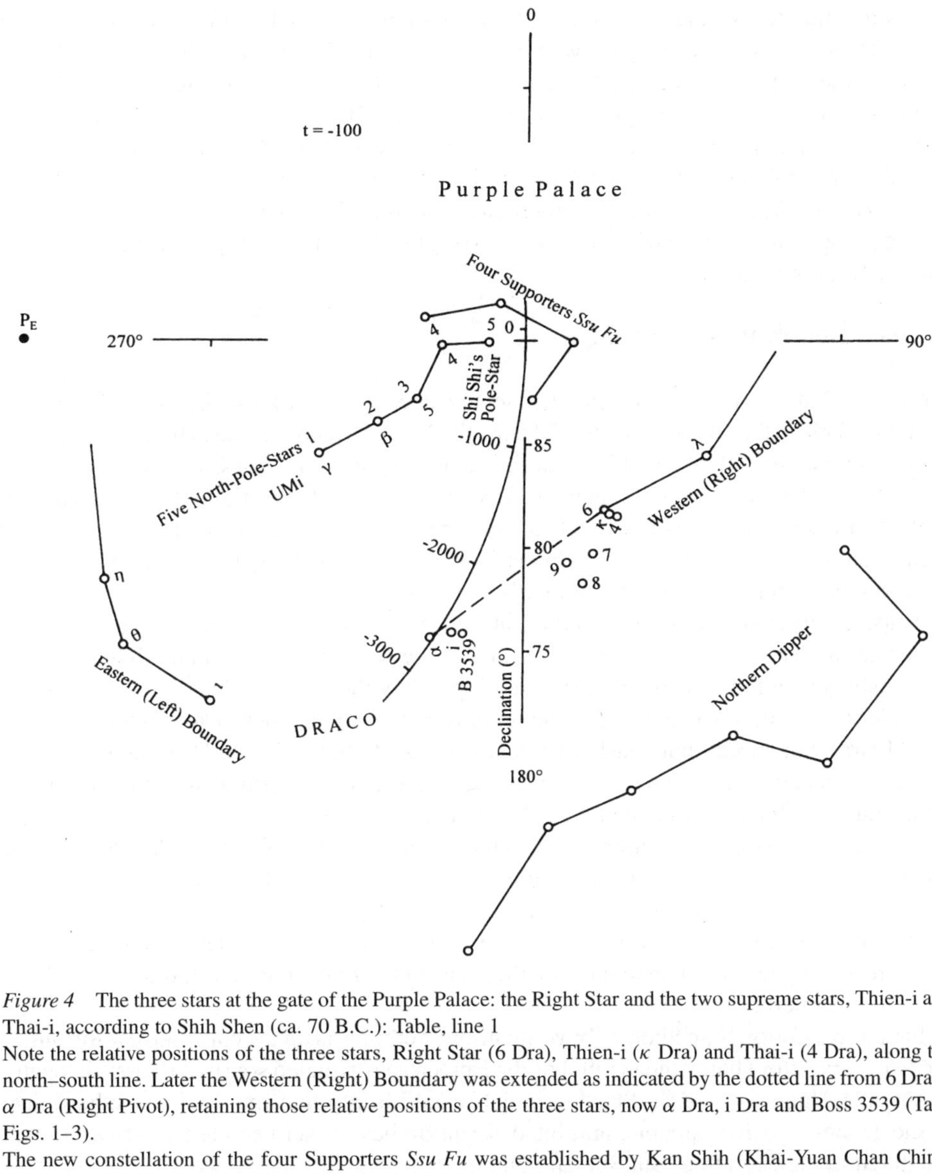

Figure 4 The three stars at the gate of the Purple Palace: the Right Star and the two supreme stars, Thien-i and Thai-i, according to Shih Shen (ca. 70 B.C.): Table, line 1
Note the relative positions of the three stars, Right Star (6 Dra), Thien-i (κ Dra) and Thai-i (4 Dra), along the north–south line. Later the Western (Right) Boundary was extended as indicated by the dotted line from 6 Dra to α Dra (Right Pivot), retaining those relative positions of the three stars, now α Dra, i Dra and Boss 3539 (Tab., Figs. 1–3).
The new constellation of the four Supporters *Ssu Fu* was established by Kan Shih (Khai-Yuan Chan Ching, ch. 69) in order to stop the farther *"precession of the North-Pole"* from the present Pole-Star [HD 117566 (Shih Shen, ca. 70 B.C.)] to another star (Maeyama, On the Pole-Star.).

consisting of the Northern Dipper, (Pei) Tou, and the then Pole-Star α Dra must have served for the Chinese as a mean for the fundamental orientation [measure (Tou)] of time and space, all according to the daily regular rotation of the celestial sphere about the Pole-Star.[8]

From the above statements it is certain that my early claim of identifying the *Ti*, the highest Deity of the Yin-people, with the Pole-Star can be considered correct.[9] It now seems justified to assume that the Pole-Star, seemingly the only unmovable point in the whole Universe, was consistently worshipped for millennia as the highest Deity; during the Yin-time (16th – 11th cent. B.C.) as *Ti* (κ Dra) and later – as this was replaced by *Thien*, Heaven, through the Chou[j]-dynasty (11th – 3rd cent. B.C.)[10] – by the name of Thien-i, Celestial Unique. It is indeed a remarkable historical reality that we find the Chou-people's probable original name, Thien-i, long after finishing its role as the Pole-Star, still in Shih Shen's Hsing Ching[k] (ca. 70 B.C.) and numerous other documents.[11]

The question arises as to what function had originally been assigned to the neighbouring star, Thai-i, Great Unique.

2. The Thai-i[b], the Great Unique

We know that the Chinese worshipped many deceased Yin-kings by the name of *Ti*, Emperor. This in fact already occurred during the Yin-time and we find in bone-inscriptions expressions such as *Ti-ting* and *Ti-chia*, Emperor Ting and Emperor Chia for *Wu-ting*[l] and *Tsu-chia*[m 12], and later we read them as *Ti Wu-ting*, Emperor Wu-ting, etc. in the History of the Yin-dynasty of the Shih Chi (ch. 3). Apparently some distinguished kings were after their death thus regarded as being almost identical to the highest Deity of the Universe, *Ti*[e], who had been in possession of men's good and evil.

Thus, for instance, the founder of the Yin-dynasty was called other names, mostly *Thai-i*[n] during the Yin-time and perhaps later also *Thien-i*[o] as Ssuma Chhien tells us (ch. 3), obviously because he was particularly worshipped by the Yin-people.[13]

Under these circumstances we came to assume that the Chinese established a special star (4 Dra), not strictly but nearly identical with the *Ti*, the Celestial Deity and Pole-Star (κ Dra), as a celestial symbol of the terrestrial Emperors, in the immediate vicinity of the Pole-Star, the Thien-i, and called it by the name of *Thai-i* (Fig. 4).[14]

The Chinese original intention seems obvious enough. A perfect identification of their Kings with the *Ti*, their ultimate Divinity for millennia, would make the authority of not legendary but real kings with their great concrete achievements even obscure. We further assume that the Chinese may have invented two similar names, Celestial Unique and Great Unique, simultaneously for the Pole-Star *Ti* and the celestial symbol of the terrestrial Emperor.[15]

In this connection Shih Shen's above-mentioned emphasis on the correlative positions of the two stars, "the Thai-i (the symbol of the Emperor) is situated south of (below, beneath, under) the Thien-i (the Pole-Star)" – which would imply that the Thai-i is subordinate to the Thien-i – is inescapable, and his description has in fact been tenaciously retained unaltered in the Chin Shu, Sui Shu and others.[16]

The question arises as to why in the Shih Chi we find the Thai-i at an entirely different position. With the Astronomical Chapter of the Shih Chi we now proceed to the problem of the Purple Palace.

3. The Purple Palace[c]

The Purple Palace is mentioned in writings such as Shih Chi (ch. 27), Huai Nan Tzu (ch. 3) and Hsing Ching of both Shih Shih and Kan Shih (Khai-Yuan Chan Ching, chs. 67, 69).

According to the frequently cited statement of Ssuma Chhien, the North-Pole is supposed to be at the second star of the so-called Five North-Pole-Stars[p] [17], β UMi, which is furthermore the permanent residence of the Thai-i. However, the Chou Pei Suang Ching tells us about the movement of the great star, certainly β UMi, in four directions near the North-Pole.[18] Undoubtedly also Ssuma Chhien knew this fact, that the star β UMi was not strictly at the North-Pole.

Long before the time of the Huai Nan Tzu and Shih Chi the Chinese must already have been increasingly uneasy about the shift of the North-Pole in connection with that of the point of the winter-solstice. Probably around 70 B.C. we find Shih Shen saying that at the winter-solstice the Sun is at the 21st degree Tou (hsiu no.8; determinative star φ Sgr), a statement that distinguishes his school's strictly scientific attitude from all others.[19] From Shih Shen's measurements of 120 star-positions in equatorial coordinates we know that his instrument was fixed to the North-Pole with an unprecedentedly high accuracy of ca. 1°. [20] Thus Shih Shen was in possession of a Pole-Star far nearer to the actual North-Pole than Ssuma Chhien and all others of his time.[21] Astronomically, Shih Shen's documents contain details of numerical measurements and can be considered the most reliable at that time.[22]

Contrary to all comments hitherto made, I claim that the Purple Palace was deliberately set up in the heavens for the eternal authority of the Emperor so as to be free of any kind of changes in that Imperial territory.

At the centre of the Palace is the residence of the Emperor and the Thai-i, the Great Unique; it is the greatest and brightest star, β UMi, which commands all in space and time. Here we finally see the Unity of the macro- and microcosm, the ultimate unification of the three most fundamental principles of the Chinese:

1. the early highest Deity of the Universe of the Yin-time, *Ti*[e] (the Pole-Star, later Thien-i, Celestial Unique),
2. the celestial symbol of the terrestrial Emperor (Thai-i, Great Unique, adjacent to the Thien-i),
3. the terrestrial Emperor himself.

The star will in fact be called Ti-wang (Imperial King) or Thien-ti (Celestial Emperor) as found in the Chin Shu (ch. 11, 13b) and others. In this regard the great bright star β UMi is no doubt more proper than any other possible Pole-Stars such as 5, 4 UMi and Shih Shen's proud one;[23] all these were at that time much nearer to the North-Pole but inevitably much smaller than β UMi (Fig. 4). As Maspero correctly says, Ssuma Chhien does not even pay much attention to the apparent tiny Pole-Star.[24]

Like Shih Shih, Kan Shih also knew that the North-Pole had changed its place and therefore he wrestled with stopping its *historical precession* by means of the Four Supporters *Ssu Fu*[q], so that the North-Pole would no longer be able to proceed to another Pole-Star, unless it broke the chained hedge of the Four Supporters, as these cannot become the Pole-Star because of the duty already imposed on them (Fig. 4).[25]

The Celestial Emperor at the centre of the Palace is surrounded by his Imperial Family-members. Within the Palace no change is allowed to occur. The whole Palace should never come into contact with the seven ever-moving luminaries – Sun, Moon and planets – nor should it sink under the horizon. It is determined to be a secluded holy territory.

How exactly the Chinese may have come to demarcate the Purple Palace from the rest of the celestial sphere, which is subject to perpetual mutations, betrayed their innermost secret and prompted our argument above.

Knowing that the present North-Pole is at, or just near, the star β UMi and far from the early Pole-Star, Thien-i (κ Dra), they drew the Western (Right) Boundary[r] immediately close to those two supreme stars, Thien-i and Thai-i, so that they, starting with the Right-Star[h], would come on the north–south line (6, κ and 4 Dra), as described by Shih Shen (ch. 67). Here the early Pole-Star of the Yin-people and the symbolized Emperor, Thien-i (κ Dra) and Thai-i (4 Dra), are excluded from the Purple Palace because they have already "changed" and are no longer those supreme stars as such. However, because of their early supremacy in the Universe they are allowed to retain their original names and to stay nearby, just outside its gate at the Right-Star (6 Dra) of the Right (Western) Boundary (Fig. 4).

The constellation of these three neighbouring stars, 6, κ and 4 Dra – its collective descriptions with their individual positions in equatorial coordinates by Shih Shih (ch. 67) – seems to me extremely significant. If we consider that sharp demarcation of the Western Boundary at the Right Star (6 Dra), critically excluding its two immediately adjacent stars, the early supreme stars of κ and 4 Dra, suspicions will be strong that Shih Shen himself (fl. ca. 350 B.C.) of the Warring States period may have been the creator of the Purple Palace.

Yet Ssuma Chhien mentions an asterism of three stars at the mouth of the Northern Dipper. We claim that the star in the north is the Right-Star of the Right (Western) Boundary, 6 Dra, and the other two stars, κ and 4 Dra, are what Ssuma Chhien calls *Yin-te*[s], adding "also called Thien-i" (ch. 27, 2a).

We further claim that these two stars had originally been called *Thien-i* and *Thai-i* as found in Shih Shen's Hsing Ching (ch. 67) and that the latter, the celestial symbol of the Emperor, was now displaced to the centre of the Purple Palace, β UMi, its permanent residence. Being thus devoid of the symbol of the Emperor, the early asterism of the two stars is now called *Yin-te*, or simply *Thien-i* as Ssuma Chhien tells us.

As we shall see below, Kan Shih also describes the asterism *Yin-te*, consisting of two stars. This name seems to have some historical significance.

The expression *Yin-te*[s] consists of shadow *yin* as opposed to *yang*[t] and virtue. Its original element seems best represented by Chavannes' expression "Vertu cachée", the Hidden Virtue.[26]

Based on it we assume that the early set of the two supreme stars – after the Thien-i was deprived of its ultimate supremacy of the Pole-Star, while the Thai-i took over its role and thus became the ultimate Divinity of the whole Universe, the highest symbol, at the centre of both macro- and microcosm – was now named *Yin-te*. These two stars are, however, allowed to retain the once-heralded Virtue, albeit "*hidden*" as a token of homage to the late Pole-Star, and to stay immediately outside the gate of the Purple Palace. Thus, the name *Yin-te* bears mute testimony to the dispossessed of the highest Divinity of the Pole-Star.

The origin of the thus newly established Purple Palace is, to my knowledge, best expressed not by Ssuma Chhien, who is eager to point to the Celestial North-Pole-Stars

(Thien-chi-hsing[n], β UMi, etc.[27]), but in the Huai Nan Tzu, extending the original concept to the whole Palace:

> "The Purple Palace is the residence of the Thai-i." (ch. 3),

and later likewise in the Chin Shu and Sui Shu:

> "...also called the Purple Tenuity (Palace), is the place of the Thai-ti (Great Emperor), it is the permanent residence of the Son of Heaven (ch. 11, 14b; ch. 19, 30b)."

Here, unlike in the Shih Chi, we find the original *Thai-i* back again in its old place south of the Thien-i in a description similar to, but more clear than, Shih Shen's:

> "The Thai-i is south of and near the Thien-i, it is also the God of the Celestial Emperor." (ch. 11, 14a; ch. 19, 30b)

That the star 6 Dra was, as we have seen above, originally assigned to the Right-Star at the gate of the Western Boundary is certain, but this star was later named *Shao-wei*[v], while that function was taken over by α Dra, now called the Right Pivot (Yu-shu[w]) (Figs. 3, 4).

If the original Boundaries had been similar to those of the Tunhuang Star-Map A (Fig. 1), the gate would actually have been very widely open and free to foreign elements. On this question the comment "Old and Now"[x] on the Shih Shen's Hsing Ching (ch. 107) is most likely to supply some valuable data. We read:

> "In the Old (records) *Thien-i* and *Thai-i* stood side by side in Chen (hsiu no. 28) but now are both in I (no. 27)."

No doubt this statement already cited above is additional support for our identification of the two stars with κ and 4 Dra. Although we subsequently read "the star Thai-i outside the gate of the Purple Tenuity Palace, south of its Right-Star", exactly the same as the *Old*, we further read:

> "In the Old, the Boundaries of the Palace were between Lou (hsiu no. 16) and Chen (no. 28), according to the new measurements they are now between Wei (no. 17) and Chen (no. 28)."

From this it follows that the last star of the Western Boundary is now in Chen (no. 28), while the two stars Thien-i and Thai-i are in I (no. 27), and therefore that the Western Boundary was already extended from the Right-Star (6 Dra) to α Dra, now the Right Pivot, at this time (Fig. 4). We are not clear when this epoch is, however, though it is very likely to be soon after the *Old* Shih Shen, and certainly some time during the Later Han.[28]

There are two star-maps, the so-called Tunhuang Star-Map A and B, probably dating from the 8th and 10th centuries (Figs. 1, 2).[29] Besides several serious inconsistencies such as the number of stars in the Boundaries and the colours of the stars, we also find some crucial ambiguities.

In A we assume the two stars above the Dipper, *Thien* and *Thai*, to be *Thien-i* and *Thai-i* (κ, 4 Dra), coloured in black, hence supposedly of Kan Shih, although we find them only in Shih Shen's Hsing Ching.

In both maps we find another set of the two stars designated as Thien-i and Thai-i above the fourth star of the Nothern Dipper, δ UMa, and these are respectively in black (Kan Shih) and in red (Shih Shen or Wu Hsien). Most remarkable is that the position of these two stars relative to the Northern Dipper is nearly the same on both star-maps, on the prolongation of $\gamma - \delta$ UMa, on which also α Dra is situated.

In A the two stars are between the Right- and the Left-Star of the Boundaries, while in B they are excluded from, and right south of, the enclosed Boundaries, consisting of presumably 12 stars like Ssuma Chhien's constellations, but contrary to Shih Shen's with 15 stars.

The Star-Map B seems likely to deal with the predecessor of the later arrangement of the three stars, α Dra (the Right Pivot of the Western Boundary) and the two adjacent tiny stars i Dra (Thien-i) and B 3539 (Thai-i), such as we find in the constellations of the Sung-time (Figs. 2–4).[30]

If this assumption is correct, the Western Boundary in B is extended from the original Right-Star 6 Dra to α Dra and together with this, the two stars Thien-i and Thai-i are displaced from their original asterism of κ and 4 Dra to their new positions mentioned above, retaining the relative positions of all these three stars "along the north-south line, near to one another", following exactly Shih Shen's original description (Fig. 4).

Despite their great similarity at a glance, the two Star-Maps differ from one another particularly at the gate of the Palace. In A the two stars Thien-i and Thai-i are isolated, as in B but, in contrast to this, are right at the wide opened gate. They are marked in black, though Kan Shih gives no stars under these names. Both Xia Nai and Deng Wenkuan considered these names to be copying errors and tried to correct them to Yin-te and Yang-te, and respectively to Yin-te alone, presumably, in the case of the former in particular, in connection with the famous 12th-century star-map (Fig. 3).[31]

Like Ssuma Chhien, Kan Shih also (ch. 69) tells us of the asterism *Yin-te* of two stars, but obviously at a different position: "The two stars *Yin-te* are west of the asterism *Shang-shu*."[y] Perhaps in this connection Deng Wenkuan was led to correct the original star-names in A to *Yin-te*.

Unfortunately, Chhen Cho's[z] (fl. A.D. 310) completion of his star-map according to the three ancient schools (Chin Shu, ch. 11, 13a), where for example no constellation is given as common to these schools, makes the early history before him completely obscure.

Our assumption about the Star-Map A (Fig. 1) goes back a long time. The Thien-i entirely isolated, only accompanied by the Thai-i, is highly likely to be identified with α Dra, the early Pole-Star around 3000 B.C.

As we have already shown, we are in possession of some increasingly compelling evidence that from those remote times of ca. 3000 B.C. onwards the Pole-Star, the only unmovable concrete point in the whole Universe, was worshipped continuously for millennia, and that by the time of Shih Shih's observations (ca. 70 B.C.) the Chinese had changed their Pole-Star at least three times: from α Dra to κ Dra to β UMi (Ssuma Chhien, etc.) to HD 117566 (Shih Shih)[32] – possibly the Kan Shih's school had its own Pole-Star near 4 UMi as seen from its constellation of the Four Supporters *Ssu-fu* (Fig. 4). Most probably, the tip of the original (Pei) Tou, as given by the 1987 excavation, is represented by the then Pole-Star α Dra, and the Yin-people's highest Deity *Ti* had been the Pole-Star κ Dra (Fig. 5).[33]

The question arises as to whether one might have named the two stars, α Dra and south of (below) it its adjacent star, by Thien-i and Thai-i, as in the later case of Shih Shen around 70 B.C. and given them the by-name *Yin-te* as found in Kan Shih, similar to what

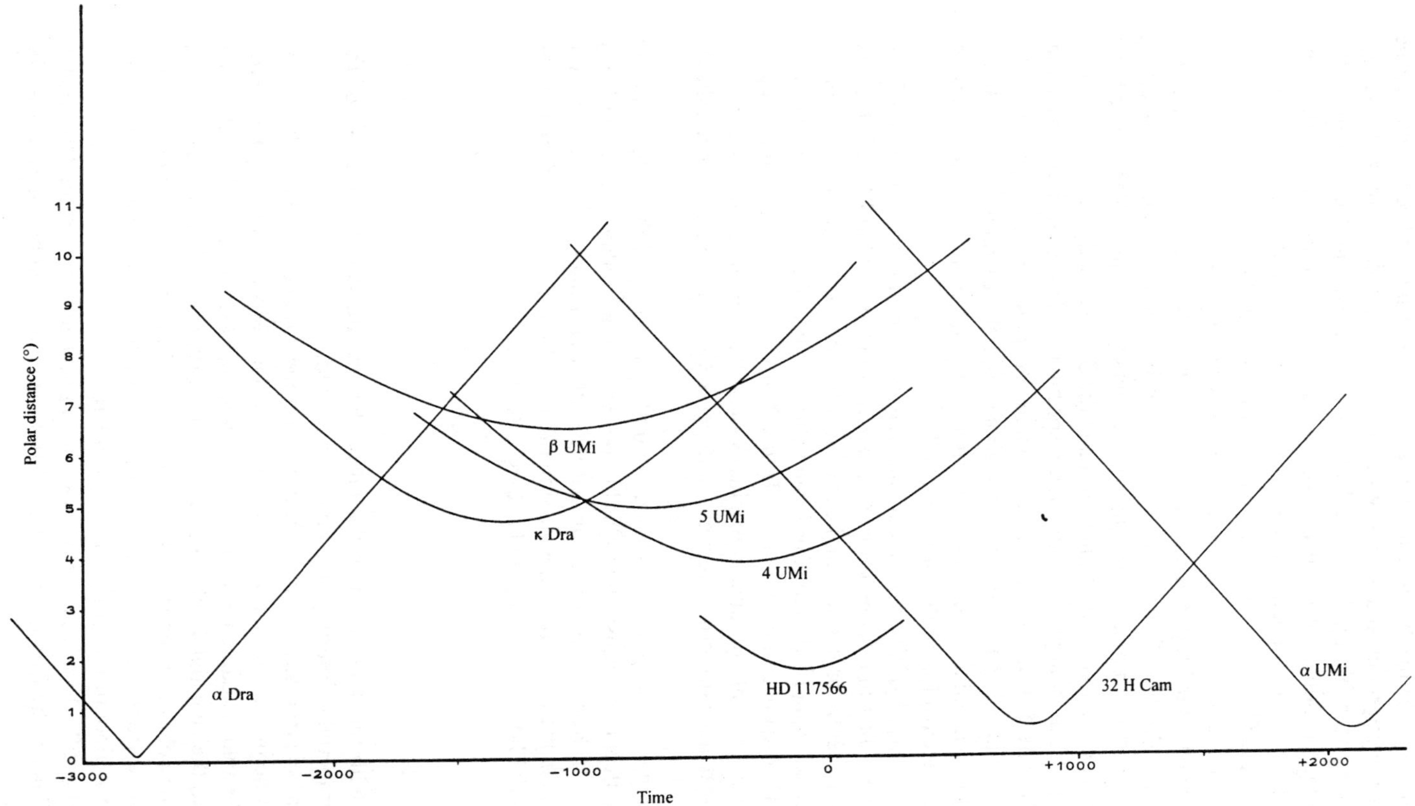

Figure 5 The polar distance of some apparent Pole-Stars in Chinese astronomy

Note: Due to the "*precession of the North-Pole*" the Chinese had at least four Pole-Stars by the Han-time:
1. α Dra [the tip of the original (Pei) Tou around 3000 B.C.; Feng Shi; Maeyama (1998)]
2. κ Dra [*Ti* (Yin), *Thien* (-i) (Chou and later); Maeyama (1992), (1995)]
3. β UMi [Ti Wang (Imperial King), Ta Ti (Great Emperor) (Ssuma Chhien, etc.)]
4. HD 117566 [Shih Shen (ca. 70 B.C.), perhaps also Kan Shih; Maeyama, On the Pole-Star.]

we have seen above in the Shih Chi, because the early Pole-Star should keep its past virtue "hidden", and whether, in order to distinguish the *older* from the *old* set of the two stars, one might have come to name the latter (κ, 4 Dra) simply *Thien* and *Thai* omitting "yi (one, unique)" (Fig. 1). Some traces of myths – as abundantly found in the Shan Hai Ching, Chhu Tzhu (ch. Astronomical questions) and others – about the earliest Pole-Star might have induced the Chinese to the circumstance mentioned above.

Table 1 shows our explanations of the three stars concerned.

Later on the Suchow Star-Map (Fig. 3) we find two stars, *Yin-te* and *Yang-te*, near the two chief stars, γ and β UMi, of Shih Shen's Five North-Pole-Stars, certainly connecting with the Yin-Yang doctrine of dualism to which these two stars, γ and β UMi, seem to be further related through their control of the Moon and the Sun respectively.[34]

Admittedly, these kinds of arguments are highly speculative. However, it is certain that for the Chinese the Pole-Star has served as the fundamental point of reference constantly since ca. 3000 B.C., and that this principle, though never explicitly attested in the bone-inscriptions, has not changed throughout the Yin-time. All this, I submit, is at least certain, thanks to the recent excavation of the "tomb no. 45".[35]

An outstanding example of the most tenacious transmission in time and space of historical elements is the fact that "*Amaterasu Oomikami*" (the Great God Illuminating the Heaven) – the fundamental principle and the mythical founder of Japan and the imperial family, the Sun-Goddess, deified in the *Ise Jingu* (Shrine) – has been worshipped – since the 7th century A.D. also under the by-name of *Thai-i* (*Tai-itsu* in Japanese), but in a highly mystical *hidden* way certainly because it originally dealt with the foreign concept – in a close connection with the Northern Dipper as the Imperial Chariot.[36] This has been shown by H. Yoshino.

The original Chinese concept of the Thai-i, the unified Divinity of the Pole-Stars and the Emperors, has thus been transmitted afar but preserved until today as a single Deity for Japanese souls. In Korea, which was at that time the transmitter of Chinese culture to Japan, such a phenomenon of the transmission of cosmological concepts founding the legacy of the imperial house seems not to have ever existed.[37]

Summary

This paper deals with the way in which the Purple Palace, the celestial symbol of the Emperor and his imperial belongings, may have originated.

I claim that this occurred strictly in connection with the old and new Pole-Stars. The Purple Palace was defined such that the centre of the Palace was occupied by the newly unified Deity of the two fundamental cencepts, the Pole-Star (Thien-i), the highest traditional Divinity, and the celestial symbol of the terrestrial Emperor (Thai-i) – now called the residence of the celestial Emperor and the Thai-i – while the Boundaries critically excluded those two supreme stars from the Palace, the old Pole-Star *Thien-i* (κ Dra) and the old star for the symbol of the Emperor *Thai-i* (4 Dra).

For this elevation of the terrestrial Emperor to the highest authority of the Universe at the centre of the Palace, the Chinese chose β UMi, the greatest and brightest star in the North-Pole region.

Table 1 The right star at the gate of the Purple Palace and the two supreme stars, Thien-i and Thai-i

	a	b	c	d	e
		Right star at the gate of the Purple Palace	Thien-i Celestial Unique	Thai-i Great Unique	Remarks
1	Shih Shih (ca. 70 B.C.)	Right Star 6 Dra	Thien-i κ Dra	Thai-i 4 Dra	Khai-Yuan Chan Ching, ch. 67, Maeyama (1977), 240–243 1-c,-d: Yin-te (Ssuma Chhien)
2	Tunhuang Star-Map A (ca. A.D. 710), Fig. 1	Right Star 6 Dra (or 7, 8, 9 Dra)	Thien κ Dra	Thai 4 Dra	The West Boundary of 8 stars, contrary to Shih Shih (7 stars)
3			Thien-i prob. α Dra, prob. also called *Yin-te*	Thai-i prob. i Dra, prob. also called *Yin-te*	3-c,-d: asterism of 2 stars *Yin-te* (Kan Shih, ch. 69)
4	B (A.D. 10th cent.), Fig. 2	Right Pivot (?) α Dra, the Right Star now called Shao-wei (?)	Thien-i i Dra	Thai-i Boss 3539	prob. predecessor of the Suchow Star-Map, the West. Boundary extended together with Thien-i and Thai-i, the Boundaries of prob. 12 stars (Ssuma Chhien)
5	Suchow Star-Map (A.D. 1193), Fig. 3	Right Pivot α Dra, the Right Star now called Shao-wei	= 4-c	= 4-d	the Boundaries of 15 stars (Shih Shih), the West. Boundary extended together with Thien-i and Thai-i

The origin of the Purple Palace is therefore ultimately attributable to the apparent "precession of the North-Pole".

This would indicate that the same fundamental concept of the Chinese might be traced back on that vast historical line of the "precession of the North-Pole", through *Thien* (Heaven of Chou's origin) and *Ti* (the Highest Deity of the Yin-people, the predecessor of *Thien* and the later Emperor, *Ti*), and finally to the original constellation of the Northern Dipper, (Pei) *Tou*, with its tip which had once been occupied by the still earlier Pole-Star, α Dra, as shown by the recent excavation of the "tomb no. 45" dating from ca. 3000 B.C.

Notes

1. For Thai-i cf. Chhien Pao-Tsung.
2. Shih Chi, ch. 28, 22a; Chin Shu, ch. 11, 14a.
3. Gaubil and many other scholars were of this opinion; cf. e.g. Saussure 495ff., Teboul. Anybody would come to this idea.
4. Maeyama (1992), 28; (1995).
5. Khai-Yuan Chan Ching, ch. 67; Maeyama (1977), 211, 242; see also Sun and Kistemaker (1997), 68 (-78).
6. Maeyama (1992), 83–87, 89, Figs. 6.1–6.5, 6.7; (1995), 390–393, Figs. 2–5.
7. Maeyama (1977), 240–243 (nos. 59, 61, 62). For my identification in detail cf. 216–218. For the three stars concerned Sun and Kistemaker chose κ, 7 and 8 Dra, which is certainly contrary to Shih Shen's descriptions.
8. See Maeyama (1998); for the (Pei) Tou in the bone-inscriptions cf. Xu, Yau and Stephenson.
9. See note 4 above.
10. For the transition from *Ti* to *Thien* see e.g. Chang 236, 239.
11. Huai Nan Tzu, ch. 3; Shih Chi, ch. 27, 2a; Han Shu, ch. 26, 2a; Chin Shu, ch. 11, 14a; Sui Shu, ch. 19, 30b.
12. E.g. Chang, 237.
13. E.g. Shirakawa, 84, 88.
14. Cf. an interesting quotation by Ho Peng Yoke, 68: "One legend says that when Huang Ti died ..., his spirit ascended up the heavens and became the deity T'ai I ...(Huang Ti Pên Hsing Chi, p. 8a)."
15. Unlike Chhien Pao-Tsung, Li Ling maintains that the history of Thai-i worship can be traced back to the Warring States period (p. 25f.). In his private communications Prof. Pankenier suspects it may be still earlier.
16. Chin Shu, ch. 11, 14a; Sui Shu, ch. 19, 30b.
17. Originally Shih Shen's designation.
18. See e.g. Cullen, 127.
19. Hou Han Shu, ch. 12, 3b.
20. Maeyama (1977), 216, 219; see also Sun and Kistemaker, 56.
21. Maeyama, On the Pole Star.
22. See also Yabuuti (1969), 52 and Maspero's erroneous statement, 281.
23. HD 117566, cf. Fig. 4 below; Maeyama, On the Pole Star.
24. Maspero, 328.
25. Maeyama, On the Pole Star.
26. Chavannes, T.3, 340.
27. This is the asterism of the Five North-Pole-Stars of Shih Shi (Fig. 4). For the erroneous interpretation of Chavannes (T.3, 339) see Herbster, 48–50.
28. For the detailed comment see Maeyama (1977), 217f.
29. See Ma Shichang; Xia Nai.
30. See Yabuuti (1936), 72, 76; (1969), 128; Maeyama (1977), 243.
31. Xia Nan, 144, 148. The lines (13) and (14) of the Star-Map B in the table should be exchanged with the line (19); Deng Wenkuan, 92.
32. Maeyama, On the Pole Star.

33. Maeyama (1998).
34. Chin Shu, ch. 11, 13b.
35. For the precession of the equinoxes as the fundamental historical problem in terms of archaic astronomy, see Dechend. For the unification of the macro- and microcosm through the terrestrial emperor, see Pankenier.
36. According to Prof. von Dechend's private communications the concept of the Northern Dipper as a chariot must have originally stemmed from Mesopotamia. See UMi /UMa as Himmelswagen/Lastwagen in Gössmann, 95–97.
37. Prof. Nha Il-Seong's private communications.

Bibliography

Chinese primary sources

Chhu Tzhu[aa]. *Elegies of Chhu.* Chhu Yuan et al. Chou, ca. 300 B.C.
Chin Shu[ab]. *History of the Chin Dynasty (265–419).* A.D. 635. Twenty-five Histories.
Chou Pei Suan Ching[ac]. *The arithmetical classic of the gnomon and the circular path of heaven.* Chou, Chhin, Han. Writers unknown.
Han Shu[ad]. *History of the Former Han Dynasty (206 B.C./A.D. 24).* ca. A.D. 100. Twenty-five Histories.
Hou Han Shu[ae]. *History of the Later Han Dynasty (A.D. 25/220).* A.D. 450. Twenty-five Histories.
Huai Nan Tzu[af]. *The book of (the Prince of) Huai Nan.* ca. 120 B.C. Repr. Taipei 1969.
Khai-Yuan Chan Ching[ag]. *Chhüthan Hsi-Ta*, 729.
Shan Hai Ching[ah]. *Classic of the mountains and rivers.* Writers unknown. Chou, Chhin, Han. Taipei 1959.
Shih Chi[ai]. *Ssuma Chhien*[aj], ca. 90 B.C. Twenty-five Histories.
Sui Shu[ak]. *History of the Sui Dynasty (581–617).* A.D. 636, 656. Twenty-five Histories.

Secondary sources

Chang, Tsung-tung: *Der Kult der Shang-Dynastie im Spiegel der Orakelinschriften. Eine paläographische Studie zur Religion im archaischen China.* Wiesbaden 1970.
Chavannes, Édouard: *Les mémoires historiques de Se-Ma Ts'ien.* 5 tomes, Paris 1895–1905, repr. Paris 1967.
Chhien Pao-Tsung: "Thai I" Khao. Investigation of the meaning of the term "Great Unique" (Chinese). *Yenching Hsüeh Pao (Yenching Univ. Journal of Chinese Studies)*, no.12 (1932), 2449.
Cullen, Christopher: *Astronomy and mathematics in ancient China: the Zhou bi suan jing.* Cambridge 1996.
Dechend, Hertha von and Santillana, Giorgio de: *Hamlet's Mill. An essay on myth and the frame of time.* Boston 1969; German translation: *Die Mühle des Hamlet.* Berlin 1993.
Deng Wenkuan: *Collected Collations of the Astronomical Texts and Calendars from Dunhuang* (Chinese). 1996.
Feng Shi: Astronomical investigation of the tomb no. 45 at the water bank west of P'u-yang, Ho Nan. *Wenwu* (1990), III, 52–60, 69.
Gössmann, P. Felix: *Planetarium Babylonicum oder die sumerisch-babylonischen Stern-Namen.* Rome 1950.
Herbster, Rainer: *Die Grundlagen der frühen chinesischen Astronomie, Bestimmung der Periode und Bahn der Sonne.* Frankfurt am Main 1986.
Ho Peng Yoke: *The astronomical chapters of the Chin Shu.* Paris 1966.
Li Ling: An Archaeological Study of Taiyi (Grand One) Worship (English transl. by D. Harper). *Early Medieval China*, 1995–96, 1–37.
Maeyama, Y. (1977): *The oldest star catalogue of China, Shih Shen's Hsing Ching.* Prismata. Festschrift für Willy HARTNER. Wiesbaden, 211–245.
——(1992): *The Four Cardinal Points of the Compass and the Chinese Lunar Mansions. The evolution of positional astronomy in ancient China.* J.W. Goethe-Univ. IGN Preprint Series No. 29.
——(1995): The stellar reference-points in ancient China and the evolution of positional astronomy. *Proc. of the 7th Intern. Conf. History of Science in East Asia, Kyoto 1994, East Asian Science: Tradition and Beyond.* K. Hashimoto et al. (eds.). Osaka 1995, 385–394.

——(1998): On the Earliest Stage of Chinese Astronomy. A new hypothesis. *Proc. of the Intern. Symp. on Ancient Astronomy and Traditional Culture of China (Nanjing, Oct. 1997)*. The Purple Mountain Observatory, Academia Sinica et al., Nanjing 1998, 6–9.

——: On the Pole Star and its Four Supporters, *Ssu fu*, in ancient Chinese astronomy. To appear in the *Proc. of The Third Intern. Conf. on Oriental Astronomy (Fukuoka, Japan, Oct. 1998)*.

Ma Shichang: Tunhuang xingtu de niandai (Chinese). On the date of the Tunhuang Star- Maps. Zhongguo gudai tianwen wenwu lunji. *Collected Papers on Astronomical Relics in China*. Beijing 1989.

Maspero, Henri: L'astronomie chinoise avant les Han. *T'oung Pao* 26 (1929), 267–356.

Needham, Joseph: *Science and civilisation in China*. vol.3. Cambridge 1959.

Nha Il-Seong: Private communications.

Pankenier, David William: *Early Chinese Astronomy and Cosmology: The "Mandate of Heaven" as Epiphany*. Stanford Univ. 1983.

Saussure, L. de: *Les Origines de l'Astronomie Chinoise*. Paris 1930.

Schlegel, Gustave: *Uranographie Chinoise*. Leyden 1875.

Shirakawa, Shizuka: *Kokotsu-bun no sekai – Kodai Yin-ocho no kozo (Japanese). The world of the bone inscriptions – The structure of the ancient Yin Dynasty*. Tokyo 1972.

Sun, Xiaochun and Kistemaker, Jacob: *The Chinese Sky during the Han. Constellating Stars & Society*. Leiden-New York-Köln 1997.

Teboul, Michel: Sur quelques particularités de l'uranographie polaire chinoise. *T'oung Pao* 71 (1985), 1–39.

Xia Nai: Another Manuscript Star-map from Dunhuang – The Dunhuang Star-map B (Chinese). *Explorations in the History of Science and Technology in China*. Li Guohao et al. (eds.). Shanghai 1982, 143–154.

Xu Zhentao, Yau Kevin K. C. and Stephenson, F. Richard: Astronomical records on the Shang Dynasty oracle bones. *Archaeoastronomy* 14 (1989), S61–S72.

Yabuuti, Kiyosi (1936): Sodai no seishuku. Descriptions of the constellations in the Sung Dynasty (Japanese). *Toho Gakuho*, Kyoto, VII, 42–89.

——(1969): *Chugoku no tenmon rekiho. Chinese astronomy and calendrical science (Japanese)*. Tokyo.

Yoshino, Hiroko: Ise-Jingu ko (Japanese). A study on Ise Shrine. *Minzoku-gaku Kenkyu*, 39 (1974), 209–232.

(a) 天一（乙） (b) 太（大）一（乙） (c) 紫宮 (d) 殷 (e) 
(f) 石申 石氏 (g) 紫微垣 (h) 右星 (i) 北斗 (j) 周 (k) 星經
(l) 武丁 (m) 祖甲 (n) 大乙 (o) 天乙 (p) 北極五星 (q) 四輔
(r) 西（右）蕃（垣） (s) 陰德 (t) 陽 (u) 天極星 (v) 少尉 (w) 右樞
(x) 古今 (y) 尚書 (z) 陳 卓 (aa) 楚辭 (ab) 晉書 (ac) 周髀算經 (ad) 漢書
(ae) 後漢書 (af) 淮南子 (ag) 開元占經 (ah) 山海經 (ai) 史記 (aj) 司馬遷
(ak) 隋書

1.2. Islamic Astronomical Tables in China The Sources for the Huihui li

Benno van Dalen
Institut für Geschichte der Naturwissenschaften, Frankfurt am Main

Islamic Astronomy in China

In the first decade of the thirteenth century, Chinggis Khan initiated the rapid expansion of the Mongol territory from the steppes on the northern border of China to a world empire that covered the largest part of the Eurasian continent. By the 1260s, the empire included both the Iranian part of the Islamic world and China, and travel between the various parts of the empire was safe and relatively quick. This not only allowed commerce to flourish, but also made possible an exchange of craftsmen and scholars and of scientific knowledge.

Thus it is known that a Chinese astronomer named Fu Mengchi or Fu Muzhai was active at the astronomical observatory in Maragha (northwestern Iran), which was founded in 1259 by the Ilkhan ruler Hülegü Khan, grandson of Chinggis. It is very probable that this astronomer provided the technical details of the so-called Chinese-Uighur calendar, a lunisolar calendar of Chinese type described in many Islamic astronomical handbooks from the Mongol period. This calendar was actively used in the Iranian world for at least a century after the Mongol conquest and has left traces on Iranian almanacs up to even the 20th century in the form of the use of the Chinese duodecimal animal cycle.[1]

On the other hand, the official annals of the Yuan Dynasty, whose first emperor was Hülegü's brother Khubilai Khan (1260–1294), inform us that a "westerner" (i.e. Muslim) named Zhamaluding[2] came to China in 1267 and offered to the emperor seven astronomical instruments and an astronomical handbook or almanac. Yamada (1980, 48–58) and Yabuuti (1997, 12) identify Zhamaluding as the astronomer Jamāl al-Dīn Muḥammad ibn Ṭāhir ibn Muḥammad al-Zaydī from Bukhara, who served the Great Khan Möngke (1251–1259) in the Mongol capital Karakorum (see also Sayılı 1960, 191–192). The instruments presented by Zhamaluding are described in detail in the Yuan annals and have been identified by Tasaka (1957), Hartner (1950) and Miyajima (1982), partially on the basis of the Chinese transliterations of their original Persian names. Zhamaluding's astronomical handbook or almanac, which is referred to in Chinese as *Wannian li* ("Ten Thousand Year Calendar" or "Eternal Calendar"),[3] is not extant, but scattered remarks concerning it in various sources make clear that it was of Islamic type and that it was officially distributed in China.

In 1271, Zhamaluding became the first director of the newly founded Islamic Astronomical Bureau, which would continue to operate parallel to the Chinese Astronomical Bureau for almost four centuries. During most of this time the results of calculations according to Islamic astronomical methods were compared with the contemporary official Chinese system. Islamic astronomy had a good reputation among the Chinese because of its theory of planetary latitudes, which did not exist in Chinese astronomical systems, and its accurate predictions of eclipses. However, the direct influence of Islamic astronomy seems to have been limited. The fundamental procedures in the official astronomical system of the Yuan Dynasty, the *Shoushi li* by GUO Shoujing[4] (1281), are all essentially of

S.M.R. Ansari (ed.), History of Oriental Astronomy, 19–30.

Chinese type. The discussion whether certain new elements, such as a pseudo-geometrical method for converting between equatorial and ecliptic coordinates, the systematic use of decimals in the underlying parameters, and the application of cubic interpolation in the calculation of the irregularity in the planetary motions, were influenced by knowledge of Islamic mathematics and astronomy, continues up to this day. Yabuuti (1997, 16–17) has claimed that various of the instruments built in China during the Yuan Dynasty, such as the Simplified Instrument (*jianyi*) by GUO Shoujing (see Needham 1959, 369–372) and the large gnomon in the Tower of the Duke of Zhou in present-day Gaocheng (Needham 1959, 294 ff.) show traces of Islamic influence.

From the fact that around 40 people were connected to the Islamic Astronomical Bureau during most of the time of its existence, we may conclude that its activities were quite significant. It is known that a large number of scientific books in "western" languages were available at the Bureau, and a list of 23 transliterated Persian titles has been analysed in detail by Tasaka (1957). Unfortunately, it seems that no direct records of observational activities by Muslim astronomers in Yuan China and no complete original works compiled by these astronomers are extant. Therefore, the scope of their activities can only be deduced from a number of sources dating mostly from the first years of the succeeding Ming Dynasty (1368–1644) and later.

The most important of these sources is referred to as *Huihui li* or *Huihui lifa* ("Islamic astronomical system"). It is a translation into Chinese of a standard Islamic astronomical handbook with tables, a so-called *zīj*.[5] *Zīj*es constitute a very important category of Islamic scientific literature. Modelled after Ptolemy's *Handy Tables*, they contain large sets of mathematical tables accompanied by instructions for their use and, in some cases, explanations of the underlying geometrical models and proofs for the calculations involved. By means of the tables, most of the common astronomical and astrological calculations could be carried out at the cost of only few simple arithmetical operations. Muslim astronomers continuously improved the values of the various astronomical parameters by conducting systematic observational programs; they increased the accuracy with which the trigonometric functions underlying most of the planetary tables were calculated, and made occasional attempts at modifying the Ptolemaic planetary models. In short, many important developments in Islamic mathematics and astronomy occurred in connection with the compilation of *zīj*es. At least 200 different Arabic and Persian *zīj*es were compiled during the Islamic Middle Ages; of these, more than one hundred are extant, but only very few have been published or extensively studied. From the researches that have been carried out, in particular by Kennedy and King, it appears that *zīj*es display a large variety in presentation, underlying astronomical parameters, layout of the tables, etc.

It seems that the above-mentioned *Huihui li* is the only Chinese translation of an Islamic astronomical handbook that is extant. The translation was started in 1382 on the order of the first Ming emperor Hong Wu. After he had conquered the Yuan capital Beijing, Hong Wu transfered the complete Islamic Astronomical Bureau with the Muslim astronomers and the large collection of mathematical and astronomical works in western languages to his new capital Nanjing. In order to make the Islamic astronomical books accessible to Chinese scholars, he assigned the task of translating some of them to a team of scholars consisting, among others, of two members of the Imperial Academy (*Hanlin*) named WU Bozong and LI Chong, and the "Islamic Great Master" Mashayihei (probably a transliteration of

Ma-sheikh (Ma being a common surname among Muslims in China), but not known from Arabic or Persian sources). It can be assumed that the original work was in Persian, not only because the Mongol contacts with the Islamic world went mainly through Iran, but also because the Chinese transliterations of month names and the names of the days of the week which are found in one version of the *Huihui li*, can be seen to be from the Persian rather than from the Arabic (the same holds for the transliterations of titles of books and names of instruments found in other sources). The translation of the *Huihui li* was finished and printed in the sixteenth year of the reign of Hong Wu, AD 1383.

The *Huihui li* now exists in three different versions, which have been studied by Japanese and Chinese scholars, most importantly by YABUUTI Kiyosi and CHEN Jiujin.[6] In each of the versions we find roughly the same extensive set of mathematical tables and brief explanations for their use. The explanatory text is generally straightforward and provides hardly any clues concerning the origin of the *Huihui li*. One of the more interesting parts of the text is the description of a type of second-order interpolation which was common among Muslim astronomers and was explained and used in many Arabic and Persian *zījes*, in particular from the Mongol period (cf. Hamadanizadeh 1979).

Although the Ptolemaic character of the astronomy contained in the *Huihui li* was recognized at an early stage, a more detailed investigation of its relation to Islamic astronomical works in Arabic and Persian has only recently been started by YANO Michio and the present author. Yabuuti (1997, 13 and 24) assumed that the *Īlkhānī Zīj* by Naṣīr al-Dīn al-Ṭūsī, compiled between 1270 and 1274 at the newly founded observatory in Maragha, was the most natural candidate for containing the material on which the *Huihui li* was based. However, our investigations (the details of which have not yet been published) have now shown that the planetary parameters in the *Huihui li* are generally different from those in around 25 Arabic and Persian *zījes* that we consulted, including the *Īlkhānī Zīj*. In fact, besides the Tibetan *Sanjufīnī Zīj* (see below) which is clearly directly related to the *Huihui li*, only the *Masᶜūdic Canon* by al-Bīrūnī has been shown to have one single table (for the first equation of Mercury, see Yano's contribution to these proceedings) in common with the *Huihui li*. Various tables in the *Huihui li* were set up in an unusual way; the best examples hereof are the double-argument tables for planetary latitudes, from which the latitudes could be taken directly without any need of calculation. Furthermore, the method of predicting eclipses is very different from the standard Ptolemaic/Islamic method, but is not typically Chinese either.[7]

From the above we conclude that the contents of the *Huihui li* derive to a large extant from original work done at the Islamic Astronomical Bureau in Yuan China. Although CHEN Jiujin (1997) argues that some tables based on the value of the geographical latitude of Nanjing derive from observations made in the early Ming Dynasty by Mashayihei, it seems safe to assume that the main part of the underlying observational work was done in what appears to have been the most active period of the Islamic Astronomical Bureau, namely the last third of the 13th century. This is confirmed by information from two Arabic sources related to the *Huihui li*, which will be discussed below.

Two of the existing three versions of the *Huihui li* contain a star table with longitudes, latitudes and magnitudes for 277 stars within a band of 10 degrees on either side of the ecliptic. The stars are indicated by their numbers within the Ptolemaic constellations, i.e. following the system used in the *Almagest*, as well as by their positions in the traditional

Chinese constellations. However, different from the fast majority of Islamic star tables, the coordinates were *not* derived from those of Ptolemy by simply adding a constant correction for precession to the longitudes, but were newly observed. CHEN Jiujin (1996, 131–41) and, most recently, SHI Yunli (to appear), have made plausible that these observations were carried out in the early Ming Dynasty rather than during the Yuan. Van Dalen (2000) contains a more complete edition of the star table than found in Yabuuti (1954) with improved identifications of stars which, often due to mistakes in the Arabic transmission of the *Almagest*, could not be correctly located by the Muslim astronomers in China.

Besides YANO Michio and the present author, extensive research on the *Huihui li* has in recent years been carried out by SHI Yunli, who concentrated in particular on the reception and adaptation of Islamic astronomy in Korea. Since in the course of these researches a number of new sources for the *Huihui li* have been located, it seemed useful to present here an overview of all sources currently available. We will concentrate on the various versions of the *Huihui li* and sources directly related to it. An influential reworking from the first half of the 15th century is briefly discussed as the last source to be treated in this article.

The Sources for the Huihui li

The Complete Books of the Four Branches of Learning

The most complete and probably most original version of the *Huihui li* that is still extant occurs in an enormous compilation of books produced under emperor Qian Long of the Qing Dynasty between 1773 and 1782. This compilation, called the *Siku quanshu* (*Complete Books of the Four Branches of Learning*), was the result of an effort to gather from all over China the most important books available at the time. A total of 10,000 works were collected and included in an index with brief descriptions; around 3500 were copied in full into the total of seven manuscript versions of the *Siku quanshu* that were prepared. Most of these have been destroyed, but facsimile editions of two extant copies are available.[8]

According to the index of works collected for the project, three copies of the *Huihui lifa*, each in four volumes, were obtained (see Shi Yunli, to appear, note 9, and Wu Weizu (1980), 54, 115 and 271). Although in each case the author is said to be BEI Lin, vice-director of the Astronomical Bureau of the Ming Dynasty in Nanjing in the second half of the 15th century, only two of the three copies are recorded under the name *Huihui lifa*. The third copy originally did not have any title at all and was named *Qizheng tuibu* (*Calculation of the Motion of the Seven Luminaries*) by the editors.[9] This copy may have been the main source for the version included in the *Siku quanshu*, since both the title and the number of chapters (seven) agree. Of one of the two versions listed in the index as *Huihui lifa*, part of the colophon is quoted, which is very similar to that included in the *Siku quanshu*. It tells us that BEI Lin restored the *Huihui li* between 1470 and 1477 in order to prevent it from being lost completely. Even though it is not its original title, for the sake of convenience we will continue to refer to the version of the *Huihui li* in the *Siku quanshu* as "Qizheng tuibu".

In early 1997, a manuscript from the Tohoku University Library in Sendai (Japan) was brought to the attention of Professor Yano by Mr. YOKOTSUKA Hiroyuki. This manuscript, no. 4226, contains a neat copy of the text of the *Qizheng tuibu* without the tables, as well as a copy of the *Mingyi tianwen shu*, a Chinese translation of an Islamic astrological work written by the Persian astronomer Kūshyār ibn Labbān (c. 1000). The latter translation was made by the same team of scholars and in the same period as the translation of the *Huihui li* (cf. above).[10] A marginal note in the Tohoku manuscript dated 1774 indicates that six paragraphs missing from the text of the *Qizheng tuibu* were supplemented (presumably from one of the two other copies listed in the index) by the editors of the *Siku quanshu*. This makes it plausible that the Tohoku manuscript, which was probably copied by Japanese scholars in either China or Korea during the Japanese occupation, is based on the so-called "rough manuscript" of the *Siku quanshu* which was presented to the Imperial Academy after the compilation had been completed (cf. Mayers 1878, 296). In fact, although the text in the Tohoku manuscript is almost identical to the *Qizheng tuibu*, it contains a few improvements over the latter.[11]

Due in particular to the efforts of SHI Yunli, four more versions of the *Huihui lifa* have recently been located. A copy in the National Library in Beijing (without tables) is indicated to have been printed in the year 1383 and hence is a candidate for containing the original text of the *Huihui li*. The first part of its preface is missing, but the remaining part is identical with part of the introduction to the *Mingyi tianwen shu*, which was written by one of the translators, WU Bozong. The main text of the Beijing copy of the *Huihui lifa* is almost identical to that of the *Qizheng tuibu*, but since the last two or three pages, including possibly the colophon, are missing, we have no means to verify whether we are not simply dealing with another version of the reworking by BEI Lin (BEI Lin's name does not appear elsewhere in the text). If the Beijing version in fact contains the original text of the *Huihui li*, we must assume that BEI Lin's restoration activities consisted mainly of collecting *Huihui li* materials and possibly recompiling some of the tables, but did not involve rewriting the explanatory text. Furthermore, it may be possible that the *Huihui li* and the *Mingyi tianwen shu* were published together with a single introduction.[12]

During the years 2000 and 2001, three more copies of the *Huihui lifa* were inspected by SHI Yunli and the present author. A copy at the Interior Department (Naikaku Bunko) of the National Archives of Japan (Kokuritsu Kōbunshokan) in Tokyo was found to contain the original printing of Bei Lin's reworking of 1477 and to differ in only very few places from the *Qizheng tuibu*. Copies in the Nanjing Library and in the *Zichuan xueshi yishu* (*Collected works of Mr. Xue from Zichuan*) by the Qing scholar XUE Fengzuo (1628–1680) (see Ding and Zhou (1956), supplement (*buyi*), 23b) contain annotations and reworkings by Qing scholars.

The Ming Dynastic history

The *Mingshi*, the official history of the Ming Dynasty, was printed in the year 1739, but the astronomical parts had been drawn up by WU Renchen in the 17th century and were corrected by the famous scholar MEI Wending (1633–1721). Besides the text of the *Datong li*, the official astronomical system of the Ming, the section on astronomy includes a work in three chapters called the *Huihui lifa*. A detailed comparison of text and tables in this work with the *Qizheng tuibu* shows that the former is basically an abbreviated version of the latter. The instructions in the *Mingshi* for using the tables can often be obtained from those in the *Qizheng tuibu* by removing repetitive parts of sentences. At some places in the *Mingshi* brief comments were added relating the Ptolemaic/Islamic terminology of the *Huihui li* to that of Chinese astronomical systems; a section concerning calendar conversion which does not occur in the *Qizheng tuibu*, was recently shown to derive from the work of the 15th-century scholar TANG Shunzhi, who made extensive studies of the *Huihui li* (Shi Yunli, to appear).

Also the tables in the *Mingshi* version of the *Huihui li* were as much as possible compressed. The tables for the mean motions of the planets, which display linear functions, were described in prose rather than given in tabular form in their entirety. The tables for the planetary equations were reduced to half their original size by making use of the fact that the tabular values in certain quadrants of the argument are identical to those in other quadrants. Furthermore, the tabular differences, used for performing linear interpolation in the tables, were left out. From the textual descriptions of the above-mentioned changes it appears that the original tables consulted by the editors of the *Mingshi* must have had precisely the same format as those in the *Qizheng tuibu*. The star table is completely missing from the *Mingshi* version, alhough it is mentioned in the section on so-called "encroachments" of stellar constellations by the moon and the planets (*ling-fan*).

The Outer Book on the Calculation of the Seven Luminaries

Under King Sejong (1419–1451) a number of astronomical works were brought to Korea from Ming China and were studied in detail by Korean astronomers. Among these works were the *Huihui li* and, as has recently been shown (Shi Yunli, to appear), the *Xiyu lifa tongjing* ("Canon of the Western Astronomical System"), an early fifteenth-century reworking of the *Huihui li* by LIU Xin (see below). In 1442, YI Sunji and KIM Tam finished the *Chiljŏngsan Oepyŏn* ("Outer Book on the Calculation of the Seven Luminaries"), which is basically an adaptation of the *Huihui li* for use in Korea.[13] The differences from the version extant as *Qizheng tuibu* are numerous. In particular, the *Chiljŏngsan* has been given the structure of a traditional Chinese astronomical system: the underlying parameters are listed at the beginning of the text, eclipses are discussed immediately after the solar and lunar theory instead of at the very end, and text and tables are intermingled instead of distributed over separate chapters. An algorithm for dealing with conversion from the Persian solar calendar to the Islamic lunar calendar was added, and the planetary mean positions were corrected for a change in epoch and in geographical longitude.

In spite of these changes, the tables in the *Chiljŏngsan* are basically identical to those in the *Qizheng tuibu* and, in fact, provide better readings for the tabular values from the original *Huihui li* than the other available sources. The *Chiljŏngsan* now exists in two different prints: as a separate work (reprinted in the series *Hanguk kwahak kisulsa charyo taigye*, vol. 4), and in chapters 159–163 of the *Sejong Sillok* (*Veritable Records of King Sejong*), the day-to-day annals of the king's reign. The *Chiljŏngsan*, which is 35 years earlier than BEI Lin's restoration of the *Huihui li*, also includes the star table contained in the *Qizheng tuibu*. The brief note attached to the title of this table describes a correction of the longitudes of the stars for precession and seems to imply that the coordinates were originally determined for the year 1391, almost ten years after the *Huihui li* was translated.

A manuscript from the Pulkovo library

In an article in the journal *Copernicus* of 1882, A. Wagner describes an Arabic or Persian manuscript in the holdings of the library of the Pulkovo Observatory near St. Petersburg (Russia). This manuscript, consisting only of numerical tables, had been obtained in China together with a Chinese manuscript. In particular the description of the layout of the tables and the sample tabular values that Wagner presents, leave no doubt that the Pulkovo manuscript contains the same tables that we also find in the *Huihui li*. Since on paleographical grounds the manuscript has been dated to the twelfth or thirteenth century, it seems probable that it contains part of the original astronomical work which was compiled by the Muslim astronomers in the service of Khubilai Khan in the early Yuan Dynasty. Unfortunately the present whereabouts of the manuscript are unknown. If it was still at the Pulkovo library at the time of the fire in February 1997, it may very well have been destroyed. However, it is also possible that it was transfered to the Academy of Sciences in St. Petersburg at an earlier stage.[14]

The same holds for the Chinese manuscript that was obtained at the same time as the Arabic or Persian one. From the German translation of the Chinese title given by Wagner ("Anleitung zur westlichen Astronomie") and two Latin comments stemming from a Jesuit, it becomes clear that also this manuscript contains material related to the *Huihui li*. Whether it was a draft translation of the Arabic or Persian work compiled in the early Yuan Dynasty, a copy of the translation prepared in the early Ming, or possibly part of the *Xiyu lifa tongjing*, can only be determined when the manuscript is rediscovered.

The Sanjufīnī Zīj

The Arabic manuscript 6040 of the Bibliothèque Nationale in Paris is a unique copy of a *zīj* compiled in 1366 by the otherwise unknown scholar al-Sanjufīnī. Hailing from the area around Samarkand in present-day Uzbekistan, al-Sanjufīnī dedicated his work to the Mongol vice-roy of Tibet. Librarian's notes and folio numbers in Chinese, translations of titles of tables into Mongolian, and transcriptions of month names into Tibetan make the so-called *Sanjufīnī Zīj* a particularly interesting topic of study. The work was already noticed by Wagner (p. 128) in his article in *Copernicus* discussed in the previous section, and Sarton included a facsimile of the title page in his *Introduction to the History of Science* (vol. 3, part 1, 1529–31). However, only in the last 15 years E.S. Kennedy has initiated more detailed studies of the work. H. Franke investigated the Mongol glosses and found Yongchang-fu (in the northeastern part of the original Tibet, close to Lanzhou) as the most probable place of compilation. Kennedy translated a procedure for predicting eclipses, and he, J.P. Hogendijk and the present author investigated technical aspects of some of the tables in the *Sanjufīnī Zīj*.[15]

During my research on the *Huihui li* at Kyoto Sangyo University (1995–97), I found that the *Sanjufīnī Zīj* has a large number of tables in common with the *Huihui li*. These include the tables for lunar and planetary equations, lunar latitude, planetary stations, the equation of time, oblique ascension (i.e. the rising time of an ecliptic arc), and a table for parallax. Furthermore, it can be shown that the tables in the *Sanjufīnī Zīj* giving planetary (mean) positions at the days of consecutive vernal equinoxes were derived from the mean motion tables found in the *Huihui li*, whereas the double-argument tables for planetary latitude in the *Huihui li* are based on the same set of planetary parameters as the latitude tables of standard-Ptolemaic type in the *Sanjufīnī Zīj*. Thus it seems that the two works had a common predecessor, which I assume was the original astronomical handbook compiled by the astronomers at the Islamic Observatory in the early Yuan Dynasty. In fact, the planetary tables in the *Sanjufīnī Zīj* are said to be based on the "Jamālī observations", which could very well refer to Zhamaluding, the first director of the observatory.

A Reworking of the Huihui li by Liu Xin (ca. 1430)

It has recently been shown (Shi Yunli, to appear) that the Korean astronomers who adapted the *Huihui li* for use in Korea had a second source at their disposal, namely a reworking called *Xiyu lifa tongjing* ("Canon of the Western Astronomical System") by a certain LIU Xin, who worked at the Astronomical Bureau of the Ming Dynasty in Beijing and died in 1449. A small number of chapters from LIU Xin's work are extant in the Beijing National Library and display various obvious differences from the *Qizheng tuibu*. In the first place, the fragment contains tables that are not part of the original work and display functions that otherwise had to be calculated. The text and tables are intermingled as in the *Chiljŏngsan*, and all intermediate results of the calculations are entered in numbered boxes in tables for easy later reference.

The existence of LIU Xin's *Xiyu lifa tongjing* may also explain some of the contradictory historical information given by BEI Lin in the colophon of the *Qizheng tuibu*. BEI Lin writes that the *Huihui li* was brought to China by a foreigner in the first years of the Ming Dynasty and then converted to Chinese computational methods by YUAN Tong, a high officer of the Astronomical Bureau. In fact, as SHI Yunli explains, YUAN Tong is the author of the *Datong lifa tonggui* ("Canon for the Datong Astronomical System", 1384), a work similar to the *Xiyu lifa tongjing* in structure, explaining the astronomical system of the Ming Dynasty. It seems plausible that BEI Lin confused the *Tonggui* and the *Tongjing* and mistook LIU Xin for the much more famous YUAN Tong. CHEN Jiujin (1997, 105) assumes that the "foreigner" mentioned above was Madeluding, father of Mashayihei and likewise an able astronomer. It seems possible that at least some parts of the translated *Huihui li* were in fact brought from the Islamic world around 1370 and were not contained in the original astronomical work compiled by the Muslim astronomers active in the Yuan Dynasty.

Concluding Remarks

After the earlier fundamental investigations of Islamic astronomy in China by YABUUTI Kiyosi and CHEN Jiujin, a new impulse has recently been given by research on the *Huihui li* carried out by YANO Michio, SHI Yunli and the present author. This research concentrates on the following three questions, which have not yet been answered satisfactorily:

(1) What was the extent of the activities of Muslim astronomers in China in the early Yuan Dynasty? Which of the various materials available in the sources described in this article can be assumed to have been part of the original astronomical handbook compiled by those astronomers?

(2) What were the new contributions of Muslim and possibly Chinese astronomers in the early Ming Dynasty to the translation now known as the *Huihui li*?
(3) Which of the materials available in the Chinese sources listed above were contained in the original translation of 1383?

It can be expected that a thorough investigation of all sources discussed in this article will provide us with much more complete answers to the above questions than can now be given. The present author expects to publish an edition and translation of the text of the *Qizheng tuibu* and a transcription of its tables in the near future. Furthermore, an investigation of the text and tables in the *Sanjufīnī Zīj* is in progress. Efforts are being made to obtain copies of the Chinese sources that have not yet been consulted and, in particular, to discover the present location of the manuscripts from the Pulkovo Library.

Acknowledgments

I would like to thank the Japan Society for the Promotion of Science (JSPS) and the Dibner Institute for the History and Technology of Science (Cambridge, Mass.) for their financial support of my research on Islamic astronomy in China. I am particularly grateful to Professor YANO Michio (Kyoto Sangyo University, Japan) for suggesting me to work on the *Huihui li*, and for his continued help on almost every aspect of the project. My frequent discussions with Dr. SHI Yunli (University of Science and Technology, Hefei, China) during our simultaneous research visits in Cambridge (Mass.) have been extremely valuable and pleasant.

Notes

1. Extensive information on the astronomical observatory in Maragha can be found in Sayılı (1960), chapter 6. The name of the Chinese astronomer who worked at the observatory is mentioned by the historian Banākatī and the astronomer al-ʿUrḍī; see Boyle (1963), 253, note 4. The use of the Chinese-Uighur calendar in the Iranian world is described in Melville (1994), its technical details are explained in van Dalen et al. (1997).
2. The Chinese characters for Chinese, Korean and Japanese names of persons and titles of historical books are listed in a separate section at the end of this article.
3. The Chinese character *li* which is here conveniently translated as "calendar", usually denotes a mathematical-astronomical system by means of which solar, lunar and planetary positions and the time and magnitude of eclipses can be calculated. Thus a *li* is in various respects similar to an Islamic work of the type "zīj" described below.
4. All Chinese, Korean and Japanese names of persons are written in their normal order, i.e. with the surname preceding the first name. In order to avoid confusion, the surnames are written in small caps, except in bibliographical references.
5. The Persian word *zīj* originally meant "thread" or "chord" and came to stand for "astronomical handbook" due to the similarity of the numerical tables in such a work to the parallel threads making up the warp in a fabric. The standard work on Islamic *zīj*es is Kennedy (1956), an update of which is currently being prepared by the present author.
6. In this article, references to the work of YABUUTI Kiyosi will mostly be to (1997), the present author's translation of the chapter on Islamic astronomy in China from Yabuuti's major work on Chinese mathematical astronomy (1969, in Japanese). In a slightly less complete form, the same chapter was published as Yabuuti (1964, in Japanese). Yabuuti (1954) contains a survey of Indian and Islamic astronomy in China plus an edition and analysis of the star table in the *Huihui li*. Yabuuti (1987) summarizes the earlier publications

and adds various historical details concerning the Islamic Astronomical Bureau in the Yuan capital. CHEN Jiujin's major work on Islamic astronomy in China is 1996 (in Chinese). He published a large number of articles on the *Huihui li*, mostly in *Studies in the History of Natural Sciences* (*Ziran kexue shi yanjiu*). The English article Chen Jiujin (1997) contains various inaccuracies.

7. The tables for planetary latitudes in the *Huihui li* were described and analysed in Yano (1999) and van Dalen (1999). The method of predicting eclipses is discussed in Yabuuti (1997, 27–30) and Chen Jiujin (1990, in Chinese).
8. The compilation of the *Siku quanshu* is described in detail in Mayers (1878, 291–299). Each of the seven manuscripts was named after the specially erected pavillion in which it was kept. The Wenyuange copy was reproduced in 1500 volumes by the Taipei Trade and Commerce Press (Shangwu yinshuguan, 1983–1986). The Wenlange copy is currently being printed by the Shanghai Ancient Book Press (Guji chubanshe). For more information about the *Siku quanshu* project and its historical and intellectual background, see Guy (1987).
9. The name *Qizheng tuibu* was presumably derived from the title of the most important section of the *Huihui li*, namely *Qizheng jingweidu fa*, "Method of [calculating] the longitude and latitude of the seven luminaries".
10. The astrological book of which the *Mingyi tianwen shu* is a translation, was identified as Kūshyār's *Madkhal fī Ṣināʿat Aḥkām al-Nujūm* by Tasaka (1957, 108) and IMAI Itaru (private publication). Yabuuti (1969, 235–42) made the first comparison of Arabic original and Chinese translation. Recently an edition and translation of the Arabic work was published together with the Chinese text by Yano (1997).
11. Since no traces of the *Mingyi tianwen shu* can be found in the index of books collected for the *Siku quanshu* or one of the catalogues deriving from the project, it seems that this work was not available to the editors. Thus only the part of the Tohoku manuscript containing the *Huihui li* could have been copied from the rough manuscript of the *Siku quanshu*.
12. The copy of the *Huihui lifa* in the National Library of Beijing has recently been reproduced in the *Xuxiu Siku quanshu* (Shanghai: Guji chubanshe, 1995–, vol. 1036), a collection of works that were not included in the *Siku quanshu* itself. The extant part of the Beijing copy has meanwhile been shown to be identical with the copy at the National Archives of Japan (cf. p. 23), i.e., it is part of the reworking by Bei Lin of 1477 rather than of the original translation of 1383. The same volume of the *Xuxiu Siku quanshu* contains another copy of the *Huihui lifa* which is kept in the library of Beijing University. It is very probably the original from which the Tohoku manuscript discussed on pp. 22–23 was made and includes the same marginal note by the editors of the Siku quanshu.
13. The *Chiljŏngsan Naepyŏn*, "Inner Book on the Calculation of the Seven Luminaries" contains a reworking of the *Datong li*, the astronomical system of the Ming Dynasty.
14. After this article had been sent to press, it was found that the two manuscripts from the library of the Pulkovo Observatory had been brought into safety at the Academy of Sciences in St. Petersburg in 1939. From there they were transferred to the Oriental Institute in that city in 1943. The Arabic / Persian manuscript, apparently dating from the time of the Chinese translation of the *Huihui lifa* (ca. 1380) is catalogued as C 2460, but the Chinese manuscript is not found in any catalogue or inventory.
15. See Franke (1988), Kennedy (1987/88), Kennedy and Hogendijk (1988), and van Dalen (1989).

References

Boyle, John A. (1963). "The Longer Introduction to the 'Zij-i Ilkhani' of Nasir-ad-Din Tusi," *Journal of Semitic Studies* 8, 244–54.

Chen Jiujin (1990), "Huili riyueshi yuanli" (The Solar and Lunar Eclipse Principle of the Islamic calendar, in Chinese), *Ziran kexue shi yanjiu* (*Studies in the History of Natural Sciences*) 9, 119–31.

Chen Jiujin (1996). *Huihui tianwenxue shi yanjiu* (Investigation of the History of Islamic Astronomy, in Chinese). Nanning: Guangxi kexue jishu chubanzhe.

Chen Jiujin (1997). "Comparative Research between the Huihui Calendar, Chiljŏngsan Oepiŏn and Qizheng tuibu," *Oriental Astronomy from Guo Shoujing to King Sejong* (eds. Il-Seong Nha and F. Richard Stephenson). Seoul: Yonsei University Press, 105–11.

Dalen, Benno van (1989). "A Statistical Method for Recovering Unknown Parameters from Medieval Astronomical Tables," *Centaurus* 32, 85–145.

Dalen, Benno van, E.S. Kennedy, and Mustafa K. Saiyid (1997). "The Chinese-Uighur Calendar in Ṭūsī's Zīj-i Īlkhānī," *Zeitschrift für Geschichte der arabisch-islamischen Wissenschaften* 11, 111–52.

Dalen, Benno van (1999). "Tables of Planetary Latitude in the *Huihui li* (II)," *Current Perspectives in the History of Science in East Asia* (eds. Yung Sik Kim and Francesca Bray). Seoul: Seoul National University, 316–29.

Dalen, Benno van (2000). "A Non-Ptolemaic Islamic Star Table in Chinese", in: *Sic itur ad astra. Studien zur Geschichte der Mathematik und Naturwissenschaften. Festschrift für Paul Kunitzsch zum 70. Geburtstag* (eds. Menso Folkerts and Richard P. Lorch), Wiesbaden (Harrassowitz), 147–76.

Ding Fubao and Zhou Yunqin (1956). *Sibu zonglu tianwen bian* (Universal Bibliography, astronomy section, in Chinese). Shanghai: Shangwu yinshuguan. New edition 1984, Beijing: Wenwu chubanshe.

Franke, Herbert (1988). "Mittelmongolische Glossen in einer arabischen astronomischen Handschrift," *Oriens* 31, 95–118.

Hamadanizadeh, Javad (1979). "Interpolation Schemes in *Dastūr al-Munajjimīn*," *Centaurus* 22, 44–52.

Hartner, Willy (1950). "The Astronomical Instruments of Cha-ma-lu-ting, their Identification, and their Relations to the Instruments of the Observatory of Marāgha," *Isis* 41, 184–195. Reprinted in: Willy Hartner (1968). *Oriens Occidens*. Hildesheim: Olms, 215–26.

Kennedy, Edward S. (1956). "A Survey of Islamic Astronomical Tables," *Transactions of the American Philosophical Society* N.S. 46–2, 123–77. Reprinted in 1989 with different pagination.

Kennedy, Edward S. (1987/88). "Eclipse Predictions in Arabic Astronomical Tables Prepared for the Mongol Viceroy of Tibet," *Zeitschrift für Geschichte der arabisch-islamischen Wissenschaften* 4, 60–80. Reprinted in: E.S. Kennedy (1998). *Astronomy and Astrology in the Medieval Islamic World.* Aldershot UK: Ashgate, chapter XIV.

Kennedy, Edward S. and Jan P. Hogendijk (1988). "Two Tables from an Arabic Astronomical Handbook for the Mongol Viceroy of Tibet," *A Scientific Humanist, Studies in Memory of Abraham Sachs* (eds. Erle Leichty, Maria de J. Ellis, e.a.). Philadelphia: The University Museum, 233–42. Reprinted in: E.S. Kennedy (1998). *Astronomy and Astrology in the Medieval Islamic World.* Aldershot UK: Ashgate, chapter XIII.

Guy, R. Kent (1987). *The Emperor's Four Treasuries. Scholars and State in the Late Ch'ien-lung Era.* Cambridge (Mass.): Harvard University Press.

Mayers, W.F. (1878). "Bibliography of the Chinese Imperial Collections of Literature," *China Review* 6, 213–23 and 285–99.

Melville, Charles (1994). "The Chinese Uighur Animal Calendar in Persian Historiography of the Mongol Period," *Iran* 32, 83–98.

Miyajima Kazuhiko (1982). "*Genshi* tenmonshi kisai no isuramu tenmongiki ni tsuite" (New identification of Islamic astronomical instruments described in the Yuan dynastical history, in Japanese), *Tōyō no kagaku to gijutsu* (Science and Skills in Asia). Kyoto: Dōhōsha, 407–27.

Needham, Joseph (1959). *Science and Civilisation in China, vol. 3: Mathematics and the Sciences of the Heavens and the Earth.* Cambridge: Cambridge University Press.

Sayılı, Aydin (1960). *The Observatory in Islam and its Place in the General History of the Observatory.* Ankara: Turkish Historical Society. Reprinted in 1981 (New York: Arno Press).

Shi Yunli (to appear). "The Korean Adaptation of the Chinese-Islamic Astronomical Tables," *Archive for History of Exact Sciences.*

Tasaka Kōdō (1957). "An Aspect of Islam Culture Introduced into China," *Memoirs of the Research Department of the Toyo Bunko* 16, 75–160.

Wagner, A. (1882). "Ueber ein altes Manuscript der Pulkowaer Sternwarte," *Copernicus* 2, 123–29.

Wu Weizu ed. (1980). *Siku caijin shumu* (A List of the Books Collected for the Compilation of the *Siku quanshu*). Shanghai: Shangwu yinshuguan.

Yabuuti Kiyosi (1954). "Indian and Arabian Astronomy in China," *Silver Jubilee Volume of the Zinbun-Kagaku-Kenkyusyo.* Kyoto: Kyoto University, 585–603.

Yabuuti Kiyosi (1964). "Kaikaireki kai" (Commentary on the Huihui li, in Japanese), *Tōhō Gakuhō* 36, 612–32.

Yabuuti Kiyosi (1969). *Chūgoku no tenmon rekihō* (Chinese astronomical calendar systems, in Japanese). Tokyo: Heibonsha. Second edition with only minor changes: 1990.

Yabuuti Kiyosi (1987). "The influence of Islamic astronomy in China," *From Deferent to Equant* (eds. David A. King and George A. Saliba). New York: The New York Academy of Sciences, 547–59.

Yabuuti Kiyosi, translated and partially revised by Benno van Dalen (1997). "Islamic Astronomy in China during the Yuan and Ming Dynasties", *Historia Scientiarum* 7, 11–43.

Yamada Keiji (1980). *Juji-reki no michi* (Way to the Shoushi li, in Japanese). Tokyo: Misuzu Shobō.

Yano Michio (1997). *Kūšyār Ibn Labbān's Introduction to Astrology*. Tokyo: Institute for the Study of Languages and Cultures of Asia and Africa.

Yano Michio (1999). "Tables of Planetary Latitude in the *Huihui li* (I)," *Current Perspectives in the History of Science in East Asia* (eds. Yung Sik Kim and Francesca Bray). Seoul: Seoul National University, 307–15.

Yano Michio (2002). "The First Equation Table for Mercury in the *Huihui Li*," *in this Volume*, p. 31 ff.

Chinese characters

Historical persons

貝琳	Bei Lin	乾隆	Qian Long
-	Fu Mengchi	世祖	Sejong
郭守敬	Gou Shoujing	唐順之	Tang Shunzhi
洪武	Hong Wu	呉伯宗	Wu Bozong
金淡	Kim Tam	呉任臣	Wu Renchen
李翀	Li Chong	薛鳳祚	Xue Fengzuo
劉信	Liu Xin	元統	Yuan Tong
馬沙亦黒	Mashayihei	李純之	Yi Sunji
梅分鼎	Mei Wending	札馬魯丁	Zhamaluding

Historical works

七政算内篇	*Chiljŏngsan Naepyŏn*	世祖實録	*Sejong Sillok*
七政算外篇	*Chiljŏngsan Oepyŏn*	授時暦	*Shoushi li*
大統暦	*Datong li*	四庫全書	*Siku quanshu*
大統暦法通軌	*Datong lifa tonggui*	通軌	*Tonggui*
回回暦	*Huihui li*	通徑	*Tongjing*
回回暦法	*Huihui lifa*	萬年暦	*Wannian li*
明史	*Mingshi*	西域暦法通徑	*Xiyu lifa tongjing*
明譯天文書	*Mingyi tianwen shu*	淄川薛氏遺書	*Zichuan xueshi yishu*
七政推步	*Qizheng tuibu*		

Modern scholars

陳久金	Chen Jiujin	呉慰祖	Wu Weizu
丁福保	Ding Fubao	藪内清	Yabuuti Kiyosi
今井溱	Imai Itaru	山田慶児	Yamada Keiji
宮島一彦	Miyajima Kazuhiko	矢野道雄	Yano Michio
石云里	Shi Yunli	横塚啓之	Yokotsuka Hiroyuki
田坂興道	Tasaka Kōdō	周云青	Zhou Yunqi

Other

簡儀	*jianyi* (Simplified Instrument)
曆	*li* (calendar)
凌犯	*ling-fan* (encroachments)
七政經緯度法	*qizheng jingweidu fa*
周公測景臺	Tower of the Duke of Zhou

1.3. The First Equation Table for Mercury in the *Huihui li*

Michio Yano*

1. Introduction

The structure of the *Huihui li*, a Chinese Islamic calendar originally compiled in A.D. 1383,[1] and the basic theory behind it were investigated by Kiyosi Yabuuti in his pioneering works.[2] However, we are not yet sure on which Islamic sources the *Huihui li* was actually based.[3]

The main body of the Arabic literature called *zīj* is astronomical tables together with the instructions how to use these tables. Very few *zīj*es give theoretical explanation of how these tables were prepared. Proofs are rarely given. In this sense we can call the *Huihui li* a typical *zīj*. What is recorded in the *Huihui li* are only astronomical tables and very brief explanations of how to use them. Nothing theoretical is found throughout the text. In order to discover the theory or the algorithm by which the tables were constructed, we can first hypothesize the theory or algorithm from the tables and, in turn, reconstruct the tables from the hypotheses. Only when the reconstructed tables agree with the actual tables to a sufficient degree we can say that the hypotheses were correct. It is only after such preliminary studies that we can guess the sources of the *Huihui li* and investigate its relation to Arabic and Persian astronomical texts.

The present author has been dealing with this problem with Benno van Dalen.[4] We read a joint paper concerning the tables of the planetary latitude at the 8th International Conference on the History of Science in East Asia, August 26–31, 1996, Seoul, Korea.[5] As van Dalen already suggested on that occasion, the *Huihui li* was very closely related to the *Sanjufīnī Zīj*[6] which was prepared by a certain Abū Muḥammad ʿAṭā ibn Aḥmad ibn Muḥammad Khwāja Ghāzī al-Samarqandī al-Sanjufīnī in 1366 for the Mongol Viceroy of Tibet.[7]

The aim of the present paper is to point out the peculiar nature of the first equation table for Mercury in the *Huihui li* and to show that this peculiarity was shared not only by the *Sanjufīnī Zīj* but also by al-Bīrūnī in his *al-Qānūn al-Masʿūdī*.[8]

The reader of this paper is assumed to be familiar with the planetary theory of the Ptolemaic system.

2. The First Equation of Planets

In the case of planets except Mercury, the geometrical model by which the equation tables of the *Huihui li* was computed seems to have been similar to that of the *Almagest*.[9] Concerning the structure of the equation tables in the *Huihui li*, the following two points are worth mentioning.

---

* This is a slightly revised version of my paper published in the *Memoirs of the International Institute for Linguistic Sciences*, Kyoto Sangyo University, No. 1 (March 1999).

S.M.R. Ansari (ed.), History of Oriental Astronomy, 33–43.

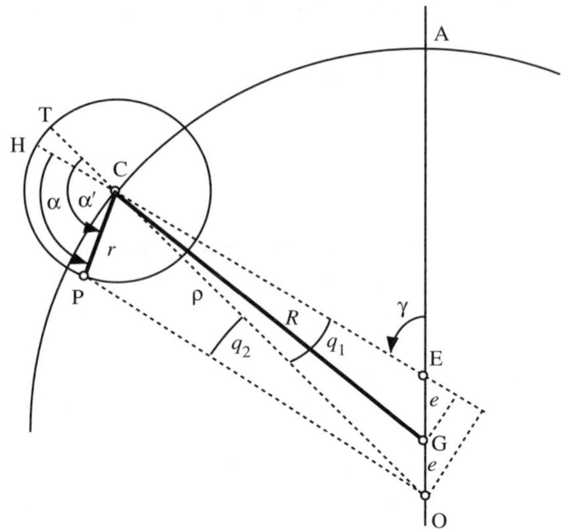

Figure 1 Planets except Mercury: q_1 and q_2.

1. The first equation (q_1 in Fig. 1), or the equation for the centre of the epicycle, was separated into two components in the *Almagest*. In the planetary equation tables the two components are arranged as column 3 and column 4. Following Neugebauer[10] I express the two sets of tabular values as c_3 and c_4, respectively.

As is shown in Fig. 2 the component c_3 ($= \angle$ODE) is the equation which would be produced if the equant (E) were the centre of the deferent, and thus the centre of the epicycle were carried on it (as the dotted circle in Fig. 2).

Actually, however, the centre of the deferent (with the radius R) carrying the epicycle C is G. The correction due to the shift from E to G is tabulated as c_4 ($= \angle$COD). Thus the first equation (q_1) is obtained by combining the two columns:

$$q_1 = c_3 + c_4.$$

Even though the first equation can be easily obtained by direct computation,[11] Ptolemy kept the two components separate 'for didactic reasons because he wanted the reader to see how c_3' (q_1 in our notation) had been obtained . . . ', as Neugebauer puts it.[12]

For the later astronomers whose aim was simply to prepare practical tables, there was no reason to provide the two separate columns, and it was natural that they tabulated only q_1. This was the case with Theon's *Handy Tables*,[13] which is originally ascribed to Ptolemy, as well as the Arabic and Persian *zījes* and the *Huihui li*.

2. The sixth column of the equation tables in the *Almagest* gives the second equation (q_2 in Fig. 1 and Fig. 3), or the equation due to anomalistic motion of the planets. This column was prepared for the special case when the centre of the epicycle is at the mean distance on the deferent, and the differences which might be produced at the farthest distance and the nearerest distance are listed in the fifth column (negative) and the seventh column (positive), respectively. For the cases when the planet is in between the mean distance and

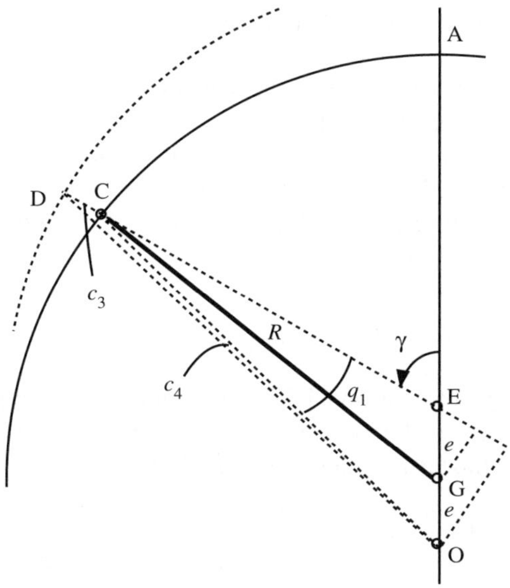

Figure 2 Planets except Mercury: c_3 and c_4.

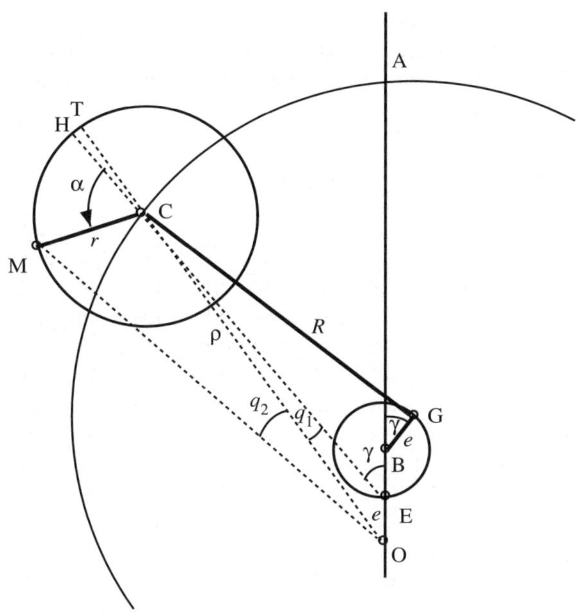

Figure 3 Mercury: q_1 and q_2.

these two extremities, Ptolemy provided interpolation function as the eighth column of the tables.

The *Huihui li*, on the other hand, gives the table of the second equation at the farthest, instead of the mean, distance. Thus what was needed was the table for the difference of equation at the farthest and the nearest distances. In Chinese this is called *yuan-jin du* (遠近度, distance degrees). For the intermediate positions the interpolation function is provided. It is remarkable that this new feature is found in the *Sanjufīnī Zīj*, but nowhere else except in Kūšyār ibn Labbān's *Zīj*.

3. The Special Case for Mercury

In the *Almagest* the geometrical model for computing the first equation of Mercury is different from that for the other four planets. The algorithm which was used in the *Almagest* can be expressed in the modern formula (cf. Fig. 3):[14]

$$\mathrm{EC} = \sqrt{R^2 - \left(2e\cos\frac{\gamma}{2}\sin\frac{3\gamma}{2}\right)^2} + 2e\cos\frac{\gamma}{2}\cos\frac{3\gamma}{2}$$

$$\rho\ (= \mathrm{OC}) = \sqrt{e^2 + \mathrm{EC}^2 + 2e\cdot \mathrm{EC}\cos\gamma}$$

$$q_1 = \arcsin\left(\frac{e\sin\gamma}{\rho}\right)$$

where γ is centrum and e is eccentricity.

As in the case of the other planets, Ptolemy decomposed q_1 into c_3 and c_4 which are shown in Table 1. Note that c_4 is negative when the centrum is less than 60° or greater than 300°. Fig. 4 is drawn for such a case.[15]

Table 1 First equation for Mercury in the *Almagest*

centrum		c_3°	c_4°	centrum		c_3°	c_4°	centrum		c_3°	c_4°
6	354	0;18	-0; 1	93	267	2;52	0;10	138	222	2; 0	0; 4
12	348	0;34	-0; 2	96	264	2;52	0;10	141	219	1;53	0; 4
18	342	0;51	-0; 4	99	261	2;51	0;11	144	216	1;46	0; 3
24	336	1; 7	-0; 5	102	258	2;50	0;10	147	213	1;38	0; 3
30	330	1;22	-0; 5	105	255	2;48	0;10	150	210	1;30	0; 2
36	324	1;37	-0; 4	108	252	2;46	0;10	153	207	1;22	0; 2
42	318	1;51	-0; 4	111	249	2;44	0; 9	156	204	1;13	0; 2
48	312	2; 4	-0; 3	114	246	2;41	0; 9	159	201	1; 5	0; 1
54	306	2;15	-0; 1	117	243	2;37	0; 9	162	198	0;56	0; 1
60	300	2;25	0; 0	120	240	2;33	0; 8	165	195	0;46	0; 1
66	294	2;34	0; 2	123	237	2;28	0; 7	168	192	0;38	0; 0
72	288	2;41	0; 4	126	234	2;23	0; 7	171	189	0;28	0; 0
78	282	2;46	0; 6	129	231	2;18	0; 6	174	186	0;19	0; 0
84	276	2;50	0; 7	132	228	2;12	0; 6	177	183	0; 9	0; 0
90	270	2;52	0; 9	135	225	2; 6	0; 5	180	180	0; 0	0; 0

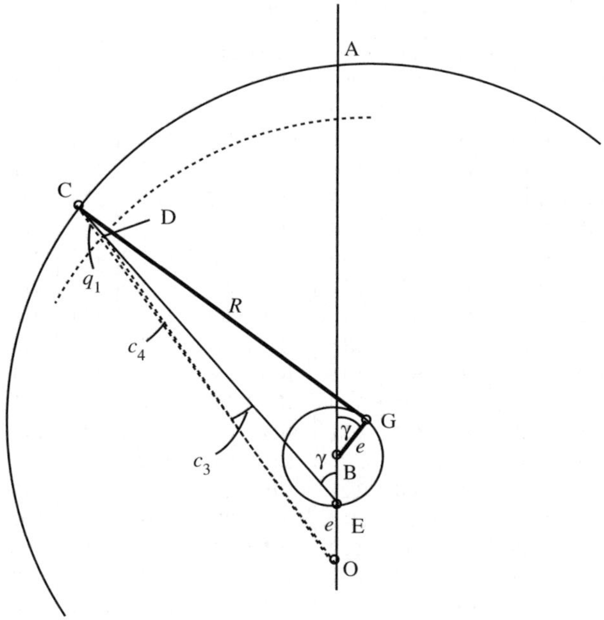

Figure 4 Mercury: c_3 and c_4.

This special treatment for Mercury was inherited by Arabic/Islamic astronomers. For instance, al-Battānī[16] tabulated q_1 for Mercury as is shown in Table 2, without, of course, decomposing it into two components. This table is practically the same as that in Theon's *Handy Tables* and the *Mumtaḥan Zīj* of Abū Manṣūr.[17] That these tabular values might have been derived from Ptolemy's geometrical model is evident when we add c_3 and c_4 of Table 1 and compare the sum with the corresponding values q_1 in Table 2.

The eccentricity (e) for Mercury used by Ptolemy, Theon, Abū Manṣūr, al-Battānī, etc. is 3 for $R = 60$.

4. Peculiarity of the *Huihui li*

The *Huihui li* does not give any separate account for the case of Mercury, but there is no doubt that the table for Mercury was constructed in a different way from that for the other planets.

Strangely enough, however, I found that the table for the first equation (q_1) for Mercury in the *Huihui li* could not be produced by the same computer program that had produced al-Battānī's table for q_1 whichever value of e was input. What is actually given in the *Huihui li* is shown in the eighth column of Table 3. After several attempts of rewriting my program in order to get the tabular values of the *Huihui li*, I happened to find that these values can be produced by subtracting Ptolemy's c_4 from c_3, instead of adding them together! In Table 3 I have shown c_3 (in degrees), c_4 (in minutes), and q_1 which I got using my program with $e = 3$.[18]

Table 2 al-Battānī's first equation for Mercury

centrum		q_1	centrum		q_1	centrum		q_1	centrum		q_1
1	359	0; 3	46	314	1;57	91	269	3; 1	136	224	2; 9
2	358	0; 6	47	313	1;59	92	268	3; 1	137	223	2; 7
3	357	0; 9	48	312	2; 1	93	267	3; 2	138	222	2; 5
4	356	0;12	49	311	2; 4	94	266	3; 2	139	221	2; 2
5	355	0;15	50	310	2; 6	95	265	3; 2	140	220	2; 0
6	354	0;17	51	309	2; 8	96	264	3; 2	141	219	1;57
7	353	0;20	52	308	2;10	97	263	3; 2	142	218	1;55
8	352	0;23	53	307	2;12	98	262	3; 1	143	217	1;52
9	351	0;25	54	306	2;14	99	261	3; 1	144	216	1;49
10	350	0;28	55	305	2;16	100	260	3; 1	145	215	1;47
11	349	0;30	56	304	2;18	101	259	3; 0	146	214	1;44
12	348	0;32	57	303	2;19	102	258	3; 0	147	213	1;41
13	347	0;35	58	302	2;21	103	257	2;59	148	212	1;38
14	346	0;38	59	301	2;23	104	256	2;59	149	211	1;34
15	345	0;40	60	300	2;25	105	255	2;58	150	210	1;32
16	344	0;43	61	299	2;27	106	254	2;58	151	209	1;30
17	343	0;45	62	298	2;29	107	253	2;57	152	208	1;27
18	342	0;47	63	297	2;31	108	252	2;56	153	207	1;24
19	341	0;50	64	296	2;33	109	251	2;55	154	206	1;21
20	340	0;53	65	295	2;35	110	250	2;54	155	205	1;18
21	339	0;55	66	294	2;36	111	249	2;53	156	204	1;15
22	338	0;58	67	293	2;38	112	248	2;52	157	203	1;12
23	337	1; 0	68	292	2;40	113	247	2;51	158	202	1; 9
24	336	1; 2	69	291	2;41	114	246	2;50	159	201	1; 6
25	335	1; 5	70	290	2;43	115	245	2;49	160	200	1; 3
26	334	1; 8	71	289	2;44	116	244	2;48	161	199	1; 0
27	333	1;10	72	288	2;45	117	243	2;46	162	198	0;57
28	332	1;13	73	287	2;47	118	242	2;45	163	197	0;54
29	331	1;15	74	286	2;48	119	241	2;43	164	196	0;51
30	330	1;17	75	285	2;49	120	240	2;41	165	195	0;48
31	329	1;20	76	284	2;50	121	239	2;39	166	194	0;45
32	328	1;23	77	283	2;51	122	238	2;37	167	193	0;42
33	327	1;25	78	282	2;52	123	237	2;35	168	192	0;39
34	326	1;28	79	281	2;53	124	236	2;34	169	191	0;35
35	325	1;31	80	280	2;54	125	235	2;32	170	190	0;32
36	324	1;33	81	279	2;55	126	234	2;30	171	189	0;28
37	323	1;36	82	278	2;56	127	233	2;28	172	188	0;25
38	322	1;38	83	277	2;57	128	232	2;26	173	187	0;22
39	321	1;40	84	276	2;57	129	231	2;24	174	186	0;19
40	320	1;43	85	275	2;58	130	230	2;22	175	185	0;16
41	319	1;45	86	274	2;59	131	229	2;20	176	184	0;13
42	318	1;47	87	273	2;59	132	228	2;18	177	183	0; 9
43	317	1;50	88	272	3; 0	133	227	2;16	178	182	0; 6
44	316	1;52	89	271	3; 0	134	226	2;14	179	181	0; 3
45	315	1;54	90	270	3; 1	135	225	2;11	180	180	0; 0

Table 3 First equation for Mercury in *al-Qānūn*, *Sanjufīnī*, and *Huihui li*

centrum		computed			texts			centrum		computed			texts		
		c_3°	c_4'	q_1	Qān.	Sanj.	Hui.			c_3°	c_4'	q_1	Qān.	Sanj.	Hui.
1	359	0; 3	0	0; 3	0; 3	0; 3	0; 3	46	314	1;59	-3	2; 3	2; 3	2; 3	2; 3
2	358	0; 6	0	0; 6	0; 7	0; 7	0; 7	47	313	2; 2	-3	2; 5	2; 5	2; 5	2; 5
3	357	0; 9	-1	0; 9	0;10	0;10	0;10	48	312	2; 4	-3	2; 6	2; 7	2; 7	2; 7
4	356	0;11	-1	0;12	0;13	0;13	0;13	49	311	2; 6	-3	2; 8	2; 9	2; 9	2; 9
5	355	0;14	-1	0;15	0;16	0;16	0;16	50	310	2; 8	-2	2;10	2;10	2;10	2;10
6	354	0;17	-1	0;19	0;19	0;19	0;19	51	309	2; 9	-2	2;12	2;12	2;12	2;12
7	353	0;20	-2	0;22	0;22	0;22	0;22	52	308	2;11	-2	2;13	2;13	2;13	2;13
8	352	0;23	-2	0;25	0;25	0;25	0;25	53	307	2;13	-2	2;15	2;15	2;15	2;15
9	351	0;26	-2	0;28	0;28	0;28	0;28	54	306	2;15	-1	2;16	2;16	2;16	2;16
10	350	0;28	-2	0;31	0;31	0;31	0;31	55	305	2;17	-1	2;18	2;17	2;18	2;18
11	349	0;31	-3	0;34	0;34	0;34	0;34	56	304	2;19	-1	2;19	2;19	2;19	2;19
12	348	0;34	-3	0;37	0;36	0;37	0;37	57	303	2;20	0	2;21	2;20	2;21	2;21
13	347	0;37	-3	0;40	0;39	0;40	0;40	58	302	2;22	0	2;22	2;22	2;22	2;22
14	346	0;40	-3	0;43	0;42	0;43	0;43	59	301	2;24	0	2;23	2;23	2;24	2;24
15	345	0;42	-3	0;46	0;46	0;46	0;46	60	300	2;25	1	2;25	2;25	2;25	2;25
16	344	0;45	-4	0;49	0;49	0;49	0;49	61	299	2;27	1	2;26	2;28	2;27	2;27
17	343	0;48	-4	0;52	0;52	0;52	0;52	62	298	2;28	1	2;27	2;29	2;29	2;29
18	342	0;51	-4	0;55	0;55	0;55	0;55	63	297	2;30	2	2;28	2;30	2;30	2;30
19	341	0;53	-4	0;57	0;58	0;58	0;58	64	296	2;31	2	2;29	2;31	2;31	2;31
20	340	0;56	-4	1; 0	1; 1	1; 1	1; 1	65	295	2;32	2	2;30	2;31	2;31	2;31
21	339	0;59	-4	1; 3	1; 4	1; 3	1; 4	66	294	2;34	3	2;31	2;32	2;32	2;32
22	338	1; 2	-4	1; 6	1; 7	1; 7	1; 7	67	293	2;35	3	2;32	2;33	2;33	2;33
23	337	1; 4	-5	1; 9	1;10	1;10	1;10	68	292	2;36	3	2;33	2;34	2;34	2;34
24	336	1; 7	-5	1;11	1;12	1;12	1;12	69	291	2;38	4	2;34	2;35	2;35	2;35
25	335	1; 9	-5	1;14	1;15	1;15	1;15	70	290	2;39	4	2;35	2;35	2;36	2;35
26	334	1;12	-5	1;17	1;17	1;17	1;17	71	289	2;40	4	2;36	2;36	2;36	2;36
27	333	1;15	-5	1;20	1;20	1;20	1;20	72	288	2;41	5	2;36	2;37	2;37	2;37
28	332	1;17	-5	1;22	1;23	1;23	1;23	73	287	2;42	5	2;37	2;37	2;37	2;37
29	331	1;20	-5	1;25	1;25	1;25	1;25	74	286	2;43	5	2;38	2;38	2;38	2;38
30	330	1;22	-5	1;27	1;27	1;27	1;27	75	285	2;44	6	2;38	2;39	2;39	2;39
31	329	1;25	-5	1;30	1;30	1;30	1;30	76	284	2;45	6	2;39	2;39	2;39	2;39
32	328	1;27	-5	1;32	1;32	1;32	1;32	77	283	2;45	6	2;39	2;40	2;40	2;40
33	327	1;30	-5	1;35	1;34	1;34	1;34	78	282	2;46	6	2;40	2;40	2;40	2;40
34	326	1;32	-5	1;37	1;37	1;37	1;37	79	281	2;47	7	2;40	2;41	2;41	2;41
35	325	1;35	-5	1;39	1;39	1;39	1;39	80	280	2;48	7	2;41	2;41	2;41	2;41
36	324	1;37	-5	1;42	1;41	1;41	1;41	81	279	2;48	7	2;41	2;42	2;42	2;42
37	323	1;39	-5	1;44	1;44	1;44	1;44	82	278	2;49	8	2;41	2;42	2;42	2;42
38	322	1;42	-5	1;46	1;46	1;46	1;46	83	277	2;49	8	2;42	2;42	2;42	2;42
39	321	1;44	-4	1;49	1;48	1;48	1;48	84	276	2;50	8	2;42	2;43	2;43	2;43
40	320	1;46	-4	1;51	1;51	1;51	1;51	85	275	2;50	8	2;42	2;43	2;43	2;43
41	319	1;49	-4	1;53	1;53	1;53	1;53	86	274	2;51	8	2;42	2;43	2;43	2;43
42	318	1;51	-4	1;55	1;55	1;55	1;55	87	273	2;51	9	2;42	2;43	2;43	2;43
43	317	1;53	-4	1;57	1;57	1;57	1;57	88	272	2;51	9	2;42	2;43	2;43	2;43
44	316	1;55	-4	1;59	1;59	1;59	1;59	89	271	2;52	9	2;42	2;43	2;43	2;43
45	315	1;57	-3	2; 1	2; 1	2; 1	2; 1	90	270	2;52	9	2;42	2;43	2;43	2;43

Table 3 (Continued)

centrum		computed			texts		
		c_3°	c_4'	q_1	Qān.	Sanj.	Hui.
91	269	2;52	9	2;42	2;43	2;42	2;42
92	268	2;52	10	2;42	2;42	2;42	2;42
93	267	2;52	10	2;42	2;42	2;42	2;42
94	266	2;52	10	2;42	2;42	2;42	2;42
95	265	2;52	10	2;42	2;42	2;42	2;42
96	264	2;52	10	2;42	2;42	2;41	2;41
97	263	2;52	10	2;41	2;41	2;41	2;41
98	262	2;51	10	2;41	2;41	2;41	2;41
99	261	2;51	10	2;41	2;41	2;41	2;41
100	260	2;51	10	2;40	2;40	2;40	2;40
101	259	2;50	10	2;40	2;40	2;40	2;40
102	258	2;50	10	2;40	2;40	2;40	2;40
103	257	2;49	10	2;39	2;39	2;39	2;39
104	256	2;49	10	2;38	2;39	2;39	2;39
105	255	2;48	10	2;38	2;38	2;38	2;38
106	254	2;47	10	2;37	2;38	2;38	2;38
107	253	2;47	10	2;37	2;37	2;37	2;37
108	252	2;46	10	2;36	2;36	2;36	2;36
109	251	2;45	10	2;35	2;36	2;36	2;36
110	250	2;44	10	2;34	2;35	2;35	2;35
111	249	2;43	10	2;34	2;35	2;35	2;35
112	248	2;42	10	2;33	2;34	2;34	2;34
113	247	2;41	9	2;32	2;33	2;33	2;33
114	246	2;40	9	2;31	2;32	2;32	2;32
115	245	2;39	9	2;30	2;31	2;31	2;31
116	244	2;38	9	2;29	2;30	2;30	2;30
117	243	2;37	9	2;28	2;28	2;28	2;28
118	242	2;35	9	2;27	2;27	2;27	2;27
119	241	2;34	8	2;26	2;26	2;26	2;26
120	240	2;33	8	2;24	2;25	2;25	2;25
121	239	2;31	8	2;23	2;23	2;23	2;23
122	238	2;30	8	2;22	2;22	2;22	2;22
123	237	2;28	8	2;20	2;21	2;21	2;21
124	236	2;27	7	2;19	2;19	2;19	2;19
125	235	2;25	7	2;18	2;18	2;18	2;18
126	234	2;23	7	2;16	2;16	2;16	2;16
127	233	2;21	7	2;15	2;15	2;15	2;15
128	232	2;20	6	2;13	2;14	2;14	2;14
129	231	2;18	6	2;12	2;12	2;12	2;12
130	230	2;16	6	2;10	2;10	2;10	2;10
131	229	2;14	6	2; 8	2; 8	2; 8	2; 8
132	228	2;12	6	2; 7	2; 6	2; 6	2; 6
133	227	2;10	5	2; 5	2; 5	2; 5	2; 5
134	226	2; 8	5	2; 3	2; 3	2; 3	2; 3
135	225	2; 6	5	2; 1	1; 1	2; 1	2; 1

centrum		computed			texts		
		c_3°	c_4'	q_1	Qān.	Sanj.	Hui.
136	224	2; 4	5	1;59	1;59	1;59	1;59
137	223	2; 2	4	1;57	1;57	1;57	1;57
138	222	1;59	4	1;55	1;55	1;55	1;55
139	221	1;57	4	1;53	1;53	1;53	1;53
140	220	1;55	4	1;51	1;51	1;51	1;51
141	219	1;53	3	1;49	1;49	1;49	1;49
142	218	1;50	3	1;47	1;46	1;47	1;47
143	217	1;48	3	1;45	1;45	1;45	1;45
144	216	1;45	3	1;42	1;43	1;43	1;43
145	215	1;43	3	1;40	1;41	1;41	1;41
146	214	1;40	2	1;38	1;38	1;38	1;38
147	213	1;38	2	1;35	1;35	1;35	1;35
148	212	1;35	2	1;33	1;33	1;33	1;33
149	211	1;32	2	1;31	1;31	1;31	1;31
150	210	1;30	2	1;28	1;28	1;28	1;28
151	209	1;27	2	1;26	1;26	1;26	1;26
152	208	1;24	1	1;23	1;23	1;23	1;23
153	207	1;22	1	1;20	1;20	1;20	1;20
154	206	1;19	1	1;18	1;17	1;17	1;17
155	205	1;16	1	1;15	1;14	1;14	1;14
156	204	1;13	1	1;12	1;11	1;11	1;11
157	203	1;10	1	1;10	1; 9	1; 9	1; 9
158	202	1; 8	1	1; 7	1; 6	1; 6	1; 6
159	201	1; 5	1	1; 4	1; 4	1; 4	1; 4
160	200	1; 2	1	1; 1	1; 1	1; 1	1; 1
161	199	0;59	0	0;58	0;58	0;58	0;58
162	198	0;56	0	0;55	0;55	0;55	0;55
163	197	0;53	0	0;52	0;52	0;52	0;52
164	196	0;50	0	0;49	0;48	0;49	0;49
165	195	0;47	0	0;46	0;45	0;46	0;46
166	194	0;44	0	0;43	0;43	0;43	0;43
167	193	0;41	0	0;40	0;40	0;40	0;40
168	192	0;38	0	0;37	0;37	0;37	0;37
169	191	0;34	0	0;34	0;34	0;34	0;34
170	190	0;31	0	0;31	0;31	0;31	0;31
171	189	0;28	0	0;28	0;28	0;28	0;28
172	188	0;25	0	0;25	0;25	0;25	0;25
173	187	0;22	0	0;22	0;22	0;22	0;22
174	186	0;19	0	0;19	0;19	0;19	0;19
175	185	0;16	0	0;16	0;16	0;16	0;16
176	184	0;13	0	0;13	0;13	0;13	0;13
177	183	0; 9	0	0; 9	0; 9	0; 9	0; 9
178	182	0; 6	0	0; 6	0; 6	0; 6	0; 6
179	181	0; 3	0	0; 3	0; 3	0; 3	0; 3
180	180	0; 0	0	0; 0	0; 0	0; 0	0; 0

I could not find out on which theory this peculiarity was based, nor was I sure whether it was based on any theory at all or whether it was a result of a sheer mistake. Therefore, I investigated the first equation table for Mercury in the several *zījes* which were at hand. The crucial point of checking was, of course, whether q_1 of each text was the result of adding c_4 to c_3 or subtracting the former from the latter. As a result, I found that only the two texts, *Sanjufīnī Zīj* and al-Bīrūnī's *al-Qānūn al-Mas^cūdī*, share this particular feature with the *Huihui li*.

It is understandable that the table of the *Sanjufīnī Zīj* (column 7 of Table 3) is almost idential to that of the *Huihui li* (column 8),[19] since these two texts show a very close relationship to each other in the other respects, too.

I was quite surprised, however, to find that al-Bīrūnī's table (column 6 of Table 3) was so close to that of the *Huihui li* and the *Sanjufīnī Zīj*[20] that there was no doubt that these three tables belonged to the same tradition. Since all the predecessors of al-Bīrūnī, including his elder contemporary Kūšyār ibn Labbān,[21] added c_4 to c_3, al-Bīrūnī seems to be the first person who subtracted c_4 from c_3 and started this strange tradition. In order to find out whether this peculiarity was based on any theoretical ground, I read the 10th part (*maqāla*) of *al-Qānūn al-Mas^cūdī*,[22] but I found nothing special concerning the particularity of Mercury's first equation table. What al-Bīrūnī says in this context was simply a summary of the *Almagest*, and no attempt at innovation could be found.

For the moment, until counter-evidence is offered to disprove my conjecture, I would call this al-Bīrūnī's mistake. It is not strange that this kind of mistake should have happened, since, in the case of the other planets, c_4 is positive when the centrum is in the first and fourth quadrants, while in the case of Mercury it is negative when the centrum is less than 60° and greater than 300°. As is seen from Table 2 and Table 3, the difference due to the difference of adding and subtracting c_4 is not so remarkable – at most 20 minutes. Observations of Mercury could not have been used for the examination of the accuracy of the table.

It is historically interesting to note that, if al-Bīrūnī made such a mistake, he must have had a separate set of tables for c_3 and c_4 at hand and simply subtracted c_4 from c_3. Then what was the case with the other Arabic/Islamic authors of astronomical texts? It is likely that some of them also had separate tables for c_3 and c_4 and simply added them together, without computing q_1 anew by the direct method which I mentioned above. Another question then arises – who was the first person to prepare a separate set of tables for c_3 and c_4 at the interval of each degree?

I have investigated the following sources in order to check the method of obtaining the first equation table for Mercury. This list is in chronological order.[23]

Appendix

List of Sources

- Abū Manṣūr: *al-Mumtaḥan Zīj* (*c.* 830), Escorial arabe 927.
- al-Ḥabash al-Ḥāsib: *K. al-Ḥabash al-Ḥāsib* (*c.* 850), Berlin 5750.
- al-Battānī: *al-Zīj al-Ṣābi'* (*c.* 900), Nallino's edition. Cf. Endnote 16 above.
- Ibn Yūnus: *al-Zīj al-Ḥākimī* (990), Leiden Or. 143.

- Kūshyār ibn Labbān: *al-Zīj al-Jāmiᶜ* (*c.* 1000), Istanbul, Vehbi 893, Fatih 3418, Berlin, Staatbibliothek, Ahlwardt 5751, etc.
- al-Bīrūnī: *al-Qānūn al-Masᶜūdī* (1030), Hyderabad ed. and British Library Or. 1997.
- al-Khāzinī: *al-Zīj al-Sanjarī* (*c.* 1120), BM Or. 6669.
- al-Ṭabarī: *Zīj-i Mufrad* (*c.* 1230), Cambridge Browne O.1.
- al-Ṭūsī: *al-Zīj-i Īlkhānī* (after 1260), Cairo DMF 1.
- al-Maghribī: *Adwār al-Anwār* (*c.* 1280), Chester Beatty 3665.
- al-Baghdādī (*c.* 1285): Paris ms., Paris arabe 2486.
- Sanjar al-Kamālī: *Zīj-i Ashrafī* (*c.* 1310), Paris Suppl. Pers. 1488.
- al-Sanjufīnī: *Sanjufīnī Zīj* (1366), Paris arabe 6040.
- Ibn Isḥāq al-Tamīmī: *Tunisian Zīj* (14th c.), Hyderabad 298.
- al-Kāshī: *Zīj-i Khāqānī* (*c.* 1420), India Office Library 430.

Notes

1. There are three different recensions of the *Huihui li* – (1) that recorded in the official *Ming Dynastic History* which was compiled during the Qing Dynasty, (2) the *Qizheng tuibu* compiled by Bei Lin in A.D. 1477, and (3) the Korean recension *Chiljong san* which forms a part of the *Sejong sillok* compiled during the reign of King Sejong (1419–1450). These recensions are considerably different, especially in the arrangement and order of the explanatory texts and tables. For the difference, see the article by Benno van Dalen in this Volume.
2. See Part 2, Chapter 3 of Yabuuti's *Chinese Astronomy and Calendrical Sciences* (*Chûgoku no tenmon rekihô* in Japanese), Tokyo (Heibonsha) 1969, 2nd ed. 1990, which was a revision of his earlier paper, published in the *Tôhô Gakuhô*, Vol. 36 (1964), pp. 611–632 with the title 'Kaikai reki kai'. This work was recently translated, with some improvements, into English by Benno van Dalen as 'Islamic Astronomy in China during the Yuan and Ming Dynasties', *Historia Scientiarum*, Vol. 7, No. 1 (1997), pp. 11–43.
3. Recently van Dalen informed me of a very interesting paper which had escaped scholarly attention for long time: A. Wagner, 'Ueber ein altes Manuscript der Pulkowaer Sternwarte', *Copernicus*, Vol. II (1882), pp. 123–129. The author of this paper happened to examine an Arabic manuscript which was brought to the library of the Pulkova observatory by a consul in China. A mere glance at the table of contents and some parameters used in this manuscript is enough to say that this text was the best candidate for the source of the *Huihui li*. Needham (*Science and Civilisation in China*, Vol. 3, 1959, p. 372, footnote e), briefly referring to Wagner's paper, just hoped that 'they were not destroyed when the Obervatory was burnt during the second world war'. It is regrettable that no historian of astronomy tried to get access to the Pulkova manuscript. Let us hope that the manuscript survived the recent fire, too.
4. I thank the Japan Society for Promotion of Science for offering scholarship to Dr. van Dalen and thus making possible our joint project.
5. My contribution was published as 'Tables of Planetary Latitute in the *Huihui li* (I)', *Current Perspectives in the History of Science in East Asia*, ed. by Yung Sik Kim and Francesca Bray, Seoul National University, 1999 (June 30), pp. 307–315, which was followed by van Dalen's paper.
6. The unique manuscript is extant in the Bibliothèque Nationale, Paris, arabe 6040. I thank van Dalen who brought a photocopy of this manuscript for my use.
7. For this very interesting *zīj*, see Herbert Franke 'Mittelmongolische Glossen in einer arabischen astronomischen Handschrift', *Oriens* 31 (1988), pp. 95–118. See also Edward S. Kennedy, 'Eclipse Predictions in Arabic Astronomical Tables Prepared for the Mongol Viceroy of Tibet', *Zeitschrift für Geschichte der arabisch-islamischen Wissenschaften* 4 (1987/88), pp. 60–80 and Edward S. Kennedy and Jan Hogendijk 'Two Tables from an Arabic Astronomical Handbook for the Mongol Viceroy of Tibet', *A Scientific Humanist, Studies in Memory of Abraham Sachs*, ed. by Erle Leichty et al., Occasional Publications of the Samuel Noah Kramer Fund, 9, Philadelphia: The University Museum, 1988, pp. 233–242.
8. I have used the printed edition in 3 vols., Osmania Oriental Publications Bureau, 1956 and a copy of the manuscript from British Library Or. 1997.

9. See Gerald J. Toomer, *Ptolemy's Almagest*, London/New York, 1984.
10. Otto Neugebauer, *Exact Sciences in Antiquity*, New York, 1969, p. 200 ff.
11. In modern expression, when the eccentricity (e) is given, q_1 is a function of centrum (γ):

$$q_1 = \arcsin\left(\frac{2e\sin\gamma}{\rho}\right)$$

where

$$\rho = \sqrt{(2e\sin\gamma)^2 + (e\cos\gamma + \sqrt{R^2 - (e\sin\gamma)^2})^2}.$$

12. Neugebauer, *op. cit.*, p. 201. See also Toomer's translation of the *Almagest*, p. 546 and footnote 48.
13. I have used the Ph.D. dissertation of Willium D. Stahlman, *The Astronomical Tables of Codex Vaticanus Graecus 1291*, submitted to Brown University in 1959.
14. See Olaf Pedersen, *A Survey of the Almagest*, Odense University Press, 1974, p. 320.
15. The angles are so small that I want to make them clear here: D is at the intersection of EC and the dotted circle of which the centre is E and the radius is R. $q_1 = \angle$OCE, $c_3 = \angle$ODE, and $c_4 = \angle$COD.
16. Carolo A. Nallino ed. *Al-Battani sive Albatenii: Opus astronomicum*, 3 vols., Milano, 1903, 1907, 1899. Reprinted from Georg Olms Verlag, Hildesheim·New York, 1977. The first equation for Mercury is in vol.II, pp. 132–137. Al-Battānī's table for $q_1$is virtually identical to that in Theon's *Handy Tables*. The difference is only 2;4 instead of 2;5 for centrum 138/222 and 1;35 instead of 1;34 for centrum 149/211.
17. I have used the facsimile edition of Escorial arabe 927 published by F. Sezgin.
18. Since the tabular values of c_3 and c_4 are rounded to the unit of minutes, some values of q_1 are different from the sum of c_3 and c_4 of this table by one minute.
19. Only two out of 180 values are different, i.e., for centrum 21/339 and 70/290.
20. Out of 180, only 11 values are different.
21. The equation tables of Kūšyār ibn Labbān have another special feature of 'displacement', in order to avoid negative values.
22. I thank Toshiaki Kashino for reading this text with me.
23. For the date of the texts, see Kennedy, *A Survey of Islamic Astronomical Tables*, Transactions of the American Philosophical Society, Vol. 46, Part 2 (1956). For the date of *al-Mumtaḥan*, al-Baghdādī, and *Īlkhānī*, I acknowledge to Benno van Dalen's personal communication.

1.4. Three Star Maps: Results of the Impact of Western Astronomy on Korean Tradition in the 18th Century*

Nha Il-Seong
Yonsei University

1. A Brief History of Star Map Makings in Korea

Star maps are known to have existed in Korea as early as in the Three-Kingdom period, 1st C. BC–10th C. AD. One of these ancient kingdoms, Koguryo (37 BC–668 AD), located in the north, had left important paintings on walls of the royal tombs, which show early groupings of stars into constellations in East Asia. The locations of these tombs are scattered along the Yalu river, now the border between China and North Korea, and in the area nearby Pyongyang, the North Korean capital city. The present survey counted nineteen such tombs. The wall painting of northern Dipper and other stars, as shown in Figure 1, is an example.

Another interesting example among many of the Koguryo period is the star map engraved on a large stone slab. This Koguryo stone star map had a large number of stars visible in the northern hemisphere in over 200 constellations. The map was sunk in the river during the war some time around 668 AD, but fortunately, a rubbed copy of this map had been preserved for over 700 years by a noble family, and it was presented to the King Taejo, the first king of the last dynasty Chosŏn (1392–1910).

A number of star maps were made in the Koryo dynasty (900–1392), after the Three-Kingdom period. Unfortunately, none of these have been preserved. During the last Chosŏn dynasty, however, a large number of star maps of various forms were made; many of them are still preserved in good condition by both public museums and private collectors around the world.

Those star maps of the Chosŏn dynasty may be grouped into two major types. The first type includes the two stone-carvings: *Chŏnsang Yŏlcha Bunya Jidos* (abbreviated as the *1395 Star Map* and the *1687 Star Map*), and those rubbings and hand-copies of these maps. *Chŏsang Yŏlcha Bunya Jido* is literally translated as

> A Chart of the Heavenly Divisions of the Celestial Bodies[1]

or as

> Positions of the Heavenly Bodies in their Natural Order, and the Regions they Govern[2].

The *1395 Star Map* is the star map engraved on a stone slab in the fourth year of King Taijo (1395) copying the rubbing of the Koguryo stone star map, which was mentioned above. This stone slab has now deteriorated rather, and is preserved in the Royal Museum in Seoul. The slab, black in colour, has a dimension 211 cm × 123 cm × 12 cm and weighs

* I.-S. Nha Observatory Contribution

S.M.R. Ansari (ed.), History of Oriental Astronomy, 45–57.

Figure 1 Seven stars in the north Dipper on walls of Koguryo grave ($\sim 5^{th}$ century AD).

about 1 ton. Both sides of the stone slab have planispheres with the same contents but in different orders (Figure 2).

In the early days, people of the Chosŏn dynasty made hand-drawn copies from the rubbings of this *1395 Star Map* for their own use. The rubbings themselves were allowed to be used mainly by a limited number of dignitaries. Many such hand-drawn copies, of various forms and accuracies, have been found in many places. In order to have uniform and accurate copies, the *Gwansang-gam*, Bureau of Astronomy of Chosŏn, made 120 copies of wood-cut prints of this map in 1571 and gave them to scholars and high-ranking officials as gifts. Of these, only one copy is extant today, this being preserved in the Library of Tenri University in Japan. In 1687, one side of the *1395 Star Map* was copied as an engraving on a white marble of 208 cm × 109 cm × 30 cm, which is well preserved today in the Museum of the King Sejong Memorial Society in Seoul. This *1687 Star Map* was made because the *1395 Star Map* had deteriorated with age lacking proper care. A detailed study of both stone slab star maps is given elsewhere[3].

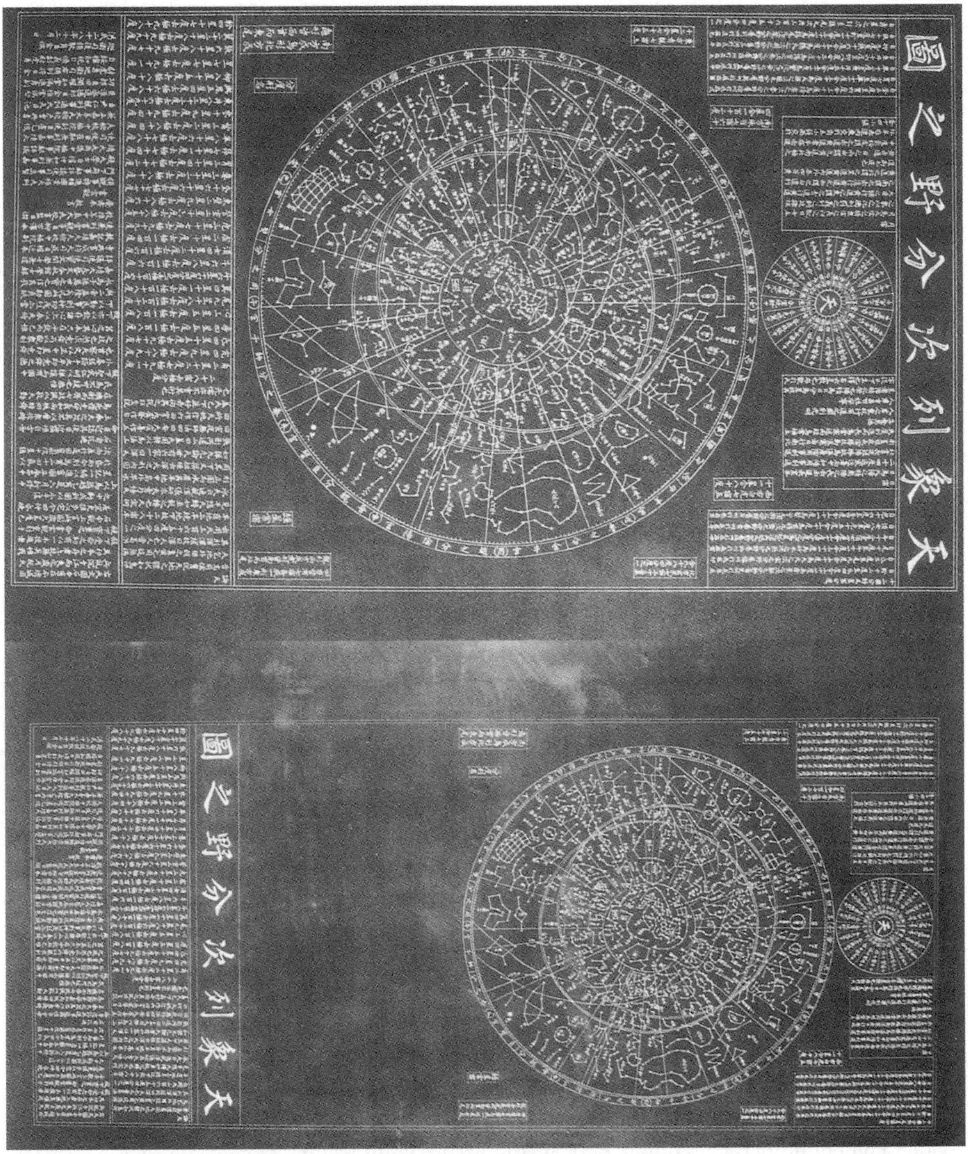

Figure 2 Both sides of a recent reproduction of the *1395 Star Map* in 1995 to commemorate the 600th anniversary of its first engraving.

The star maps of the Chosŏn dynasty of the second type are maps made in the later period of the dynasty. These maps are the results of Korean astronomers interacting with Jesuit astronomers in Beijing. In the mid-18th century, the astronomers of the Bureau of Astronomy constructed three modified star maps with the aid of occidental knowledge.

The first of these is called the *Honchŏn Jŏndo* (abbreviated as *HJ*), which appears in wood-cut prints. The other two are large folding screen star maps, which consist of 8 panels each. Each of these three star maps has explanatory texts in appropriate areas. These texts provide information on the transmission of western astronomy to Korea by the Royal astronomers of the Bureau of Astronomy.

2. Three Star Maps in the 18th Century

The three star maps under discussion were made by the astronomers of the Bureau of Astronomy of the Chosŏn dynasty in the 18th century. Now we introduce them in chronological order.

2.1. Honchŏn Jŏndo (HJ), the star map of wood-cut prints

No clear information has been found in Korean history regarding when this map was made and by whom.

The print is in black and white with the name *Honchŏn Jŏndo* on the upper part of the frame, which is 86 cm high and 59 cm wide. The central part of the print is a circular star map 57.6 cm in diameter. The circle is divided into 12 equal regions with 12 straight lines drawn from the north celestial pole, which is the centre of the circular map. The angular separation between adjacent lines is thus 30 degrees. The stars on this map were plotted in the equatorial coordinates and have different symbols regarding their brightnesses. The symbols for stars are as follows: double circles (◎) for the 1st magnitude stars, circles with a point at the center (⊙) for the 2nd magnitude stars, larger open circles (◯) for the 3rd magnitude stars, smaller open circles (∘) for the 4th magnitude stars, and the asterisks (∗) probably for the 5th magnitude stars, globular clusters and nebulae. The number of stars plotted on the *HJ* is 2,034 with 3 additional symbols.

We have counted the number of stars plotted on *HJ*. The numbers of stars in different ranges of the sky are tabulated in Table 1. The listing is in the order as given in Tables 1–12 of the *Yixiang Kaocheng* (abbreviated as *YK*) given by a Jesuit astronomer Ignatius Kögler, S. J. (1716–1746 in Qing) and his collaborators in 1752. The number of stars on *HJ* is significantly smaller than those listed in *YK*.

There are inscriptions on both the upper and the lower parts of *HJ*, which show the astronomical knowledge known to Korean astronomers at the time when *HJ* was being made. The contents of the upper part, written from right to left, are as follows:

(1) Sketches and the astronomical data of the Sun, the Moon, and 5 planets
(2) Diagrams of the solar and lunar eclipses
(3) Tables of sunrise and sunset times.

Of these three items, the first one is worthy of detailed discussion. The seven celestial bodies are described below:

The Sun: Darkspots are aligned horizontally on the solar disk. The number of these spots exceeds 60, which appears to be unusual in view of modern observations.

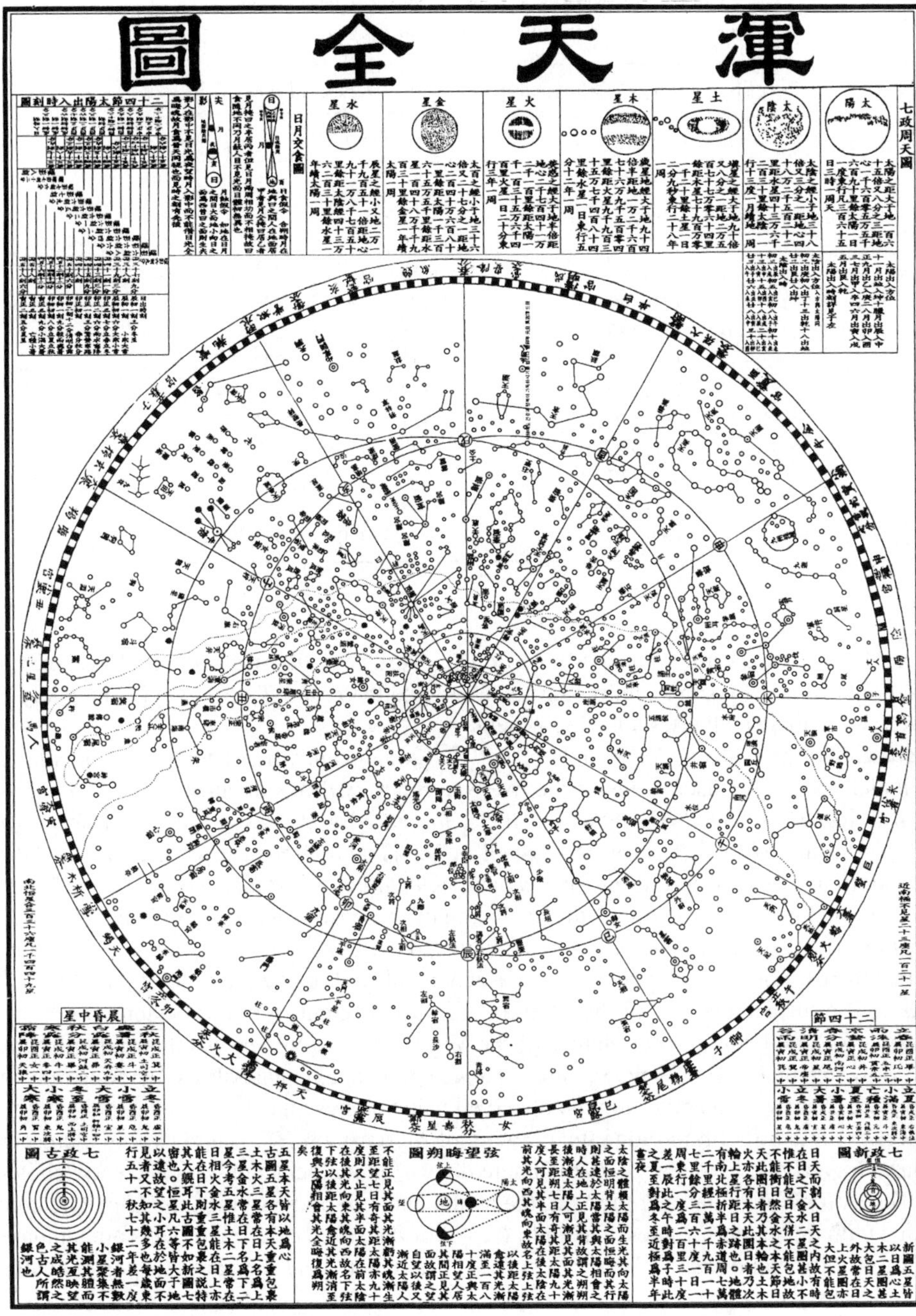

Figure 3 A wood-cut print of the *Honchŏn Jŏndo* (*HJ*) by the Bureau of Astronomy in the 18th c.

Table 1 Number of stars plotted in the *Honchŏn Jŏndo* with those listed in tables of the *Yixiang Kaocheng (YK)*.

No.	Range of Sky (in degrees)	Number of stars	
		HJ (star map)	*YK* (tables)
1	1–30	170	269
2	31–60	177	301
3	61–90	159	259
4	91–120	154	254
5	121–150	187	236
6	151–180	235	296
7	181–210	179	237
8	211–240	153	262
9	241–270	151	241
10	271–300	143	227
11	301–330	164	246
12	331–360	162	255
	Total	2,034 (+3)	3,083

The Moon: Craters are well distributed on the surface of the full moon. However, one might question why some significant features, like the crater Tycho which can easily be observed with a small telescope, are absent.
Saturn: Saturn is surrounded by an unusually wide ring. Only five satellites are positioned to its left side. The discoveries of the five Saturnian satellites in chronological order are as follows: *Titan* (by C. Huygens in 1655), *Iapetus* (by D. Cassini in 1671), *Rhea* (by D. Cassini in 1672), *Tethys* (by D. Cassini in 1684), and *Dione* (by D. Cassini in 1684). The sixth and seventh satellites, *Mimas* and *Enceladus*, were discovered much later by W. Herschel in 1789. Therefore, the exclusion of these two satellites in *HJ* is obvious.
Jupiter: The disk of Jupiter has two wide bands, one each in the northern and the southern hemispheres. Four Jovian satellites are clearly displayed on the left side of the planet.
Mars: The surface features of the Martian disk are different from our present views.
Venus: Although the present picture of the Venusian environment is rather complicated, the drawing in *HJ* shows no features. However, it does successfully demonstrate the crescent phase which could only be observed using a telescope.
Mercury: It is difficult to judge the drawing of Mercury on *HJ*. It is shown in a crescent phase. Nevertheless, the bright hemisphere due to the incident sun light is clearly shown.

The astronomical data for the Sun, the Moon and five planets given in *HJ* can be compared with the modern values. The radii (sizes) of the seven celestial objects are tabulated in Table 2 for comparison. The discrepancies between the *HJ* data and the present values are obviously very large.

The contents of the lower part, arranged counter-clockwise, are as follows:

(1) Table of the meridian stars at dusk and dawn in the 24 fortnightly periods
(2) The old solar system

Table 2 The sizes of the Sun, the Moon and five planets

Object	Diameter (Earth unit)	
	HJ	Present value
Sun	165 + 3/8	109.00
Moon	1/(38 + 1/3)	0.27
Saturn	90 + 1/8	9.01
Jupiter	94 + 1/2	10.77
Mars	1/2	0.53
Venus	1/(36 + 1/27)	0.95
Mercury	1/21,951	0.38

(3) The diagram of the phase variations for the Moon
(4) The new solar system.

The inscribed meridian stars at dusk and dawn in the 24 fortnightly periods characterize *HJ* as a unique star map, because they provide valuable information used for the comparative study with another important document of the mid-18th century, the *Dongguk Munhŏn Bigo* (abbreviated as *DMB*, literally means the Complete Serviceable Study of the History of Korean Civilisation), published in 1770. The *DMB* has a precise table of the meridian stars in the *Shinbŏb Jungsŏng-gi* (which literally means the *New Record of Meridian Transits*). The date of this *New Record of Meridian Transits* is 1755.

2.2. *Screen Star Map Type 1 (SSM1)*

This is the first type of star map which is on an 8-panel folding screen. We shall name this the *Screen Star Map Type 1 (SSM1)*, which shall be distinguished from the star map in the next section.

The *SSM1* contains two different star maps. The one on the first three panels, Nos. 1–3, is a copy, 147 cm in diameter, of the *1395 Star Map*. The other circular maps, 107 cm in diameter, shown on panels Nos. 4–7, is the *General Map of the Stars in Northern and Southern Hemispheres in Eclipstic Coordinates*. The last panel, No. 8, contains the diagrams of the Sun, the Moon and five planets. Each panel is about 160 cm high and 56 cm wide, which makes the total width 4.48 m for this *SSM1*. There are two inscriptions on the panels Nos. 4–7. The one in the top part is a copy of the statement made by Ignatius Kögler and Fernando Bonaventura Moggi in 1723, and the one in the bottom part represents the revisional work on the number of stars, which is given in *Jungbo Munhon Bigo*[4].

Three copies of this *SSM1* are known to exist, one each in England, Japan and Korea. Comprehensive investigations of the copy in England were made in 1966 by Needham and Lu[5] and later by Needham *et al.*[6], and of the copy in Korea by Nha and Lee[7]. These three copies of *SSM1* are shown in Figure 4. The top one, the Japanese copy, was made using

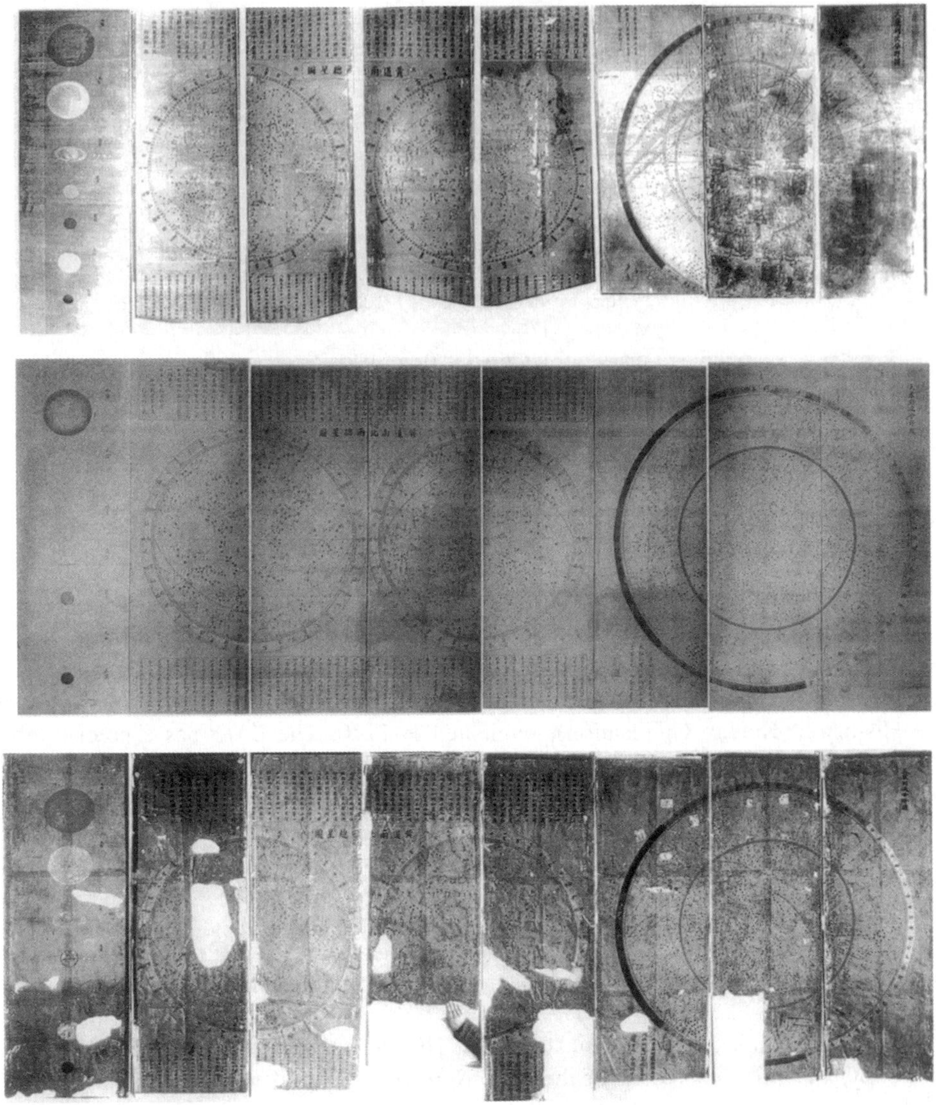

Figure 4 Three copies of the *SCREEN STAR MAP TYPE 1 (SSM1)*. The Nanbang Bunkakang in Osaka, Japan (top), the Whipple Museum in Cambridge, UK (middle), and the National Folk Museum in Kyŏngbok Palace in Seoul, Korea (bottom).

several photos taken at different angles and of different sizes with the use of computer scanning techniques.

Needham *et al.* considered that the map was a product of the Jesuit–Korean contacts, particularly of Ignatius Kögler and Andre Perreira with two Koreans, An Kuk-bin and Pyŏn Chung-wha. They postulated that the *SSM1* appears to have been painted sometime between 1755 and 1760. However, this author has discovered a statement on the picture of an additional panel photographed by Miyajima[8], saying that it was made in the fifth year King Sejo, or in the 44th year of King Yŏngjo. Accordingly, if this 9th panel is indeed a part of the original *SSM1*, this screen map would have been made in 1708 or 1768. Unfortunately, this additional panel is not available for examination at this time; hence the dating of *SSM1* is still in doubt.

2.3. *Screen Star Map Type 2 (SSM2)*

This is a different type of folding screen star map, which seems to be a reformed version of the *SSM1*. The more widely used name of *SSM2* is *Sinbŏb Chŏnmun-do*, which has been listed as a national treasure No. 848. Each panel is 183.5 cm high and 57 cm wide, with the total width of 4.51 m for the screen stretch. Panel No. 1 contains the inscription of the related astronomical knowledge on the right side and the diagrams of the Sun, the Moon and five planets on the left side. The next three panels, Nos. 2–4, contain a map of stars in the northern hemisphere in ecliptic coordinates and panels Nos. 5–7 contain a map of stars in the southern hemisphere in ecliptic coordinates. The last panel, No. 8, contains the names of officials who have engaged in the map making.

SSM2 was discovered by accident in a storage room of Bŏbju-sa Temple in Korea in 1961 by the late Professor Lee Yongbom, who later reported in 1966 a detailed study of this map[9]. It is remarkable that two Korean screen star maps, *SSM1* and *SSM2*, which were made by the same institute at about the same time two centuries ago, became known to the world in the same year, 1966 though from two different places.

The date of making of *SSM2* is not clearly stated on the screen. Therefore, we shall look for clues to determine this in two ways. The first clue is the exclusion of the *1395 Star Map* in *SSM2*. This may suggest to us that it was made later than *SSM1*, since the necessity of using the *1395 Star Map* as reference had diminished with time. The size of the star maps for both hemispheres is larger on the 6 panels of *SSM2* than on the 4 panels of *SSM1*. The second clue is the six names of the government officials listed on panel No. 8. The first three high-ranking officials are the Prime Minister, Kim Jae-ro, the Minister of Department *Yijo*, Yi Ki-jin, and the Minister of Department *Hojo*, Seo Jong-gub. These three officials are well documented in *Chosŏn Wangjo Silloks* (abbreviated as *Sillok*), and we have discovered their dates of office listed on the last panel.

Kim Jae-ro was the prime minister four times in 18 years from 1740 to 1758; Yi Ki-jin was the minister of *Yijo*, a department related to the internal affairs, for the period from 1741 to 1744; and Seo Jong-gub became the minister of *Hojo*, a ministry related to the taxation and agriculture, in 1741. Although it is not clear how long Seo Jong-gub held this office, we can determine that *SSM2* was made between 1741 and 1744 from the overlapping period when these three officials held their respective offices.

The other three were of a lower rank, and their names are not recorded in *Silloks*, except that of An Kuk-bin. An Kuk-bin was one of the most active observers of Halley's comet in 1759, when he was already retired from the head of local government. Needham *et al.* gave credit to him as one of the makers of *SSM1*. An's office, listed on the last panel of *SSM2*, was a branch of the ministry of defence, which apparently had no connection with astronomy. Therefore, one may guess that he might have been interested in astronomy long before his new assignment to the Bureau of Astronomy.

3. Comments on the Dates of Honchŏn Jŏn-do

Unlike the two screen star maps *SSM1* and *SSM2, Honchŏn Jŏn-do* has no name that can be used for reference. At least five identical copies of wood-cut prints of this star map have been found to date. This implies that *HJ*s were printed in large quantities and distributed widely by the Bureau of Astronomy, *Guansang-gam*, which was the only place where this kind of work took place.

It is very important to obtain an estimate of the date for these *HJ* prints, so that we might trace back the roots of interactions between Jesuit astronomers in Beijing and Korean envoys to Beijing in the 18th century. Although a tentative conclusion has been reached that *SSM1* precedes *SSM2*, because of the inclusion of the *1395 Star Map* in the *SSM1*, the chronological order of *HJ* in star-map making is uncertain. In order to estimate the dates of *HJ* the following information from the inscriptions on *HJ* are investigated:

(a) Surface features of Jupiter and Saturn

Among sketches of the seven celestial objects in *HJ* and *SSM1*, Jupiter and Saturn with their satellites are interesting in terms of comparison between the two maps. The Japanese copy of panel No. 8 of *SSM1* is adopted for comparison of its superior quality over the other two.

Markings on the surface of Jupiter in *HJ* are slight different and so are those of Saturn, as shown in Figure 6. For Saturn, the inside of the major ring is occupied by a shade and dots in *HJ*, but these markings are absent in *SSM1*. For Jupiter, two wide bands dominate the disk of Jupiter in *HJ*, but they seem to be lacking in *SSM1*. With these differences on the star maps, one may postulate that observations made at the time of making *HJ* were superior to those of *SSM1*. This fact may lead us to a conclusion that the date of *HJ* was later than *SSM1*. On the other hand, it could simply be that these were made by superior map makers. Therefore, the time difference between the two maps may not be found with these surface features.

(b) Satellites of Saturn

As was mentioned earlier, the chronological order of the discovery of Saturnian satellites could serve as a good indicator in estimating the map-making dates. There are five satellites shown in *HJ*, as in *SSM1* and in *SSM2*. This may suggest to us that *HJ* was made some decades after 1684, the year of discovery of the 4th and 5th satellites by Cassini. Because of the communication delay between the two Worlds in those days, it would have taken time for the news of the discovery in Europe to reach Korea through the Jesuit astronomers in Beijing.

Five Saturnian satellites in *SSM1* are marked on the right side of the planet. The left edge of the first panel of *SSM2* is heavily damaged and the right side of Saturn on this panel has an erosive feature. These make the identification of Saturnian satellites in *SSM2* impossible. In *HJ* and the *General Map of Stars in Ecliptic Coordinates* (1723), these satellites are on the left side of Saturn. Thus, the most probable date of making *HJ* is sometime after 1723 and before 1741–1744 (for *SSM2*).

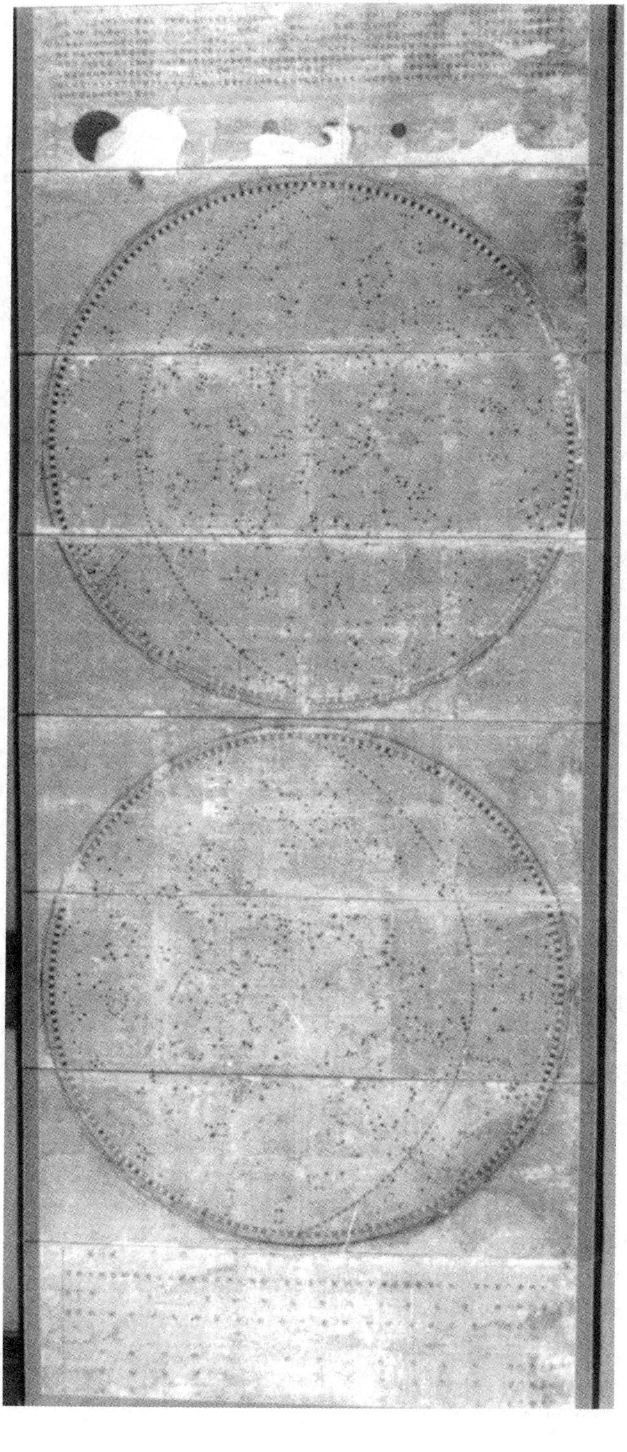

Figure 5 Sinbŏb Chŏnmun-do, *SCREEN STAR MAP TYPE 2 (SSM2)*, preserved in the Bŏbju-sa Temple. (Treasure No. 848.)

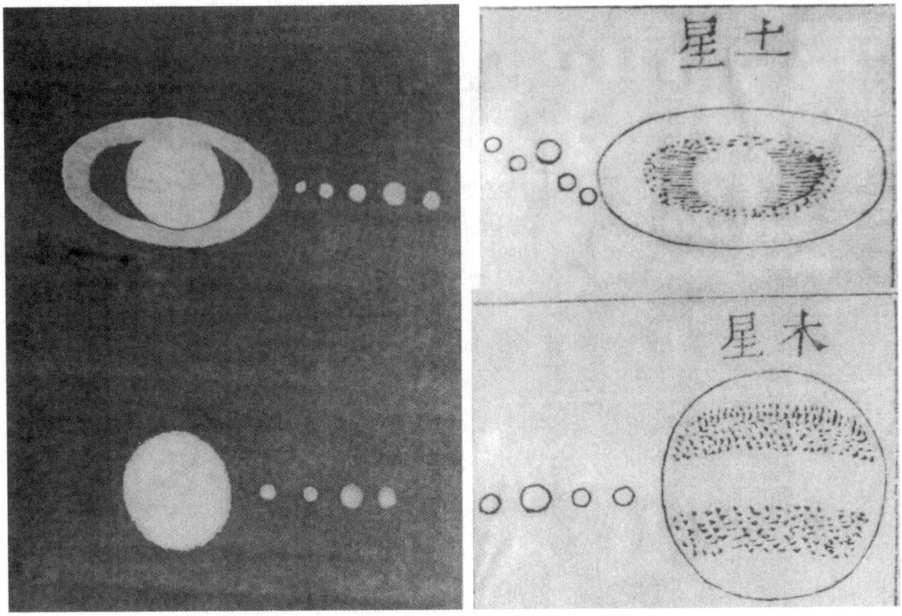

Figure 6 Surface features and satellites of Jupiter and Saturn. From left to right are *SSM1* and *HJ*, respectively.

In *SSM1*, the fourth satellite, which is marked with a bigger symbol than other satellites, correctly represents Titan. However, the biggest marking in *HJ* is assigned to the third satellite, that is Rhea. It is unfortune that no direct comparison of the satellites in the *HJ* and *SSM1* is possible, because *HJ* has information only on Saturn.

(c) Coordinate systems

The stars in *HJ* were plotted in equatorial coodinates, while those in the *SSM*s were in ecliptic coordinates. In ancient Korea, and also China, the equatorial coodinates were adopted for all known star maps. This is a strong indication that the *HJ* precedes the *SSM*s. One might also think that in earlier days *HJ* was easier to produce because of its small size.

Conclusion: With the information currently available, a tentative conclusion may be derived that the prints of the *HJ* were made in the early or mid-18th century, followed by the two screen star maps.

Notes

1. Rufus, W. C., 1913, 'The Celestial Planisphere of King Yi Tai-jo', *Transactions of the Korean Branch, The Royal Asiatic Soc.*, Vol. IV, p. 23.
2. Needham, J. and Lu, Gwei-Djen, 1966, 'A Korean Astronomical Screen of the Eighteenth Century from the Royal Palace of the Yi Dynasty', *Rivista Internationale Storia Della Scienza*, Vol. VIII, Fasc. 2.
3. Nha Il-Seong, 1996, 'Stone carved star map, Chonsang Yolcha Bunya-Ji-Do, and its 600th memorial reconstruction', *The Journal of Korean Studies*, Vol. 93, pp. 41–132 (in Korean).

4. *Jungbo Munhon Bigo* (Comprehensive Study of [Korean] Civilisation, revised and expanded), 1959. (Koso Kanhaeng-hoe: Seoul), vol. 1: p. 23.
5. Joseph Needham and Lu Gwei-Djen, 'A Korean Astronomical Screen of the Mid-Eighteenth Century from the Royal Palace of the Yi Dynasty (Choson Kingdom, 1392 to 1910)', *Physis*, 1966, vol. 8(2): pp. 137–162.
6. Joseph Needham, Lu Gwei-Djen, John H. Combridge and John S. Major, *The Hall of Heavenly Records, Korean Astronomical Instruments and Clocks 1380–1780* (Cambridge University Press: Cambridge), pp. 153–179, 1986.
7. Nha Il-Seong and Lee Yong-Sam, *Palpok Byŏngpungsik Chŏnmun-do* (National Folk Museum), Seoul, 1955 (in Korean).
8. private communication.
9. Lee Yongbom, 1966. 'On the New Star Map Preserved in the Bŏbju-sa Temple', *Yoksa Hakhoi*, vol. 31: pp. 1–66 (in Korean).

1.5. Projection Methods in East Asian Star Maps*

Kazuhiko Miyajima
Doshisha University, Kyoto

The celestial map of the Chunyou (淳祐) stone inscription in Suzhou (蘇州) and that contained in the *Xin yixiang fayao* (新儀象法要), both belonging to the Song (宋) Dynasty, are among the oldest of the extant celestial maps in China. The former has a circular form, and the latter is in two forms, circular and rectangular. I have been wondering which projection methods were used in these star maps, or rather, whether they were ever drawn accurately enough for us to be able to discuss their projection methods. While the stone inscription in Suzhou remains original without any change, we are not sure to which extent the star map in the *Xin yixiang fayao* maintains its original form, because the text underwent frequent reimpressions.

The rectangular celestial map of the *Xin yixiang fayao* consists of two parts, namely, *Hunxiang tongbeifang zhongwaiguan xingtu* (渾象東北方中外官星圖, 'the figures of inner and outer bureaus in the east and north on the celestial globe') and *Hunxiang xinang-fang zhongwaiguan xingtu* (渾象西南方中外官星圖, 'the figures of inner and outer bureaus in the west and south on the celestial globe').[1] The former is drawn on two pages (7 verso and 8 recto) and the latter on the next two pages (8 verso and 9 recto). Therefore, the four successive pages cover roughly 12^h to 18^h, 18^h to 0^h, 0^h to 6^h, and 6^h to 12^h, respectively.[2] We call these pages the third, fourth, first, and second quadrant, respectively.

Since the existing text of the *Xin yixiang fayao* is based on the recension of the *Shoushange congshu* (守山閣叢書), I photocopied a text of this recension belonging to the Research Institute for Humanistic Studies, Kyoto University. I used a macro lens which yields less distortion and I took great care so that the optical axis of the camera was perpendicular to the map. I used a negative film of the photograph in order to analyze the projection method and measured the positions of the *juxing* (距星, determinative stars) in each constellation. I compared the measured positions of the determinative stars with the observational values which are preserved in ancient documents. I used $y = 0$ for the equator line in the map and $x = 0$ for the right edge of each quadrant.

Observations of fixed stars were carried out four times during the Song Dynasty, namely, in the Jingyu (景祐) period (1034–1038), the beginning of the Huangyou (皇祐) period (1049–1054), the Yuanfeng (元豐) period (1078–1085), and the Chongning (崇寧) period (1102–1106). We have only a fragment of the record on the first set of observations. As for the last two observations, the records we have are limited to those of difference in right

* I have chosen this title for the following reason. I had a chance of analyzing two celestial maps, one from Japan and the other from Korea. The Japanese one is a rectangular star map in the *Tenmon seishō zu* (天文成象図, 'Map of the arrangement of stars and constellations'), a work of the astronomer Shibukawa Harumi (渋川春海) and his son Hisatada (昔伊), published in 1699; that from Korea is the circular star map *Ch'ōnsang yōlch'a punyajido* (天象列次分野之圖, 'Chart of the constellations and the regions they govern'), originally engraved on a stone in 1395. As a result of my analyses I found that the celestial maps in the Chinese cultural area in East Asia were based on the same projection method.

S.M.R. Ansari (ed.), History of Oriental Astronomy, 59–65.

ascension of the twenty-eight lunar lodges. For the second set of observations, however, records of a total of 358 observed stars survive in the *Wenxian tongkao* (文献通考), *Lingtai biyuan* (靈臺秘苑), and *Kuankui jiyao* (管窺輯要). In Yabuuti (1) and (2) attempts were made to identify these stars and it was demonstrated that the star map in the *Xin yixiang fayao* was based on the observations carried out during the Yuanfeng period. However, since no observational records from the Yuanfeng period are extant, I had to use the record belonging to the Huangyou period for my present purposes. I converted Chinese degrees into modern degrees by multiplying the former by 0.9856. Since the values relevant to the right ascension are only given in *ruxiudu* (入宿度, the distance in right ascension from the determinative stars of the twenty-eight lunar lodges), Yabuuti simply added them to the right ascension of the determinative stars in A.D. 1050 in order to obtain the observational value of the right ascension (α) of each star. I followed the same method, but used the polar distance ($p = 90° - \delta$) instead of the declination (δ).

There were some difficulties and problems in my analyses, which are summarized as below.

(1) Errors might have occurred in the process of transmitting the records of the observational data.

(2) In the *Xin yixiang fayao*, it is written that '1464 stars are recorded in the map',[3] but actually only the determinative star of each constellation was observed. It is difficult for us to decide which star in a constellation was determinative. In our sources very ambiguous expressions are used such as 'a star in north-west', or 'a big star'. Sometimes there is more than one star in 'the north-west', and the 'big star' is not always the brightest star in the constellation.

(3) Errors might be possible because the observational data I used were not those in the Yuanfeng period, but those in the Huangyou period. Therefore, two types of possible errors must be considered:

- The observational errors of the two sets of records are different. Besides the ordinary observational errors, there might have been errors due to misreading of scales (eg. errors of 10 degrees or those due to the wrong direction of counting) as well as those due to misrecording.
- Different stars might have been observed in the two sets of observations. It is possible that, although the selection of such determinative stars had been established, a wrong star was caught in the sighting tube (望筒) of the armillary sphere and was regarded as determinative.

(4) There might have been errors which occurred when the original figures were drawn or when they were transferred to printing blocks.

(5) Deformation might have occurred at the time of reimpression.

If the result of my analyses is reliable enough, however, we can detect these errors and make corrections, and decide which is the determinative star.

It seems that a cylindrical projection was applied for the rectangular maps.[4] Thus, for the right ascension, we have a relation of the type: $\alpha = A \cdot x + B$. As for the polar distance, since there are various types of cylindrical projections, different formulas might be possible. For instance, $y = C \cdot \log\tan(p/2)$, according to Mercator's method, or

$\delta = C \cdot y$ (i.e., $p = 90° - C \cdot y$), according to the cylindrical equidistant projection (正距円筒図法).

I applied the method of least squares for the right ascension of the stars in the third quadrant. The result was:

$$\alpha = -8.2933 \cdot x + 180.23.$$

The constant of this formula represents the right ascension of the right edge of the figure. Although the ecliptic drawn in the star map does not cross the equator at the right edge of the figure, it is evident that this figure was drawn making the autumnal equinox the starting point, since the constant is nearly 180°.[5]

When we express the relation between y and the polar distance $p = 90° - \delta$ in the graph, the plotted dots are almost on a straight line. In other words, the cylindrical equidistant projection must have been used. When we compute the coefficient of a formula of the type $p = C \cdot y + D$ by means of the method of least squares, we obtain

$$p = -8.0030 \cdot y + 89.91.$$

The fact that the constant of this formula, i.e. the polar distance of the equator line in the map, is very close to 90° means that this figure was drawn accurately. The upper edge of the figure turns out to be 33.49°, which is slightly less than the latitude of Kaifeng (開封). This means that the figure was drawn northward to the circle of constant visibility and a little further. In the case of the circular celestial maps, the drawing is made outward to the constant visibility circle and a little exterior. Thus they must have been made in such a way that the two areas slightly overlapped each other.[6]

I applied the same method to the other quadrants and found that the errors in them are slightly larger than those in the third quadrant. Nevertheless, the results were rather consistent.

The circular celestial map in the *Xin yixiang fayao*[7] is divided into a right half and a left half which are drawn on two separate pages. If the circular map was drawn by the azimuthal equidistant projection the polar distance p must be proportional to the distance r from the center of the figure. If, on the other hand, it were drawn by a stereographic projection, as in the Hellenistic period, a relation such as $p = 2 \cdot \arctan(r/E)$ should have been assumed. When I plotted the relation between the measured values and the polar distances, I found that they formed nearly a straight line. Therefore I concluded that the azimuthal equidistant projection was used in this map.

But there remain some problems. For example, the horizontal radius and the vertical radius are not equal. This means that the center of the figure is not necessarily the center of mapping (the north pole). So I thought it necessary to compute the coordinates (x_0, y_0) of the north pole in comparison to the center temporarily set at the time of measurement. Assuming that the relation is not proportional but a constant value is involved, and that the distance from the center can be expressed by the formula like $p = E \cdot r + F$, I applied the method of least squares. As a result I found that all the values x_0, y_0, and F are close to zero and, therefore, that the map was drawn by azimuthal equidistant projection rather accurately.

I have compared the positions of stars measured in the figure with the positions expected from the observational data in the literature. Even though we admit that the figure was drawn rather accurately, we notice a considerable number of errors. This is probably due to frequent reimpressions.

Since the two star maps, that of the *Ch'ōnsang yōlch'a punyajido* and that of the *Chunyu tianwen tu*, are both circular, we can apply the same method to them as we already did to the *Xi yixiang fayao* mentioned above. But unlike the *Xi yixiang fayao* which consists of two hemispheres, these two maps contain only one figure which represents the whole sky (excepting the area near the south pole). Therefore we must take a special care at the place crossing over the line of 360°. Since the observational data of the positions of the stars in the *Chunyu tianwen tu* seem to belong to the Yuanfeng period, i.e., the observation time of the *Xi yixiang fayao*, we substituted them by those in the Huangyou period. Since, again, there are no such observational data in the case of the *Ch'ōnsang yōlch'a punyajido*,[8] I used the theoretical value in A.D. 1400 which is close to A.D. 1395, the date of the inscription itself. As a result, I confirmed that the both maps were drawn by means of the azimuthal equidistant projection, where the polar distance p of the stars can be obtained by

$$p = E \cdot r + F.$$

In this case F was almost 0, and thus I could confirm both the accuracy of the map and the correctness of the method of analysis. Now that we have realized that the maps were drawn by the azimuthal equidistant projection, what we should do is only to determine E by means of the least square method with the formula:

$$p = E \cdot r.$$

Since E depends on the size of the map, the value we get from the photographs are different from the actual value.

As for the right asension α of the stars in the circular maps whose center is the north pole, it should be of course in the form of

$$\alpha = \theta + H$$

where θ is the positional angle from the arbitrary line (a parallel to y axe is convenient) drawn from the center of the figure (x_0, y_0), and H is the right ascension of this line and therefore constant. Here again, however, we first hypothesized the formula:

$$\alpha = G \cdot \theta + H$$

in order to scrutinize the acurateness of the drawing and the correctness of our method and, using the least square method, I found that G was almost 1. Therefore we used the formula:

$$\alpha = \theta + H$$

and we obtained H. In this case H depends on the direction of the standard line by which to measure the positional angles.

The polar coordinates (r, θ) of the stars in the maps should be measured with reference to the center of the drawing, i.e., the north pole (x_0, y_0), but at first we can not judge where it is, because errors are inevitable in drawing figures and because we cannot say that the reference point was the center of the circle drawn in the map. In fact the centers of three circles, namely, the inner (constant visibility) circle, the equator, and the outer (constant invisibility) circle, are slightly different, and they are not perfectly circular. Moreover the boundary lines of the twenty-eight lunar lodges do not converge at one point, i.e. a point which is theoretically the north pole. Therefore I used an arbitrary point as the hypothesized origin (0, 0) and one of the boundary lines as the reference direction, and I set the relation between the polar coordinates (r, θ) of the determinative star and the orthogonal coodinates (x, y) in the form:

$$x = x_0 + r \cdot \sin\theta$$

$$y = y_0 + r \cdot \cos\theta$$

and applying the least square method, I determined E, H, x_0, and y_0. If the map were accurate, (x_0, y_0) should be identical with the center of the three circles, but actually they were slightly different.

With reference to the original point (x_0, y_0) of the drawing (i.e., the correct position of the north pole) thus determined, and using the relations:

$$r = p/E$$

$$\theta = \alpha - H$$

we plotted the positions of the determinative stars of the maps which were obtained either from the recorded observational data or from the theoretical value, and compared them with the positions of the determinative stars actually drawn on the map. As a result we found that as far as the determinative stars are concerned the maps were drawn almost correctly, with a few exceptions. The *Chunyou tianwen tu* was very good, while some errors were found in the *Ch'ōnsang yōlch'a punyajido*.

Now it became evident that the circular star maps in the East Asia were drawn by means of the azimuthal equidistant projection. Since the relation $p = E \cdot r$ is applicable not only to stars but also to the size of three circles, we can estimate the value of the geographical latitude used for the drawing by means of the polar distance or declination corresesopnding to the inner and the outer circles. Since, further, the polar distance of the equator is about 90° if the maps were drawn with a certain accuracy, we can safely assume that

$$r_i : r_e : r_o = \phi : 90^\circ : (180^\circ - \phi)$$

where r_i, r_e, and r_o are radii of the inner circle, the equator, and the outer circle, and ϕ is the geographical latitude where the map was supposedly used. Applying this relation to the two star maps, I found from one the latuitude of Seoul, the capital of Korea, and from the other that of Kaifeng, the capital of the Northern Song.[9]

I do not go into further details, but the ecliptic circle drawn in the East Asian star maps are not accurate and therefore we cannot use it in order to fix the date by means of the position of the vernal equinox. It is possible, however, that we can approximately determine the year by checking the difference of the posisions of stars actually drawn and those theoretically computed for various times. Then, the year when the difference of the mean squares is minimum can be regarded as the year of the observation. I have not used this method in the two maps mentioned above, but I tried this method when I examined the ceiling star map of the Kitora tumulus which was discovered in 1998, and I could estimate the latitude of its locality. However, I refrain from discussing this problem, because this would be more than an additional remarks to my presentation at the IAU general assembly in 1997.

Compared with these maps, the *Tenmon seishō zu*, based on Shibukawa Harumi's observation and published by his son, is quite accurate. This is because the printing block was original. When we plot in a graph the relation between the polar distance and the *y* coordinate position, the scarcity of dispersions supports my assumption. The recorded position and the observed position agree quite well.[10] These observational values are recorded in Shibukawa's *Tenmon keitō* (天文瓊統) and they were studied in Watanabe (1963) and (1987).

From these we can conclude:

(1) Needham's interpretation of the projection method of the *Xin yixiang fayao* as Mercator's projection is wrong.

(2) The most Western star maps before early modern times are circular and they are based on stereographic projection. This method is quite geometrical, while the star maps in the Chinese cultural area are based on 'numerical projection'. Thus we can say that the different projection methods reflect the difference in character of astronomy and mathematics.

(3) The star maps in China, Korea and Japan appear as if they were drawn without good care, but, in fact, they can be analyzed quantitatively. Some of the star maps, in particular, retain the original accuracy even after frequent reimpressions as in the case of the *Xin yixian fayao*.

(4) This analysis is applicable only to the determinative stars, but the result which I obtained through this study will give some help for the identification of the other stars in Chinese constellations.

(5) The determinative stars are mostly located near the western end of constellations, but sometimes they are in other places. Sometimes they are less bright. We do not know what criterion was used for the selection. The distribution of the determinative stars in the *Xin yixiang fayao* shows that there are many cases where the determinative stars are aligned vertically, that is, stars with approximately the same right ascension were chosen. This is probably for the sake of effectiveness of observation – one has only to move the sighting tube along the north–south direction with less motion in the east–west direction.

Notes

1. *Hun xiang* (渾象) means celestial globe. The names of the four cardinal directions were used in order to designate the four parts of the wide belt which has a width of about 55° on both sides of the equator. The starting point of the 'east' is at the first point of *jiaoxiu* (角宿), i.e. about 12^h, and the remaining directions are counted to the direction of increasing right ascension. The 'inner' and 'outer' mean north and south of the equator, respectively. Following the Chinese tradition of designating constellations after the bureaucracy, the stars and constellations were called *guan* (官, 'bureau').
2. The verso pages are a little narrower than the recto pages.

3. In fact the number is 1463, because one star was dropped by mistake.
4. In other words, the azimuthal projection method like central projection or stereographic projection was not used.
5. The right ascension of the left edge of the figure turned out to be 261.82°. The reason that this is less than 270° is that the horizontal length of the figure in the right hand page (verso) is shorter than the figure in the left hand page (recto), because Chinese characters are placed at the right edge of the right hand page.
6. The coefficients of x and y are of course variable according to the size of the figure. The values shown here are those in the negative film.
7. There are two kinds of maps, one is only around the north pole and the other is around both the north pole and the south pole making the equator outermost circle. I made my analyses of the first type only.
8. Probably independent observation was never carried out, but I put aside this problem.
9. The *Chunyou tianwen tu* was made in the Southern Song, but the observational data are those of the Northern Song period.
10. I followed Shibukawa's identification in the present research, although there are several differences from the identifications in the *Yixiang kaocheng* 儀象考成 and Yabuuti's paper.

References

1. Needham, J. (1959). *Science and Civilisation in China*, Vol. 3. Cambridge.
2. Pan Nai & Wang Dechang (1983). 'The Star Observation at Northern Song Dynasty and Hunag You Star Catalogue' (北宋的恒星観測及《宋皇祐星表》上), *Kejishi Wenji* (科技史文集) 10, pp. 98–121.
3. Pan Nai (1989). *A History of Fixed Star Observations in China* (中国恒星観測史), Shanghai.
4. Watanabe Toshio (1963). 'A Report on Yasui Harumi's Star map' (in Japanese), *Journal of the University of Mercantile Marine (Natural Science)* 14.
5. Watanabe Toshio (1987). *A History of Modern Japanese Astronomy* (in Japanese), Vol. 2. Tokyo.
6. Yabuuti Kiyosi (1936). 'Sōdai no Seishuku', *Tōhō gakuhō* 7, pp. 42–89.
7. Yabuuti Kiyosi (1969). *Chūgoku no tenmon rekihō*. Tokyo.

1.6. Gnomon Measurements and the Obliquity of the Ecliptic

K.-Y. Chen
University of Florida

1. Introduction

This study was originally motivated from the discussion of the obliquity of the ecliptic, ϵ, in the work of Needham (1959) and his collaborators, and the article, "An 8-th Century Meridian Line," by Beer *et al.* (1961). A search in the historical records yields values, published and unpublished, of ϵ.

The gnomon-shadow lengths of the sun at noons on the solstitial days as recorded in the Chinese Annals were collected. Each summer-winter pair was used to compute the latitude of the observing site and the value of ϵ. The values of ϵ, determined in the Orient and in the Occident, are used to show the variation of ϵ in the three thousand years of recorded history. The result is compared with those of the modern investigations.

2. The Gnomon Observations

The heights of the Chinese gnomon were given as 8 *chi*; except the 9-*chi* gnomon used for the first listed set of observations at Jiankang, and the 40-*chi* gnomons used in Beijing. The calculated values are not corrected for the effects of atmospheric refraction and local height from the sea level. In Table 1, the years in which the observations were made are listed in column 1, the places of observation in column 2, the calculated values of latitudes and obliquity in column 3, the altitudes of the sun at noon of the summer solstice in column 4. The latitudes and longitudes of the cities in the fifth column were copied from the *Gazetteers* of the U. S. Office of Geography, except those of Gaocheng which were measured in the twentieth century for the gnomon of Zhou Gong, the Duke of Zhou (*d* 1036 *BC*). This monumental stone gnomon was elected by the astronomer royal Nangong Shuo in 723 *AD*. In Table 1, the angles are given in degrees, and the latitudes and longitudes are measured to the north and to the east, respectively.

The land of the State of Linyi has been known to China since the Han dynasty. Its capital is said to be near the ancient city of Hue. The value of 0.57 *chi*, which is recorded in the *Tang Annals* for gnomon-shadow length as measured by Da Xiang and Yuan Tai in Linyi on the summer-solstice day, does not appear to be correct. Instead, the value of 0.91 *chi*, which was obtained after the military campaign of Liu Song against Linyi, under the command of Tan Huo-zhi, in 433 *AD*, is used here. The land of Tiele was known as the land of Xiongnu in the Han dynasty. The calculated latitude of the place is not as far north as that of Lake Baykal. On his journey to Samarkand, the Taoist Qui Chu-ji measured the gnomon shadow of the sun on the summer-solstice day along the shore of the Kerulen river. The longitude of the place of the observation is estimated to be about 112°E according to the event written in the book, *The Journey West*, by Li Zhi-chang.

S.M.R. Ansari (ed.), History of Oriental Astronomy, 67–73.

Table 1 Chinese records with Gnomon measurements

Year	Place	Latitude/Obliquity	Altitude ⊙	City	Lat./Long.	Dynasty
ca 1100 *BC*	Yangcheng	34.51/23.89	79.38	Gaocheng	34.41/113.14	Zhou
442 *AD*		–	79.38			Liu Song
724		34.15/23.68	79.52			Tang
822		34.16/23.69	79.53			
1024		34.15/23.68	79.53			Bei Song
442	Jiaozhou	–	87.85	Hanoi	21.03/105.85	Liu Song
442 *AD*	Linyi	–	83.51	–	–	
ca 300 *BC*	Luoyang	35.05/23.73	78.69	Luoyang	34.68/112.47	Zhan Guo
115–120 *AD*		34.51/23.89	79.38			Dong Han
508		–	78.83			Bei Wei
581–584		34.24/23.76	79.52			Sui
596–597		33.94/23.69	79.73			
544	Jiankang	31.90/23.55	81.67	Nanjing	32.05/118.78	Liang
544		–	81.66			
721–725	Linyi	17.14/23.63	85.93	–	–	Tang
	Jiaozhou	21.21/23.57	87.64	Hanoi	21.03/105.85	
	Langzhou	29.14/23.64	84.50	Changde	29.03/111.68	
	Caizhou	33.24/23.74	80.35	Runan	33.02/114.37	
	Xuzhou	33.83/23.62	79.80	Fugou	34.07/114.38	
	Bianzhou	34.46/23.63	79.17	Kaifeng	34.85/114.35	
	Huazhou	34.75/23.65	78.90	Huaxian	35.58/114.50	
	Weizhou	39.63/23.65	74.03	Weixian	39.80/114.40	
	Tiele	51.01/23.70	62.70	–	–	
	Henanfu	34.18/23.63	79.45	Luoyang	34.68/112.47	
1049–1052	Junyi	34.60/23.48	78.88	Kaifeng	34.85/114.35	Bei Song
1064		34.60/23.50	78.90			
1092		34.60/23.50	78.90			
1106		34.55/23.51	78.97			
1221	Kerulen R.	–	65.47	–	–	Yuan
1277	Da Du	39.86/23.54	73.68	Beijing	39.93/116.40	
1368 *AD*	Beijing	39.85/23.54	73.68			Ming

The correlation between the calculated latitutes and the geographic latitutes of the modern cities is an indication of the accuracy of the calculated values, including ϵ. Using the method of least square, one finds that $Latitude_{calc} = 0.574 + 0.977 \times Latitude_{geo}$, The quantity, β, defined as *the unexplained variation/the number of n degrees of freedom*, is equal to 0.08048. Similarly, one obtains that $Latitude_{calc} = 22.798 + 1.115 \times zenith \odot -0.00396 \times zenith\odot^2$, where $zenith\odot$ is the zenith distance of the sun at noon of the summur solstice (measured toward south) with $\beta = 0.12838$. Fig. 1 depicts the plot of the calculated *vs.* the geographic latitudes, and Fig. 2 that of the calculated latitutes *vs.* the zenith distances of the sun. The lines in these diagrams represent the respective relations.

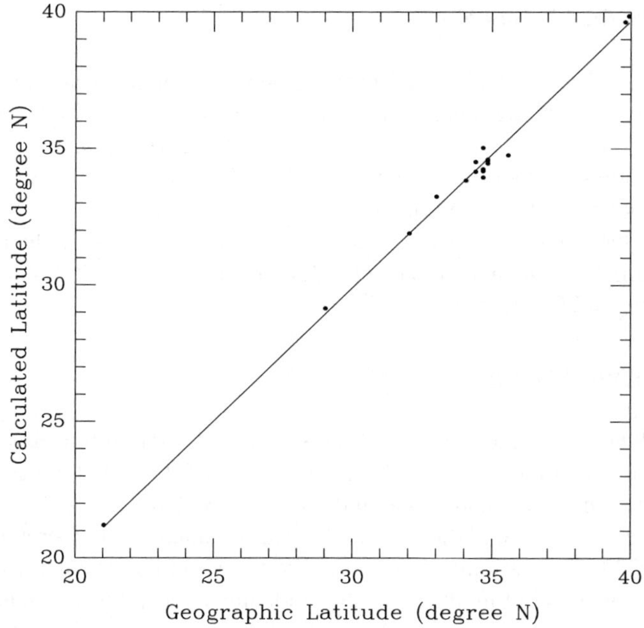

Figure 1 Latitudes of Chinese places; calculated from gnomon measurements vs. geographic values.

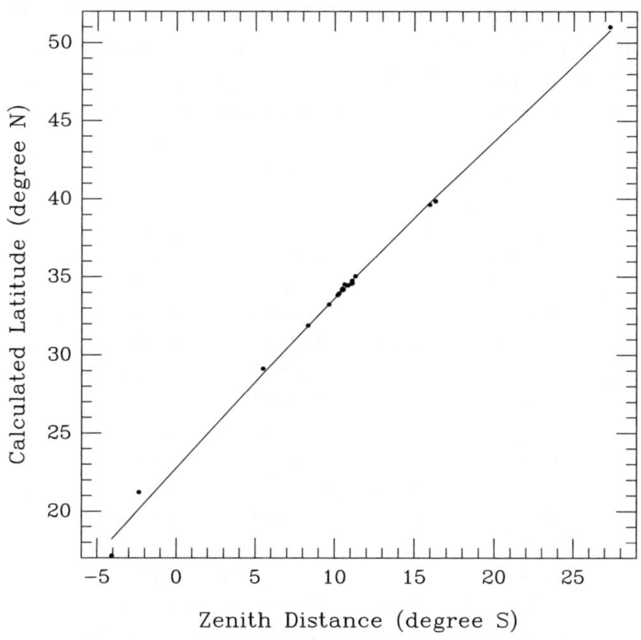

Figure 2 From gnomon observations in China, calculated latitudes vs. zenith distances at noon of summer solstices.

3. Other Published Values of ϵ

Other published values of ϵ observed in the Orient are collected and listed in Table 2. Again, the latitudes and longitudes in the last column are from the *Gazetteers*, except that of the first entry, in which the values were derived from *Artha-sastra* (ref. Y. Ohashi's article in this Volume). The other exception is the latitude of the Sun temple at Modhera, India, which is given by Rao (1998).

The list of values of ϵ from Occidental observations is given in Table 3. These values were copied from the book by La Lande (1792) and the article of Wittmann (1984), except those of Ptolemy and Copernicus.

4. The Saeculum of Decline

A 2-degree polynomial is used to express the secular variation of the values of ϵ as listed in Tables 1 and 2, together with the 1976 IAU value of 23.439° for the epoch 2000.0. Using the method of least squares, one find that $\epsilon = 23.73768151 - 0.00014884 \times T - 0.00000001 \times T^2$, where T is the number of year counted in the Common Era. In this case, β is 0.00333. The variation of ϵ with time based on the oriental observations and the IAU value is shown in Fig. 3. If the 38 occidental values, the 38 oriental values, and the IAU value are used together, one obtains the 2-degree polynomial with the use of the method of least squares, $\epsilon = 23.74583283 - 0.00016342 \times T$, which is a linear function representing the secular decline of the value of ϵ at the rate of 0.01634° per century. Here, β is 0.00208. The variation of ϵ based on all 77 data points is shown in Fig. 4.

Determination of the value of the secular decrease of ϵ in modern time was first given by Newcomb (1898) at the rate of $-46.845''/cy$ at epoch 1900.0. In the twentieth century, numerous observational and theoretical investigations, *e.g.* Lieske (1970), Laubscher

Table 2 Other oriental records

Year	Obliquity	Place	Lat./Long.
ca 300 *BC*	23.7	India	23.7/–
230 *BC*	23.86	Alexandria	31.20/29.90
830 *AD*	23.56	Baghdad	33.30/44.60
832	23.55	Damascus	33.50/36.25
869	23.55	Baghdad	33.30/44.60
880	23.58	ar Raqqah	36.00/39.00
970	23.58	Shiraz	29.60/52.53
970	23.58	Cairo	30.05/31.25
987	23.58	Baghdad	33.30/44.60
994	23.54	Shahr Rey	35.58/51.42
1009	23.58	al-Mugattam	30.03/31.28
1026	23.57	Modhera	23.58/–
1259	23.50	Maragha	37.33/46.38
1424 *AD*	23.52	Samarkand	39.40/66.58

Table 3 Obliquity Data of the Occident

Year	Observer	Obliquity
150 *BC*	Hipparchus	23.856
140 *AD*	Ptolemy	23.839
1079	Arzachel	23.567
1460	Peurbach	23.467
1490	Walther	23.496
1525	Copernicus	23.475
1560	Wilhelm	23.525
1570	Danti	23.483
1587	Brahe	23.525
1589	van Lansberg	23.528
1610	Wendelin	23.504
1636	Gassendi	23.516
1644	Linemann	23.508
1650	Riccioli	23.500
1660	Hevelius	23.483
1661	Mouton	23.500
1670	Picard	23.508
1672	Richer	23.482
1686	de Lahire	23.485
1690	Flamsteed	23.482
1700	Wurtzelbauer	23.498
1700	Manfredi	23.477
1703	Bianchini	23.476
1709	Romer	23.480
1719	Horrebow	23.480
1725	Manfredi	23.476
1736	Bouguer	23.473
1738	Cassini	23.472
1750	de Lacaille	23.472
1750	Cassini	23.472
1750	Cassini	23.469
1750	Bradley	23.472
1759	Mayer	23.471
1769	Maskelyne	23.469
1780	de Lalande	23.470
1788	Cassini	23.467
1814	Gauss	23.464
1817 *AD*	Gauss	23.464

(1972), Morrison (1982), Wittmann(1984), Jodi and Rossello (1987), and references quoted in their papers, yield values around $-0.3''/cy$ for the correction of Newcomb's value. However, the values derived from this study, $-53.58''/cy$ and $-58.83''/cy$, are significantly different from the modern values. One might speculate that the discrepancy could be caused by a yet undetermined acceleration or a cyclic change in the variation of ϵ.

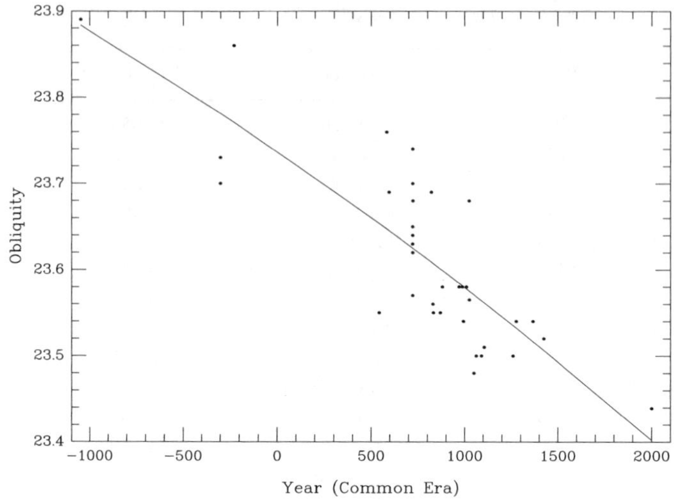

Figure 3 Variation of the obliquity of the ecliptic with time, Oriental data.

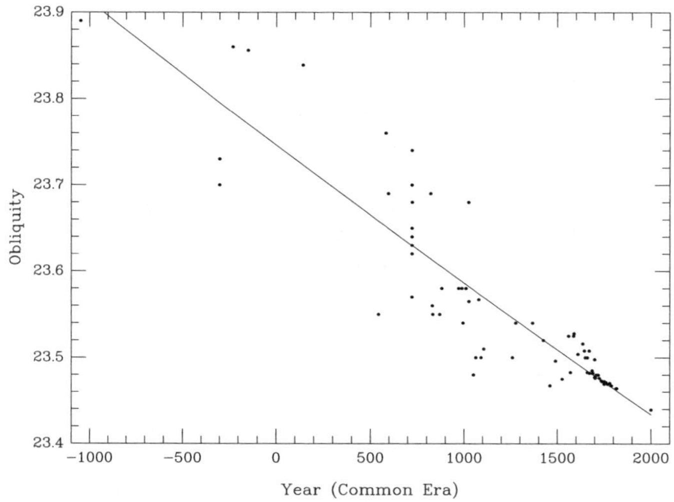

Figure 4 Variation of the obliquity of the ecliptic with time.

References

Beer, A., Ho, P-Y., Lu, G-D., Needham, J., Pulleyblank, E.G., and Thompson, G.I. (1961). An 8th-Century Meridian Line, *Vista in Astronomy*, **4**, pp. 3–28.

Jordi, C., and Rossello, G. (1987). Corrections to the FK5 Reference Frame, *Mon. Not. R. Astr. Soc.*, **225**, 723.

La Lande, J. (1792). *Astronomie*. Johnson Reprint Corp., New York (1966).

Laubscher, E. (1972). Determination of the Motion of the Ecliptic, *Astron. Astrophys.*, **20**, 407.

Lieske, J. H. (1970). On the Secular Change of the Obliquity of the Ecliptic, *Astron. Astrophys.*, **5**, 90.

Morrison, L. V. (1982). Analysis of Lunar Occultations, *Mon. Not. R. Astr. Soc.*, **198**, p. 1119.
Needham, J. (1959). *Science and Civilisation in China*. University Press, Cambridge.
Newcomb, S. (1898). *Astron. Pap. Washington*, **6**.
Rao, N. K. (1998). Sun Temple at Modhera, *Bull. Astron. Soc. India*, **26**, p. 1.
Whittmann, A.D. (1984). On the Variation of the Obliquity of the Ecliptic, *Astronomische Gesellschafts-Mitteilungen*, Nr. 62, pp. 201–204.

Glossary of Chinese Names

Beijing	北京	Bei Song	北宋
Bei Wei	北魏	Bianzhou	汴州
Caizhou	蔡州	Changde	常德
Dadu	大都	Da Xiang	大相
Dong Han	東漢	Fugou	扶溝
Gaocheng	告成	Han	漢
Hanoi	河内	Henanfu	河南府
Huaxian	滑縣	Huazhou	滑州
Jiankang	建康	Jiaozhou	交州
Junyi	浚儀	Kaifeng	開封
Langzhou	朗州	Li Zhi-chang	李志常
Liang	梁	Linyi	林邑
Liu Song	劉宋 (Song 宋)	Luoyang	落陽
Ming	明	Nangong Shuo	南宮説
Nanjing	南京	Qiu Chu-ji	丘處機
Runan	汝南	Sui	隋
Tang	唐	Tan Huo-zhi	檀和之
Tiele	鐵勒	Weixian	蔚縣
Weizhou	蔚州	Xiongnu	匈奴
Xuzhou	許州	Yangchang	陽場
Yuan	元	Yuan Tai	元太
Zhan Guo	戰國 (Warring States)	Zhou	周
Zhou Gong	周公		

1.7. The Legends of Vasiṣṭha – A Note on the Vedāṅga Astronomy

Yukio Ôhashi

1. Introduction

In this paper, I would like to discuss some topics about the Vedāṅga astronomy, which is a kind of ancient Indian astronomy. Before proceeding to the discussion, I would like to explain the position of the Vedāṅga astronomy in the whole history of Indian astronomy.

The history of Indian astronomy can roughly be divided into the following periods.[1]

(1) Indus valley civilization period.
(2) Vedic period (ca. 1500 BC – ca. 500 BC).

- (a) Ṛg-vedic period (ca. 1500 BC – ca. 1000 BC).
- (b) Later Vedic period (ca. 1000 BC – ca. 500 BC).

(3) Vedāṅga astronomy period.

- (a) Period of the formation of Vedāṅga astronomy (the 6/5/4th century BC (?)).
- (b) Period of the continuous use of Vedāṅga astronomy (up to the 2/3/4th century AD (?)).

(4) Period of the introduction of Greek astrology and astronomy.

- (a) Period of the introduction of Greek horoscopy (the 2nd – 3rd century AD).
- (b) Period of the introduction of Greek mathematical astronomy (ca. 4th century AD (?)).

(5) Classical Siddhānta period (the end of the 5th century AD – the 12th century AD).
(6) Coexistent period of Hindu astronomy and Islamic astronomy (the 13/14th century AD – the 18/19th century AD).
(7) Modern period (coexistent period of modern astronomy and traditional astronomy), (the 18/19th century AD).

The Vedāṅga (limb of the Veda) is a class of works regarded as auxiliary to the Veda. It consists of six divisions, one of which is astronomy (*jyotiṣa*).

2. Indian Originality of the Vedāṅga Astronomy[2]

The fundamental text of the Vedāṅga astronomy is the *Jyotiṣa-vedāṅga*, of which two recensions, Ṛg-vedic and Yajur-vedic, are extant.[3]

S.M.R. Ansari (ed.), History of Oriental Astronomy, 75–82.

The main structure of the Vedāṅga astronomy is as follows.

1 *yuga* = 5 years,
= 60 solar months (one solar month is 1/12 of a year),
= 61 *sāvana*-months (one *sāvana*-month is 30 *sāvana*-days),
= 1830 *sāvana*-days (civil days),
= 62 synodic months,
= 1860 *tithi*s (one *tithi* is 1/30 of a synodic month),
= 67 sidereal months,
= 1835 sidereal days,

1 year = 2 *ayana*s (half-years),
= 6 *ṛtu*s (seasons),
= 12 solar months,
= 366 *sāvana*-days,
= 372 *tithi*s.

The Vedāṅga calendar is a luni-solar calendar, and there are two intercalary months in a *yuga* (five years). One *sāvana*-day (civil day) is from sunrise to sunrise.

David Pingree argued that Vedāṅga astronomy was formed under Mesopotamian influence during the Achaemenid occupation of the Indus valley.[4] Pingree's argument is, however, definitely wrong. I shall show that Vedāṅga astronomy is based on the actual astronomical observations in North India.

Firstly, let us examine the length of a year. Pingree argued that one year in the Ṛg-vedic recension of the *Vedāṅga-jyotiṣa* was 366 sidereal days and not 366 *sāvana*-days. Although Yajur-vedic recension of the *Vedāṅga-jyotiṣa* states that one *yuga* is 61 *sāvana*-months (= 1830 *sāvana*-days) and the number of sidereal days in a *yuga* is the number of *sāvana*-days plus five, that is one year is 366 *sāvana*-days or 367 sidereal days, Pingree argued that this is due to the misunderstanding of the compiler of the Yajur-vedic recension. Pingree concluded that one year of the original Vedāṅga astronomy was 365 civil days, which is the same as the Egyptian–Persian year, and that it was introduced into India through Persia. This Pingree's argument is, however, completely wrong. I shall show that one year of the Vedāṅga astronomy was definitely 366 civil days. According to the *Vedāṅga-jyotiṣa* itself, the purpose of the Vedāṅga astronomy was to determine the proper time of sacrifices. Vedic sacrifices include the new and full moon sacrifices, seasonal (four monthly) sacrifices etc. At the time of the new and full moon sacrifices, the date of the new and full moon was fairly accurately determined. For example, the *Śāṅkhāyana-śrauta-sūtra* (I.3.5) states that the two days of full moon are the day on which the moon appears full about the setting of the sun and its succeeding day.[5] The day of full moon can be determined fairly accurately by this method, because the time of moonrise changes by about 49 minutes on the average per day. On the contrary, the change of season cannot be determined so accurately by naked eye observation. So, we can suppose that Vedāṅga astronomy could predict the date of the new and full moon for at least five years with accuracy, if it could not predict the season with the same accuracy. Now, the modern exact value of 62 synodic months is 1830.90 days, and that of 67 sidereal months is 1830.55 days. Therefore, one *yuga* of the Vedāṅga astronomy could not be different from 1830 days or so. If Pingree's argument is true, one *yuga* becomes 1825 days, and it makes nearly 6 days' error of the new and full moon days, which is evidently inadmissible at the time of the new and full moon sacrifices. Now, it is clear that one year of the Vedāṅga astronomy was 366 civil days. There is no similar calendar in ancient West Asia. So, the Vedāṅga astronomy must be Indian original astronomy.

Secondly, let us examine the seasonal change of the length of daytime and nighttime. The *Vedāṅga-jyotiṣa* states that the length of daytime is given by the following zig-zag function.

$$\text{The length of daytime} = \left(12 + \frac{2}{61}n\right) \textit{muhūrtas},$$

where n is the number of days after or before the winter solstice. One *muhūrta* is 1/30 of a day. According to this formula, the proportion of daytime and nighttime at the solstice becomes 2:3 which is observed at the latitude 35°N or so. This is the latitude of Kashmir area, and much north of the basin of the Ganga River which was the central area in Post-Vedic period. So, Pingree conjectured that this value was borrowed from Mesopotamia, of which the central area is at the latitude 35°N or so. This Pingree's conjecture is also wrong. I shall show that the above mentioned formula is based on the actual observations in North India. The seasonal movement of the sun was well noticed by Vedic people. For example, the *Kauṣītaki-brāhmaṇa* (XIX.3) states that the sun goes north for six months and stands still being about to turn southwards, and then goes south for six months and stands still being about to turn northwards.[6] This statement probably refers to the change of the position of sunrise or sunset. It changes much around the equinox, but does not change much around the solstice. So, the sun looks standing still around the solstice. This fact must have produced an idea that the seasonal change of certain phenomenon should be obtained from the observations around the equinox, and not from those around the solstice. So, the above mentioned formula of the Vedāṅga astronomy must have been obtained by the extrapolation from the observation of the change of the length of daytime around the equinox, and not by the interpolation from the observation around the solstice. Practically, there are two possibilities. If we assume that the formula was extrapolated from one *muhūrta*'s difference of the length of daytime during one solar month after the equinox, the most suitable latitude for this observation becomes 27°N. If we assume that the formula was extrapolated from two *muhūrta*'s difference during two solar months after the equinox, the most suitable latitude becomes 29°N. In any case, it is clear that this formula is based on the observations in North India. In order to make it intelligible, the actual length of daytime at 35°N, 29°N, and 27°N, and the *Vedāṅga-jyotiṣa*'s linear zig-zag function are together graphed in Fig. 1.

3. Continuous Use of the Vedāṅga Astronomy

The five-year cycle of the Vedāṅga astronomy was used in the *Arthaśāstra* (a political work attributed to Kauṭilya),[7] the *Śārdūlakarṇa-avadāna* (a Buddhist work),[8] the *Sūriya-pannatti* (a Jaina work),[9] etc. And also, the *Paitāmaha-siddhānta* quoted in the *Pañca-siddhāntikā* (XII) of Varāhamihira (the 6th century AD)[10] is a text of the Vedāṅga astronomy.

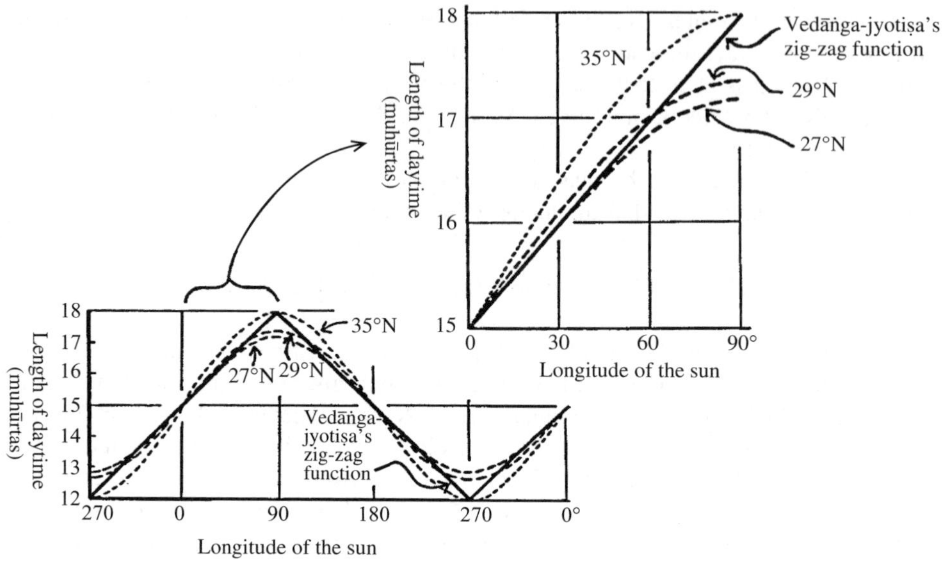

Figure 1 Vedāṅga-jyotiṣa's annual variation of the length of daytime.

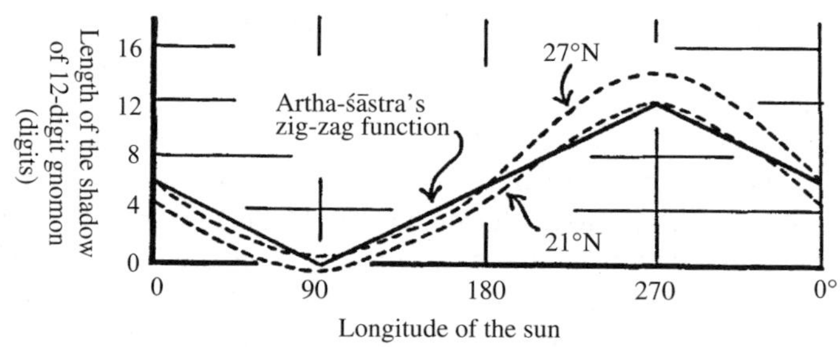

Figure 2 Artha-śāstra's annual variation of the midday shadow.

The epoch of this *Paitāmaha-siddhānta* is AD 80. So, the Vedāṅga astronomy must have been used in the 1st century AD. The *Śārdūlakarṇa-avadāna* was translated into Chinese in the 3rd century AD, and the *Sūriya-pannatti* is included in the canon of Śvetāmbara sect of Jainism which is said to be edited in the 5th century AD. It may be that the Vedāṅga astronomy was still used in their time.

What is interesting is that the annual variation and diurnal variation of the gnomon-shadow are mentioned in some of those texts, and they show that they are also based on the observations in North India.

The *Artha-śāstra* (II.20.41–42) gives the annual variation of the gnomon-shadow. It is graphed in Fig. 2 together with the actual variation at 27°N and 21°N. From this figure, it is clear that the data in the *Artha-śāstra* is based on the observation in North India. Similar data are found in the *Śārdūlakarṇa-avadāna*, the *Sūriya-pannatti*, etc.

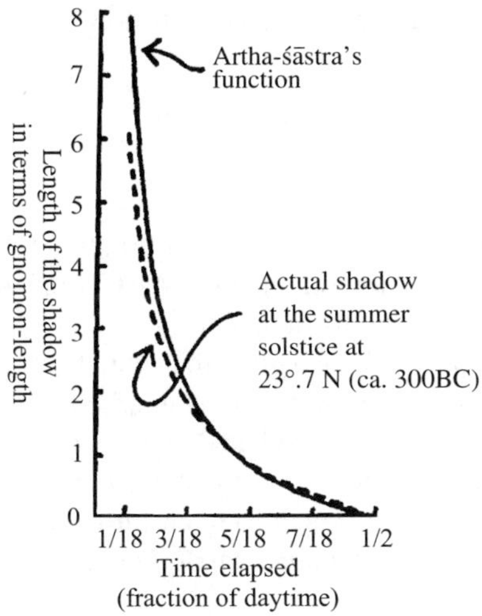

Figure 3 Artha-śāstra's diurnal variation of the gnomon-shadow.

The *Artha-śāstra* (II.20.39–40) gives the diurnal variation of the gnomon-shadow. As George Abraham rightly pointed out, it follows the following formula:[11]

$$\frac{d}{2t} = \frac{s}{g} + 1, \tag{1}$$

where t/d is the fraction of daytime which has elapsed since sunrise or is remaining until sunset, and s is the length of the gnomon of length g. It is graphed in Fig. 3 together with the actual variation at the summer solstice at the Tropic of Cancer (23°.7N in ca. 300 BC). It is again clear that this formula is based on the observation in North India. The *Sūriya-pannatti* also gives similar data.

From the above discussions, it is now clear that the Vedāṅga astronomy was produced in North India without explicit foreign influence.

4. Vedāṅga Astronomy under Greek Influence

It seems that Vedāṅga astronomy was still used when Greek astrology was introduced into India, and records of the Vedāṅga astronomy under Greek influence are found in the *Yavana-jātaka* of Sphujidhvaja and the *Pañca-siddhāntikā* of Varāhamihira.

The *Yavana-jātaka* (AD 269/270) of Sphujidhvaja is a Sanskrit work on Greek horoscopy.[12] Its last chapter (chapter 79) is devoted to mathematical astronomy. The text tells that "the instruction of the Greeks" (*Yavana-upadeśa*) is explained there, but also mentions the name of "the sage Vasiṣṭha" (*Vaśiṣṭha-muni*), who seems to be an Indian traditional astronomer. In the *Yavana-jātaka* (LXXIX.32), the diurnal variation of the gnomon-shadow is given. It can be expressed as follows.

$$\frac{d}{2t} = \frac{s - s'}{g} + 1, \tag{2}$$

where s' is the noon shadow.

The *Pañca-siddhāntikā* (IV.48–49) of Varāhamihira (the 6th century AD) also gives the same diurnal variation of the gnomon-shadow. As George Abraham pointed out,[13] the formula (1) is a special case of the formula (2).

The *Pañca-siddhāntikā* (II.9–10) also gives the annual variation of the gnomon-shadow, which is the same as that of the *Artha-śāstra* as graphed in Fig. 2. However, the position of the sun is given with reference to zodiacal signs, which must have been introduced into India along with Greek astrology, in the *Pañca-siddhāntikā*. So, this must be the remnant of the Vedāṅga astronomy under Greek influence. Varāhamihira tells that this annual variation is from the *Vāsiṣṭha-samāsa-siddhānta*. Here again appears the name of the sage Vasiṣṭha.

Now, let us examine the formula (2). David Pingree argued that the diurnal variation mentioned in the *Artha-śāstra* etc. is an adaptation of a Mesopotamian method.[14] I shall show that Indian formula is original.

According to Otto Neugebauer, Mesopotamian shadow table in the *mul Apin* can be obtained from the following equation:[15]

$$t = \frac{c}{s}, \tag{3}$$

where t is the time after sunrise, counted in time degrees (1 day $= 360°$), s the length of the shadow in terms of cubits, and c a constant, 60 at the winter solstice, 75 at the equinoxes, and 90 at the summer solstice. Strangely, the noon shadow is always 5/6 cubit.

Let us take up the gnomon-shadow after a quarter of daytime since sunrise. According to the Indian formula (2), the shadow becomes one gnomon-length longer than the midday shadow. According to the Mesopotamian formula (3), the shadow becomes double the length of the midday shadow. The shadow-length of a gnomon of length 1 at a quarter of daytime after sunrise according to the formulae (2) and (3) as well as the actual shadow-length for the latitude 35°N and 23°.7 N for a half year is graphed in Fig. 4. Here, we can see that the Mesopotamian formula (3) is good around the equinox at the latitude 35°N, while Indian formula (2) gives wrong value throughout an year, and that the Indian formula (2) is excellent around the summer solstice at the latitude 23°.7 N, while the Mesopotamian formula (3) is of no use there. So, we can conclude that the Indian formula (2) must have been originated in the observational data at the summer solstice in North India (particularly at the Tropic of Cancer), while the Mesopotamian formula (3) may be originated in the observational data at the equinox in Mesopotamia.

From the above discussion, it is now clear that the formula (1) was produced in North India, and it developed into the formula (2).

Mention may be made here that the *Yavana-jātaka* (LXXIX.31) and the *Pañca-siddhāntikā* (II.8) give the annual variation of the length of daytime which is the same as the *Vedāṅga-jyotiṣa* as graphed in Fig. 1. And also, the *Pañca-siddhāntikā* (II.11) gives

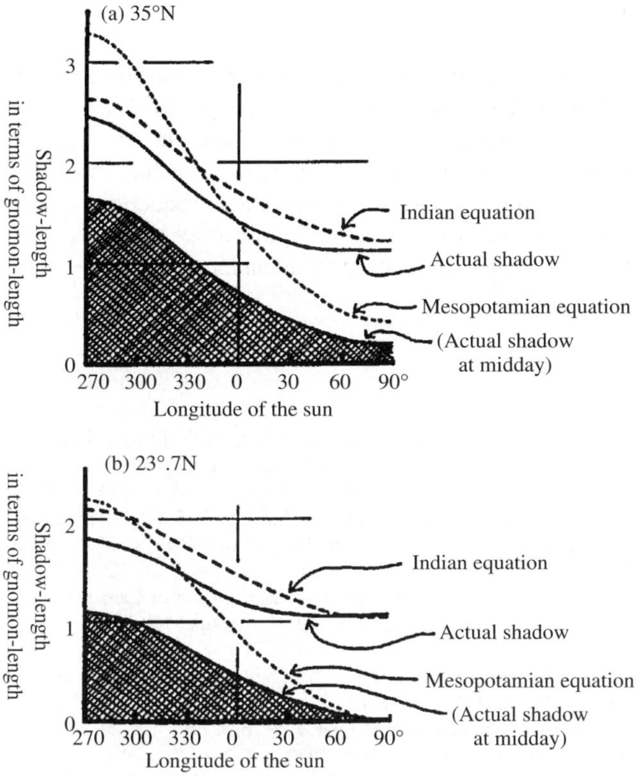

Figure 4 Shadow at the end of the first quarter of daytime.

the method to obtain *lagna* (rising point of the ecliptic) as follows.

$$lagna = \frac{36}{12 + s - s'} + L, \tag{4}$$

where L is the longitude of the sun, and the height of the gnomon is 12 *aṅgula*s (digits). As *lagna* minus L roughly corresponds to the time elapsed since sunrise, and 6 signs to the length of daytime, this formula can be considered to be the same as the formula (2). This seems to be a mixture of the Vedāṅga astronomy and the elements of Greek horoscopy like *lagna*. The *Pañca-siddhāntikā* (II.8) and (II.11) are also attributed to the *Vāsiṣṭha-samāsa-siddhānta*.

It is quite possible that the astronomical system of Vasiṣṭha was basically the Vedāṅga astronomy but received certain influence of Greek astrology. Greek influence of mathematical astronomy was not much at this stage.

The Vedāṅga astronomy under Greek influence is also found in the *Dafangdeng-dajijing, Ricangfen* (大方等大集經・日藏分), Chinese translation of a Buddhist text translated by Narendrayaśa in 586 AD.[16] The annual variation of the length of daytime and of midday shadow which is the same as Vedāṅga astronomy is recorded there, but the position of the sun is given with reference to zodiacal signs. From this text, we know that the Vedāṅga astronomy under Greek influence was widely used in India for certain period.

5. Conclusion

The Vedāṅga astronomy is Indian original astronomy based on the actual astronomical observations in North India. After the introduction of Greek astrology into India, the Vedāṅga astronomy was still used with some modifications, such as the use of zodiacal signs, and was connected with the name of the sage Vasiṣṭha. This astronomical system was widely used for a certain period. Then, after the introduction of Greek mathematical astronomy, the Vedāṅga astronomy gradually gave away to new astronomical systems, and finally the Hindu classical astronomy (Siddhānta astronomy) was established at the end of the 5th century AD.[17]

Notes

1. For the outline of the history of Indian astronomy, see Daqiao Youjifu (大橋由紀夫) (= Yukio Ôhashi): "Yindu tianwenxueshi cuoyao" (印度天文学撮要) (Introduction to the History of Indian Astronomy, in Chinese), *Xibei Daxue Xuebao (Ziran-kexue-ban)* (西北大学学報(自然科学版)) (*Journal of Northwest University (Natural Science Edition)*), Xi'an, China, Vol. 27, No. 3, (1997), pp. 185–189.
2. For the Vedāṅga astronomy in detail, see Ôhashi, Yukio: "Development of Astronomical Observation in Vedic and Post-Vedic India", *Indian Journal of History of Science*, Vol. 28, No. 3, (1993), pp. 185–251.
3. Kuppanna Sastry, T. S. (ed. by K. V. Sarma): "Vedāṅga Jyotiṣa of Lagadha with the Translation and Notes", *Indian Journal of History of Science*, Vol. 19, No. 3 (Suppl.), (1984), pp. 1–32, and No. 4 (Suppl.), pp. 33–74.
4. Pingree, David: "The Mesopotamian Origin of Early Indian Mathematical Astronomy", *Journal for the History of Astronomy*, Vol. 4, (1973), pp. 1–12.
5. Caland, W. (tr.): *Śāṅkhāyana Śrautasūtra*, Nagpur, 1953, p. 5.
6. Keith, A. B. (tr.): *Rigveda Brahmanas*, London, 1920, p. 452.
7. Kangle, R. P.: *The Kauṭilīya Arthaśāstra*, 3 parts, Bombay, 1965–1972.
8. It is the 33rd story of the *Divyāvadāna* (P. L. Vaidya ed., Buddhist Sanskrit Texts, Vol. 20, Darbhanga, 1959).
9. Kohl, J. F.: *Die Sūryaprajñapti versuch einer textgeschichte*, Stuttgart, 1937.
10. Thibaut, G. and Sudhākara Dvivedī: *The Pañcasiddhāntikā*, Benares, 1889; and Neugebauer, O. and D. Pingree: *The Pañcasiddhāntikā of Varāhamihira*, 2 parts, Copenhagen, 1970–71; and Kuppanna Sastry, T. S.: *Pañcasiddhāntikā of Varāhamihira with Translation and Notes*, Madras, 1993.
11. Abraham, George: "The Gnomon in Early Indian Astronomy", *Indian Journal of History of Science*, Vol. 16, No. 2, pp. 215–218.
12. Pingree, David: *The Yavanajātaka of Sphujidhvaja*, 2 parts, Cambridge, Mass., 1978.
13. See note (11).
14. See note (4).
15. Neugebauer, Otto: *A History of Ancient Mathematical Astronomy*, Part 1, Berlin, 1975, pp. 544–545.
16. *Taishō-shinshū-daizōkyō* (Taishō edition of the Chinese Tripitaka), Vol. 13, Tokyo, 1924.
17. For the observational astronomy in Classical Siddhānta period, see Ôhashi, Yukio: "Astronomical Instruments in Classical Siddhāntas", *Indian Journal of History of Science*, Vol. 29, No. 2, (1994), pp. 155–313.

1.8. Spherical Trigonometry and the Astronomy of the Medieval Kerala School

Kim Plofker
Brown University

1. Introduction

Geocentric astronomy in the Graeco-Islamic tradition is practically synonymous with spherical trigonometry: the chief physical structures underying its measurement systems are great circles on the sphere.[1] Although the methods of plane trigonometry inspired by the Greek chord function became the cornerstone of classical Indian mathematical astronomy, the corresponding techniques for exact solution of triangles on the sphere's surface seem never to have been fully developed within this tradition. Numerous rules nevertheless appear in Sanskrit texts for finding the great-circle arcs representing various astronomical quantities; these were presumably derived not primarily by spherics *per se* but from plane triangles inside the sphere or from analemmatic projections, and were supplemented by approximate formulas assuming small spherical triangles to be plane.[2]

The activity of the school of Mādhava (originating in the late fourteenth century in Kerala in South India) in devising, elaborating, and arranging solutions of celestial triangles, as well as in refining formulas or interpretations of them that depend upon approximations, has received a good deal of notice.[3] This paper presents two such rules from the *Tantrasaṅgraha* of Mādhava's student's son's student, Nīlakaṇṭha Somayājin, and examines them in comparison with corresponding formulas from other Indian treatises and from Islamic spherical astronomy.

2. Finding True Declination

2.1. *Pre-Islamic Indian formulas for declination*

Though the primary system of celestial coordinates (see Figure 1) in Sanskrit siddhāntas is that of ecliptic longitude λ and latitude β (as well as the important ecliptic declination here called $\delta_1(\lambda)$ from its Arabic/Persian designation as the "first" declination), other coordinates for celestial positions are sometimes required in these astronomical treatises. Among these, the "true" declination δ (the north or south distance perpendicular to the equator) of bodies other than the sun is needed for the risings and conjunctions of stars and planets, and in the case of the moon also for its *pātas*, ominous astrological events determined by the sum of the longitudes of the moon and the sun, and sometimes modified by constraints upon their true declinations.[4] True declination can be found from $\delta = \beta_* \pm \delta_1(\lambda_*)$, the sum of a body's polar latitude and the ecliptic declination of its polar longitude; but as the distinction between ecliptic and polar coordinates is often not made, the quantity is approximated by $\delta \approx \beta \pm \delta_1(\lambda)$.[5]

S.M.R. Ansari (ed.), History of Oriental Astronomy, 83–93.

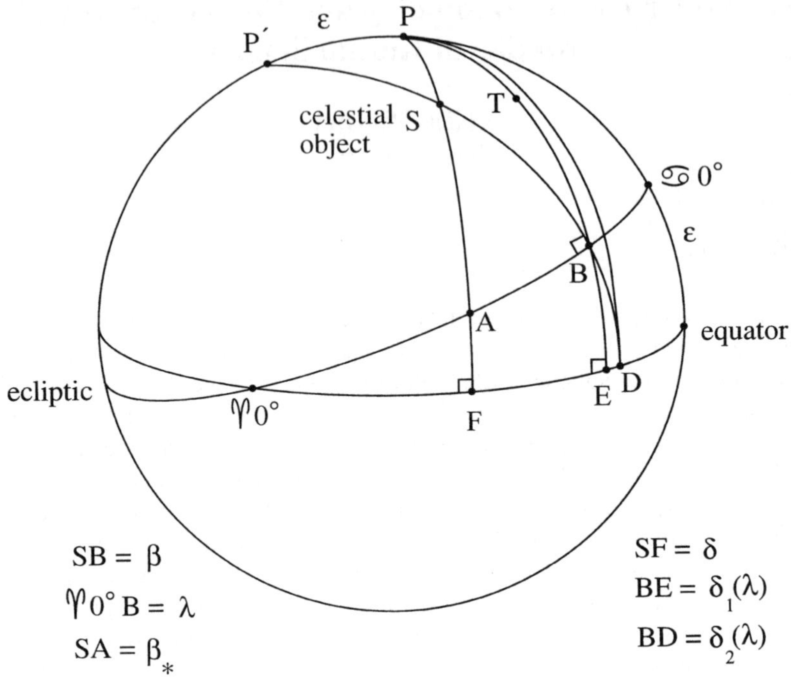

Figure 1

The first significant refinement of the calculation of δ appears to be given by Bhāskara (b. 1114) in calculating planetary risings and settings. He prescribes finding not the δ-arc through the body itself (arc SF in Figure 1), but an arc TB along PB that can be combined with BE $= \delta_1(\lambda)$ to give a value equal to SF. This is dependent upon the ecliptic deviation or *ayanavalana* ($\angle$ASB $=$ $\angle$PSP′), which we may derive by the Law of Sines from the spherical triangle PSP′ as follows (using the convention of capitalized trigonometric functions, e.g., Sin for $R \cdot \sin$ when the trigonometric radius $R > 1$):

$$\begin{aligned} \operatorname{Sin} \angle \mathrm{ASB} &= \operatorname{Sin} \angle \mathrm{PSP'} \\ &= \frac{\operatorname{Sin} \epsilon \cdot \operatorname{Sin} \angle \mathrm{SP'P}}{\operatorname{Cos} \delta} = \frac{\operatorname{Sin} \epsilon \cdot \operatorname{Cos} \lambda}{\operatorname{Cos} \delta} \\ &= \frac{\operatorname{Sin} \epsilon \cdot \operatorname{Sin}(\lambda + 90^\circ)}{\operatorname{Cos} \delta}. \end{aligned} \tag{1}$$

A common approximation to this *ayanavalana* rule,[6] which is not far off if δ is small (and has the advantage of being expressible simply in terms of the first declination δ_1, from spherical triangle Aries0°BE), is the following:

$$\operatorname{Sin} \angle \mathrm{ASB} \approx \frac{\operatorname{Sin} \epsilon \cdot \operatorname{Sin}(\lambda + 90^\circ)}{R} = \operatorname{Sin}(\delta_1(\lambda + 90^\circ)). \tag{2}$$

Bhāskara's prescribed method for finding arc TB allows for the use of either (1) or (2), as follows:[7]

> When one has subtracted the square of the Sine of the *ayanavalana* [$\angle$ASB] from the square of the Radius, the square-root is the "perpendicular" (*yaṣṭi*). The latitude of the planet is multiplied by the perpendicular and divided by the Radius, or else multiplied by the "day-sine" [Cos δ_1] of the planet increased [in longitude] by three signs, and divided by the Radius. [The result] is indeed the accurate [additive correction] appropriate for the correction of the declination.

In other words:

$$\delta = \delta_1 + \frac{\beta \cdot \sqrt{R^2 - \mathrm{Sin}^2 \angle \mathrm{ASB}}}{R} \approx \delta_1 + \frac{\beta \cdot \mathrm{Cos}(\delta_1(\lambda + 90^\circ))}{R}. \tag{3}$$

Clearly, this is essentially an approximation on the analogy of plane triangles: if one draws in the arc ST of a small circle parallel to the equator, and considers the resulting figure STB to be a plane right triangle,

$$\frac{\mathrm{TB}}{\beta} = \frac{\sqrt{R^2 - \mathrm{Sin}^2 \angle \mathrm{ASB}}}{R} \tag{4}$$

and the above equation follows.

2.2. *Declination-rules and Sanskrit astronomy in imperial India*

Versions of Islamic rules for δ proliferate in later Sanskrit astronomical works,[8] beginning with two formulas in the *Yantrarāja* of Mahendra Sūri, written in 1370. Although, as I have shown elsewhere,[9] this earliest adoption reveals some misunderstanding of the transmitted rules, this did not deter subsequent efforts at transmission and adaptation, often influenced by the authors' familiarity with existing Indian declination-rules. Eventually, some of the Islamic techniques were blended into the Sanskrit canon.

The beginning of such a synthesis is illustrated in the *Yantraratnāvalī* of Padmanābha in 1423. In one chapter of this work[10] he supplies formulas for the geometrical projections needed in the construction of the Islamic astrolabe, including the computation of δ for the stars to be marked in its rete. But his rule for the polar longitude β_*,[11] from which δ can be computed as explained in the previous section, is

$$\mathrm{Sin}\beta_* = \frac{\mathrm{Sin}\ \beta \cdot \sqrt{R^2 - \mathrm{Sin}^2 \angle \mathrm{ASB}}}{R}, \tag{5}$$

the right side of which is just a variant of Bhāskara's approximation in equation (4) to arc TB, *not* arc SA (that is, β_*): Padmanābha or his post-Bhāskara source appears to have merely substituted Sines for the corresponding arcs. Here it seems that a traditional Sanskrit rule has been transferred to the context of Islamic-inspired astronomy and pressed into service to solve a slightly different problem.

The reverse effect—the "Indianization" of a traditional Islamic rule—is seen in, for example, the *Sarvasiddhāntarāja* of Nityānanda, written in 1639 under the patronage of the

Mughal emperor Shāh Jahān to explain Islamic astronomy to readers of Sanskrit.[12] Among Nityānanda's rules for various coordinates for celestial positions is included the Islamic second declination δ_2, the distance perpendicular to the ecliptic between the ecliptic at the object's longitude and the equator:

> Whatever [ecliptic] declination is [derived] from the complement [of the longitude], that is called "reverse." The Radius is divided by the Cosine of the reverse declination and multiplied by the Sine of the [ecliptic] declination. The arc produced from the result is called the "second declination."[13]

That is,

$$\text{Sin}\,\delta_2 = \frac{\text{Sin}(\delta_1(\lambda)) \cdot R}{\text{Cos}(\delta_1(90 - \lambda))}, \tag{6}$$

a rule appearing in, for example, the *Ḥākimī Zīj* of ibn Yūnis in 1003.[14] This formula is true by the Law of Sines if $\angle$BDE is the complement of $\delta_1(90-\lambda)$; and since $\angle$BDE is the complement of $\angle$PDB, whose Sine by that Law is equal to $(\text{Sin}\,\epsilon \cdot \text{Sin}\,\angle\text{SP}'\text{P})/\text{Sin PD} = (\text{Sin}\,\epsilon \cdot \text{Sin}(90 - \lambda))/R = \text{Sin}\,\delta_1(90 - \lambda)$, the truth of the statement is confirmed.

Here Nityānanda is openly accepting Islamic techniques and terminology, in fact, the entire structure of Muslim astronomy. By contrast, a slightly later work, the 1646 *Siddhāntasārvabhauma* of a Benares astronomer named Munīśvara, strongly opposes the Indian adoption of most aspects of Islamic cosmology or astronomical theory;[15] yet even in this work, a blend of traditional Indian and Indianized Islamic trigonometry is found. Discussing planetary conjunctions, Munīśvara first directs the reader to compute δ by a method almost the same as Bhāskara's in equation (3), except that the Sine of arc TB is specified by an expression identical to Padmanābha's for the Sine of β_* in equation (5).[16] After criticizing this approach on geometrical grounds Munīśvara then remarks:[17]

> Because of the crudeness of the previously-stated rule for the declination, [the author] states a suitable rule for the true declination by another method, in two verses.
>
> The product of the [ecliptic] declination of the planet and the Radius is divided by the "day-sine" [Cos δ_1] of [the longitude of] the planet plus three signs. The arc of the result is the "other declination" of the planet, beginning with degrees; the sum [or] difference [of that] with the latitude in its own direction, when they are in the same or different directions respectively, is [computed]. The Sine of that arc times the "day-sine" of [the longitude of] the planet plus three signs is divided by the Radius. The arc of that [result], beginning with degrees, is the true declination.

This comprises a rule for the second declination δ_2 equivalent to equation (6), and the following expression:

$$\text{Sin}\,\delta = \frac{\text{Sin}(\beta + \delta_2(\lambda)) \cdot \text{Cos}(\delta_1(\lambda + 90))}{R}. \tag{7}$$

an expression similarly derivable by the Law of Sines from triangle SDF, and similarly originating in Islamic trigonometry.[18]

Munīśvara's contemporary and rival Kamalākara,[19] in his *Siddhāntatattvaviveka* of 1658, repeats the rules from equations (6) and (7), and supplements them with the following:[20]

> Or, the Sine of the "other declination" corrected [by adding the latitude] and divided by the Sine of the "other declination" is multiplied by the Sine of the [ecliptic] declination of the planet; the arc [from that result] is the true declination of the [planet's] disk.

Equivalently,

$$\text{Sin}\,\delta = \frac{\text{Sin}(\beta + \delta_2(\lambda)) \cdot \text{Sin}\,\delta_1(\lambda)}{\text{Sin}\,\delta_2(\lambda)}, \tag{8}$$

as appears from triangles SFD and BED via the Law of Sines (but which I have not so far seen in an Arabic or Persian treatise).

Islamic declination rules, then, have by the middle of the seventeenth century become so naturalized within Indian techniques that even a traditionalist such as Munīśvara has no objection to using them or even to preferring them on the basis of superior accuracy to a more orthodox siddhānta rule. This suggests that there must have been more activity in "Sanskritizing" Islamic trigonometry in the two and a half centuries following Mahendra Sūri's efforts in 1370 than is presently known to any detailed extent.[21]

3. Declination-rules in South India's Kerala school

A similar increase in complexity over the earliest Indian δ-rules is visible in those of the *Tantrasaṅgraha* of 1500 mentioned above; it is natural to ask whether spherical trigonometry in the school of Mādhava also continued a tradition affected by Islamic sources. In contrast to the Sanskrit tradition of the north described above, the overt evidence here for transmitted textual content currently seems to be nil.[22] That is, there are in the Keralese works as studied so far no references to foreign sources, no apparent borrowings of technical vocabulary, and no explicit adaptations of Islamic rules; there are, however, some interesting parallel results, none of which are implausible as independent achievements by such remarkable thinkers as Mādhava and his followers.[23] The following discussion, therefore, merely attempts to begin an examination of some of these parallels, in the hope that it may be useful for more thorough comparative analyses of the two traditions.

In chapter 6 of the *Tantrasaṅgraha*, Nīlakaṇṭha discusses the computation of the ominous *pātas*. After defining these and prescribing the calculation of the ecliptic declinations of sun and moon, he explains the calculation of the moon's latitude β from its maximum value $\beta_{\max}$ (owing, as his commentator Śaṅkara points out, to "the [dependence of the] accuracy [of the lunar declination] on correction by the moon's latitude"), and prescribes two rules for obtaining its true declination δ. The first of these is as follows:[24]

> When one has multiplied the Sine of the [lunar] latitude by the Cosine of the maximum declination [ϵ, the obliquity of the ecliptic] and the [Sine of the] given declination [δ_1] by the Cosine of that [latitude], both [products] are divided by the Radius. They are to be added or subtracted.
>
> [Their] sum (when in the same direction) or difference (when in different directions) is [the Sine of] the true declination [δ]. The Cosine of the true declination should remain [like] a "day-sine" [r, radius of a "day-circle" parallel to the equator] [or like the radius] in a [small] circle of latitude [parallel to the ecliptic].

> [This declination] should be taken for the true declination by those most knowledgeable in the calculations of the sphere.

Śaṅkara comments in the *Laghuvivṛti*:[25]

> Now, when one has multiplied the Sine of the given latitude of the moon by the complementary Sine of twenty-four degrees, [namely] the greatest [ecliptic] declination, and multiplied the Sine of its given [ecliptic] declination by the Cosine of that given latitude, and has made the sum of those two products when in the same direction [or] their difference in different directions, one should divide [that result] by the Radius. Then the quotient is the Sine of the true given declination of the moon. When one has subtracted its square from the square of the Radius, the square-root of the remainder should remain [like the radius] in a latitude-circle, and it should be [like] a day-sine: thus it is incidentally said. The true declination derived by the method thus stated should be used by those most knowledgeable in the calculations of the sphere, in *vyatīpāta* and other [matters] pertaining to the visible celestial sphere.

In other words,

$$\frac{\mathrm{Sin}\beta \cdot \mathrm{Cos}\,\epsilon}{R} \pm \frac{\mathrm{Sin}\,\delta_1(\lambda) \cdot \mathrm{Cos}\,\beta}{R} = \mathrm{Sin}\,\delta. \tag{9}$$

Typically, Nīlakaṇṭha's verses say nothing about how he derived or demonstrated the rule, and the brief commentary is of little help. (He may, of course, owe it to one of his predecessors or colleagues among those "most knowledgeable in the calculations of the sphere," or "*golavittama*;" this may in fact be a reference to Mādhava himself, who is sometimes styled "the *Golavid*" by later members of his school.)[26] Both in its competence and its conciseness, his treatment is reminiscent of his complete solution of the astronomical triangle in chapter 3 of the *Tantrasaṅgraha*.[27]

Essentially the same formula as Nīlakaṇṭha's is included among the δ-rules of Ibn Yūnis;[28] but here the solution is expressed as a combination of two terms he calls "Sines," as follows:

$$\begin{aligned} S_1 &= \frac{\mathrm{Sin}\,\delta_1(\lambda) \cdot \mathrm{Cos}\beta}{R}, \\ S_2 &= \frac{\mathrm{Sin}\,\beta \cdot \mathrm{Cos}\,\epsilon}{R}, \\ \mathrm{Sin}\,\delta &= S_1 \pm S_2. \end{aligned} \tag{10}$$

Nīlakaṇṭha's next few verses contain another and a remarkable rule for δ, for which I have neither seen nor produced any thoroughly satisfactory explanation:[29]

> Or, the declination can also be deduced by means of the "highest declination" of the moon.
>
> One should divide the Sine of 24 degrees, multiplied by the Cosine of the maximum latitude, by the Radius: the so-called "elevation of the center" [n] of the latitude-circle is obtained.
>
> Now both the Sine and Cosine of [the longitude of] the node [of the lunar orbit] corrected for precession, multiplied by [the Sine of] the maximum latitude and divided by the Radius, become the two results of that.

One should divide the Cosine[-result], multiplied by the final day-sine [i.e., Cos $\delta_{1_{max}}$ = Cos ϵ], by the Radius. It is applied positively or negatively to the elevation of the center when it is produced from the node [located in the half of the ecliptic] beginning with Capricorn or Cancer [respectively].

The square root of the sum of the squares of that [above result] and the Sine-result is the "highest declination" of the moon. The Sine-result multiplied by the Radius and divided by that is the "motion of the *ayana*."

That is applied positively or negatively to [the longitude of] the moon corrected for precession when the node is in [the half of the ecliptic] beginning with Libra or Aries [respectively]. The Sine of the arc of that [result] is then multiplied by its own "highest declination" and divided by the Radius.

[Thus] the declination of the moon at that time is made correct.

Śaṅkara recapitulates the procedure without explaining much more about it:

Now, he states [a rule] for the purpose of calculating its true declination by another method: Or else, let the true declination of the moon be derived from its "highest declination" [D], as follows: when one has multiplied the Sine of twenty-four degrees by the Cosine of the maximum latitude measured by zero–twenty-seven [270; $\beta_{max} = 270' = 4;30°$], one should divide [the product] by the Radius. Then the quotient is named the "elevation of the center" [n] of the circle of radius [determined by] the maximum latitude, [with its] center [upon] a string tied to two sign-projections [?].

Now when one has computed both the Sine and Cosine of the [longitude of the] node corrected by precession, and multiplied [them] by [the Sine of] the maximum latitude, one should divide [them] by the Radius. The two quotients from that should be the Sine- and Cosine-results. Having multiplied [one] of those two, the Cosine-result, by the final day-sine, one should divide [the product] by the Radius. Then one should apply the resulting quotient to the elevation of the center positively or negatively according to the [location] of the precession-corrected node in [the six signs] beginning with Capricorn or with Cancer. The square-root of the sum of the squares of the elevation of the center thus modified and the Sine-result is called the "highest declination" of the moon.

Then, having multiplied the previously-computed Sine-result by the Radius, one should divide [the product] by that "highest declination." The quotient from that is called the "motion of the *ayana*." Again, that is also called the "motion of latitude" because of its being computed by means of the latitude. And that is to be applied positively or negatively to the precession-corrected moon according to the [location] of the precession-corrected node in [the six signs] beginning with Libra or Aries. Then having computed the Sine of [the longitude of] the moon made accurate by the correction using the previously-stated and derived "motion of the *ayana*," and then having multiplied [that] by its own computed "highest declination," one should divide [the product] by the Radius. The quotient from that is the given declination of the moon [that] is accurate, as [he says].

Using the above notation and λ_M and λ_N for the precession-corrected longitudes of the moon and its node respectively, we may express these instructions as

$$\frac{\text{Sin}\,\epsilon \cdot \text{Cos}\,\beta_{\max}}{R} = n,$$

$$\left(n \pm \frac{\text{Cos}\,\lambda_N \cdot [\text{Sin}]\beta_{\max}}{R} \cdot \frac{\text{Cos}\,\epsilon}{R}\right)^2 + \left(\frac{\text{Sin}\,\lambda_N \cdot [\text{Sin}]\beta_{\max}}{R}\right)^2 = [\text{Sin}]^2 D,$$

$$\frac{\text{Sin}\,\lambda_N \cdot [\text{Sin}]\beta_{\max}}{R} \cdot \frac{R}{[\text{Sin}]D} = [\text{Sin}]\Delta\lambda, \tag{11}$$

$$\frac{\text{Sin}(\lambda_M \pm \Delta\lambda) \cdot [\text{Sin}]D}{R} = [\text{Sin}]\delta.$$

The simplest case of this procedure is illustrated in Figure 2, where the lunar node N is assumed to be at the intersection of the ecliptic and the equator (here considered to coincide with the zero-point of the zodiac, meaning that longitudes are precession-corrected). The circle of maximum latitude with radius Cos $\beta_{\max}$ = VH is drawn parallel to the ecliptic Aries0°J. Then the obliquity of the ecliptic $\epsilon = \angle$JNQ is equal to $\angle$WHV, and from the first relation in (11), we see that in triangle HWV, WV $= n$. Since $\lambda_N = 0$, the term to be added to n reduces to (Sin $\beta_{\max}$) $\cdot$ (Cos ϵ)$/R$, which from triangle VUN is seen to be VU. And since Sin $\lambda_N = 0$, the Sine of the "highest declination" D is simply WV+VU = WU, which will indeed be equal to the Sine of the moon's maximum true declination, namely arc HQ. But the third step of the required procedure is puzzling. Geometrically, we ought to derive from the longitudinal arc Aries0°L the arc Aries0°M along the lunar orbit, so that by the Law of Sines we would correctly have (Sin δ)/(Sin D) = (Sin Aries0°M)$/R$, analogous to the typical calculation of δ_1 from λ and ϵ. But we are actually instructed to

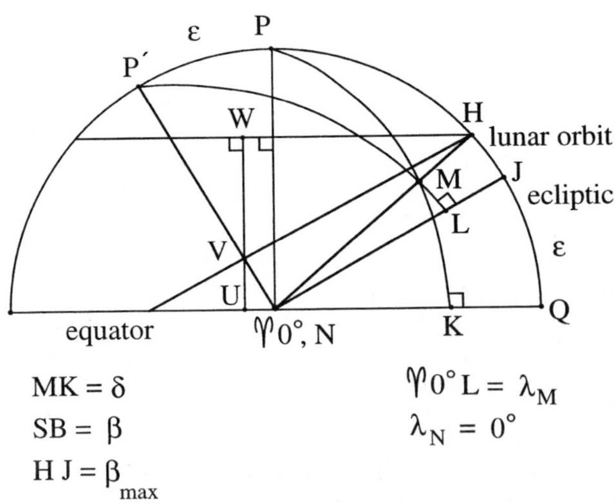

Figure 2

calculate not Aries0°M but an additive correction to the longitude Aries0°L, which in this special case with $\lambda_N = 0$ is zero; so the fourth and final step of the procedure substitutes an arc of the ecliptic for what should be an arc of different length along the lunar orbit. Practically speaking, however, the smallness of β_{max} makes the resulting inaccuracy in δ no more than a few minutes, at least in this case.

In the late medieval period, then, we see a continued growth in complexity in standard Indian trigonometrical methods for finding astronomical quantities, both in the northern schools and courts (where this growth was stimulated by encounters with Islamic techniques) and in the Keralese tradition. In neither case, though, do the more sophisticated methods for deriving spherical arcs appear to bud off into a separate topic on the solution of spherical triangles in general; and in neither case do the demonstrably exact solutions eliminate the simultaneous use of the ingenious approximations characteristic of the Indian astronomical tradition.

Notes

1. All the spherical trigonometry we will actually use in this paper is the Law of Sines for spherical triangles: i.e., for a triangle on the sphere composed of three great-circle arcs a, b, and c with opposing angles α, β, and γ respectively, $\sin\ a/\sin\ \alpha = \sin\ b/\sin\beta = \sin\ c/\sin\ \gamma$.
2. The role of spherics in Indian astronomy is also discussed in Plofker (2000).
3. See, e.g., Gupta (1974) and (1985), and presumably Hariharan (1988), which I have not yet been able to consult.
4. The *pātas* are defined by longitudes only in, e.g., *Pañcasiddhāntikā* III, 20; see Neugebauer-Pingree (1970), part I, pp. 44–45. In the Brāhmapakṣa founded on the *Paitāmahasiddhānta*, however, the true declination of the moon at a *pāta* is supposed to equal δ_1 of the sun; see, e.g., *Paitāmahasiddhānta* V, 9, in Pingree (1967), p. 498.
5. For example, in the *Brāhmasphuṭasiddhānta* of Brahmagupta (7, 5); see Dvivedi (1901), p. 101. Al-Bīrūnī notes that combining $\delta_1(\lambda)$ and β to give δ is standard practice for both Indians and Persians; see Debarnot (1985), pp. 212–213.
6. The exact rule for $\angle$ASB is given by Bhāskara in *Siddhāntaśiromaṇi* I, 5, 21; see Chaturvedi (1981), p. 247. The approximation shown is a standard formula for *ayanavalana* in the Brāhmapakṣa (as in *Brāhmasphuṭasiddhānta* 4, 17; see Dvivedi (1901), p. 77); the Āryapakṣa generally substitutes for the Sine of $(\lambda + 90)$ its Versine (as in *Śiṣyadhīvṛddhidatantra* 5, 25; see Chatterjee (1981), part I, p. 102).
7. *Siddhāntaśiromaṇi* I, 7, 3: see Chaturvedi (1981), pp. 276–277.
8. For a discussion of the evidence for the use of particular Arabic and Persian works in India, see Ansari (1995).
9. See Plofker (2000).
10. See Ohashi (1997), p. 216.
11. Ohashi (1997), p. 265.
12. See Pingree (1978), pp. 323–326.
13. Verses I, 4, 49–50ab of the *Sarvasiddhāntarāja* as they appear on f. 29r of Rajasthan Oriental Research Institute (Alwar) 2619, which reads as follows: yā koṭitto bhavetkrātiḥ sā vilomā nigadyate vilomakrāṃtikoṭijyoddhṛtā trijyā guṇā punaḥ 49 krāṃtijyā phalajaṃ cāpaṃ dvitīyā krāṃtirucyate.
14. See King (1972), pp. 293–294.
15. See Pingree (1978), pp. 321–322.
16. *Siddhāntasārvabhauma* bhagrahayuti, 31: see Ojhā (1978), p. 413. This suggests that Padmanābha did not originate the modification of Bhāskara's TB-rule replacing arcs with Sines, but simply reinterpreted this modified expression to refer to β_* instead.
17. *Siddhāntasārvabhauma* bhagrahayuti, 41–42: Ojhā (1978), p. 420.
18. This rule was known, e.g., to Abū Naṣr in the eleventh century; see Debarnot, p. 212.

19. See Pingree (1978), pp. 322–323.
20. *Siddhāntatattvaviveka* 8, 24cd–25ab: Dvivedi (1925), fasc. 3, p. 388.
21. Known examples of Sanskrit works treating Islamic astronomy in this period include the 1615 *Yantraśiromaṇi* of Viśrama (which contains no coordinate conversions), an earlier work of Nityānanda's, the 1628 *Siddhāntasindu*, one or more translations of the *Zīj* of Ulugh Beg, and possibly the anonymous *Hayatagrantha* (which correctly identifies δ, δ_1, and δ_2, but does not derive them mathematically); for more details on these works, see Pingree (1978). This conclusion also requires the modification of the closing statement of Plofker (2000), which deduces from Mathurānātha Śukla's repetition in 1782 of some errors in the *Yantrarāja* of 1370 that the methods of spherical trigonometry made little impression on Indian mathematical astronomy in the intervening 400 years. Mathurānātha's perpetuation of the early inaccuracies (which I have not seen reproduced in any of the late medieval treatises discussed here) was apparently more the exception than the rule.
22. However, the probable influence of Islamic lunar theory on the work of a later member of this school, Acyuta Piṣāraṭī, is mentioned in Pingree (1978), p. 319.
23. For an overview of some of the mathematical and astronomical accomplishments of this school, see Sarma (1972).
24. The passages quoted from Nīlakaṇṭha and Śaṅkara, as well as the ones described above, appear in Sarma (1977), pp. 311–313.
25. The concise *Laghuvivṛti* is the only commentary on this portion of the *Tantrasaṅgraha* currently published; see Sarma (1977), p. lxxv, and Sarma (1972), p. 126.
26. See Sarma (1972), p. 51.
27. See Gupta (1974).
28. See King (1972), pp. 290–293.
29. Sarma (1977), pp. 313–314.

Bibliography

Ansari, S. M. R. (1995): "On the Transmission of Arabic-Islamic Astronomy to Medieval India," *Archives intern. d'histoire des sciences*, Vol. 45, No. 135, pp. 273–295.

Chatterjee, Bina (1981): *Śiṣyadhīvṛddhidatantra of Lalla*, 2 volumes, Indian National Science Academy, New Delhi.

Chaturvedi, Murali Dhara (1981): *Siddhāntaśiromaṇi of Bhāskarācārya*, Sampurnanand Sanskrit University, Varanasi.

Debarnot, Marie-Thérèse (1985): *Kitāb maqālīd 'ilm al-hay'a: la trigonométrie sphérique chez les Arabes de l'Est à la fin du Xe siécle*, Institut Français de Damas, Damas.

Dvivedi, S. (1901): *Brāhmasphuṭasiddhānta of Brahmagupta* (1901/1902), Benares.

Dvivedi, S. (1925): *Siddhāntatattvaviveka, Benares Sanskrit Series* fasc. 1, 2, 3, 6, and 14, 2nd ed. rev. by Muralīdhara Jhā (1924–1935), Benares.

Gupta, R. C. (1974): "Solution of the Astronomical Triangle as Found in the *Tantra-Saṃgraha* (A.D. 1500)," *Indian Journal of History of Science*, Vol. 9, No. 1, pp. 86–99.

Gupta, R. C. (1985): "Mādhava's Rule for Finding Angle between the Ecliptic and the Horizon and Āryabhaṭa's knowledge of It" in *History of Oriental Astronomy*, G. Swarup et al., eds, Cambridge University Press, Cambridge, pp. 197–202.

Hariharan, S. (1988): "Declination in Indian Astronomy and the Approach of Kerala Astronomers," *Journal of the Asiatic Society*, Vol. 30, pp. 39–49.

King, David A. (1972): *The Astronomical Works of ibn Yunus*, PhD dissertation, Yale University.

Neugebauer, O., and D. Pingree (1970): *Pañcasiddhāntikā of Varāhamihira*, 2 volumes, Kongelige Danske Videnskabernes Selskab, København.

Ohashi, Y. (1997): "Early History of the Astrolabe in India," *Indian Journal of History of Science*, Vol. 31, pp. 199–295.

Ojhā, Mīṭhālāla (1978): *Siddhāntasārvabhauma by Munīśvara*, Part III, Sampurnanand Sanskrit University, Varanasi.

Pingree, David (1967): "The *Paitāmahasiddhānta* of the *Viṣṇudharmottarapurāṇa*," *Brahmavidyā* 31–32 (1967–68), pp. 472–510.
Pingree, David (1978): "Islamic Astronomy in Sanskrit," *Journal for the History of Arabic Science*, Vol. 2, pp. 315–30.
Plofker, Kim (2000): "The Astrolabe and Spherical Trigonometry in Medievel India," *Journal for the History of Astronomy*, Vol. 31. pp. 37–54.
Sarma, K. V. (1972): *A History of the Kerala School of Hindu Astronomy*, Vishveshvaranand Institute, Hoshiarpur.
Sarma, K. V. (1977): *Tantrasaṅgraha of Nīlakaṇṭha Somayāji*, Vishveshvaranand Institute, Hoshiarpur.

1.9. Astronomical Dating and Statistical Analysis of Ancient Eclipse Data

Kevin D. Pang, Kevin K. Yau and Hung-hsiang Chou

Abstract

All 13 Shang dynasty oracle bone eclipse records have been uniquely matched to six solar and seven lunar eclipses in the 14th to 12th centuries B.C. The King Zhong Kang fifth year autumnal (October 16, 1876 B.C.) and King Yu third year "double sunset" (September 24, 1912 B.C.) eclipses confirm the accuracy of the revised *Bamboo Annals* Xia dynasty chronology (Nivison and Pang, *Early China* **15**, 1990, 87–95). The eclipse dates are plotted against the number of generations before 841 B.C. (the earliest accurate historical date), the respective kings ruled. The curve of bestfit has both the strengths of verified royal genealogy – continuity – and eclipse dating – accuracy. It is 99% accurate, and can be confidently used as a foundation for building a detailed absolute chronology for the Xia, Shang and Zhou dynasties, an important project in China's current Five-Year Plan (Song Jian, *Sci. Tech. Daily*, May 17, 1996; Wehrfritz, *Newsweek*, July 7, 1997).

1. Introduction

Astronomical dating involves retroactively computing and matching the circumstances of a celestial phenomenon observed by ancient people, but recorded without a precise date. When applied to ancient Oriental astronomical records such dating can give the double benefit of improving historical dates, and our knowledge of the motions of Solar System bodies, particulary cometary orbits and the Earth's past rotation rate. Let us first introduce a recent compilation of these records.

The systematic observation of heavenly bodies and celestial phenomena began at the dawn of civilization in China (Pang, 1985; Chen and Xi, 1993; Pang and Yau, 1996). Both the quantity and quality of astronomical records kept for 4,000 years are quite impressive. This vast treasure trove of information has recently become readily accessible to users after a ten-year literature search and compilation effort. Beginning in 1974 hundreds of scholars from the Beijing Observatory and cooperating institutions scanned 150,000 volumes of Chinese historical texts and local gazettes, and excerpted 1.2 million words pertaining to astronomy. The fruit of this labor is the *Union Compilation of Ancient Chinese Records of Celestial Phenomena* (Zhuang and Wang, 1988). It lists 1,600 + solar eclipse, 1,100 + lunar eclipse, 1,000 + cometary, and 200 + occultation records.

Past analysts have found these records to be generally reliable. For example of the past 30 apparitions of Halley's Comet all, with possibly two exceptions, have been faithfully observed and recorded by Chinese astronomers (Yeomans and Kiang, 1981). One of us (K.K.Y.) has had similar success in reconstructing the historical orbit of Comet Swift-Tuttle back to 703 B.C., using these records (Yau *et al.*, 1994). These results have, in turn, refined the ephemerides, important tools for historians of astronomy. The lack of clock error (delta T) values before 700 B.C. (defined in Eq. 1) had left the timescale of the

S.M.R. Ansari (ed.), History of Oriental Astronomy, 95–119.

ephemerides unconstrained (Morrison, 1992). To make optimum use of the ephemerides we need to determine delta T as far back in time as possible (Pang and Yau, 1996).

The analyses of Chinese historical eclipse and occultation records by many researchers have also given consistent results for the Earth's past rotation rate, or delta T as a function of time (Han *et al.*, 1984; Pang *et al.*, 1988, 1995, 1998; Liu, 1988; Hilton *et al.*, 1988, 1992; Liu and Yau, 1990; Zhang, 1994; Zhang and Han, 1995). It is our purpose to both extend these results further back into the High Antiquity, as well as make them more precise. The oldest of these records have turned out to be especially valuable for the following reason.

The longterm time dependence of the clock error or delta T can be well-approximated by a parabolic equation (due to a linear lengthening of the day, expected from constant tidal braking),

$$\text{delta T} = \text{AT} - \text{UT} = c\text{T}^2, \tag{1}$$

where UT is Universal Time (based on the Earth's rotation), and AT is atomic (cesium clock) time. T is conveniently chosen to be, say, the number of centuries before January 0, 1800. Therefore the weight of a record grows to the second power of its age. The constant a and the first power bT can be neglected for this reason. Dawn and dusk observations are especially valuable, as sunrise and sunset times can be retrospectively computed. We will analyze 13 such records in the following 11 sections, beginning with two fascinating sunrise eclipse records from the Western Zhou (*ca.* 1100 to 771 B.C.) and Eastern Jin (A.D. 317 to 420) dynasties, observed from the *same* place.

2. The April 21, 899 B.C. and April 4, A.D. 368 "Double Sunrise" Eclipses Over Zheng, China

The *Bamboo Annals*, entombed in 299 B.C. and unearthed in A.D. 281, states that "in the first month of spring in the first year of King Yi of the Western Zhou dynasty, he ascended the throne, the day dawned twice at Zheng (34.5°N, 109.8°E)," now called Hua District, in Shaanxi Province (Herrmann, 1966). *Kaiyuan zhanjing*, compiled by Gautama Siddhartha in A.D. 714–724, cites this passage and *adds* that "in the 2nd (*actually 12th*) year of Sheng Ping reign period of King Shang (*actually King Xi*) the day (*also*) began twice at Zheng (italic ours)." Siddhartha was an Indian astronomer who worked in the court of Tang dynasty emperor Xuan Zhong (A.D. 712–756). A reign period is a subdivision of a monarch's rule. These events have been uniquely matched with the April 21, 899 B.C. and April 4, A.D. 368 sunrise eclipses at Zheng (Fig. 1). See Pang *et al.* (1988) and Pang *et al.* (1995) for detailed discussions.

The 899 B.C. and A.D. 368 solar eclipses had respective magnitudes of 0.95–0.97 and 0.991–0.998. The brightness changes were thus greater than observed during the January 4, 1992 sunset eclipse over Southern California, USA, which had a magnitude of 0.91–0.92. The 1992 event was indeed perceived as a "double dusk" by many observers. For example David H. Levy (1992) noted that "... as annularity ended ... Sunset had come and gone, but the sky began to brighten not darken ... For almost 15 minutes it continued to brighten until the onrushing shadow of Earth took over and darkness fell again." In a "double dawn" the sequence of events would be reversed, as it is the symmetrical opposite of a "double dusk."

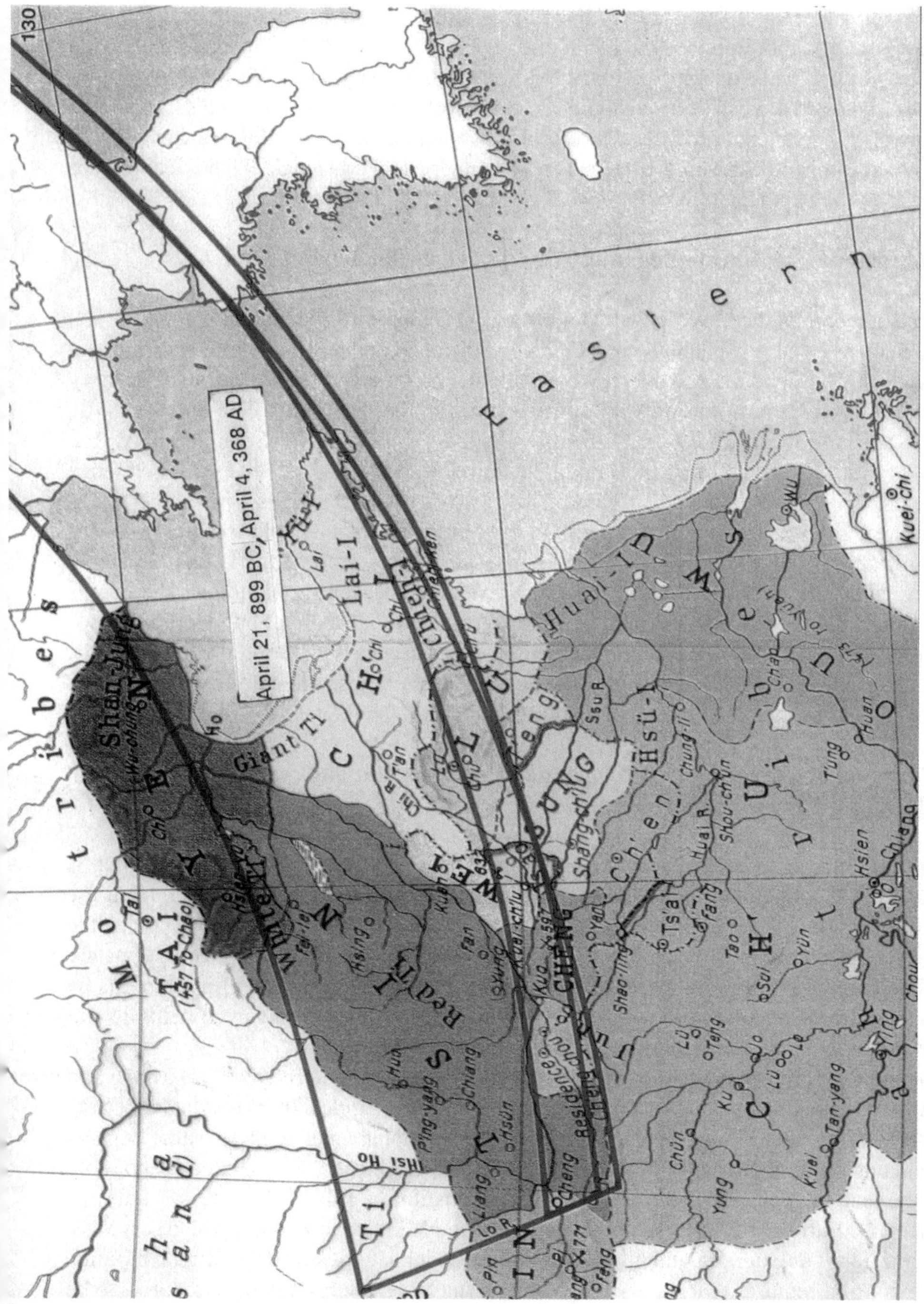

Figure 1 The paths of annularity of the solar eclipses of April 21, 899 B.C. (wide), and April 4, A.D. 368 (narrow). The paths were truncated at Zheng or Cheng (small circle) by the morning terminator. "Double dawns" had been observed there on both dates. Reproduced from Albert Herrmann's *Historical Atlas of China* (1966) with the permission of Edinburgh-Aldine Press.

The recurrence of a central solar eclipse at the same site under almost identical circumstances, after 1,266 years, accurately links up an ancient delta T value (5.8 ± 0.15 hr for 899 B.C.) with a more precise medieval one (1.7 ± 0.1 hr for A.D. 368), and makes the statistics of such early data more robust. Moreover, the first year of King Yi, heretofore known only to be between 966 and 895 B.C., can now be firmly fixed at 899 B.C. Medieval records of a "double sunset," described with as much detail as the 1992 sunset eclipse, have recently been discovered in a search of Chinese local gazettes. These records are analyzed in the next section.

3. The May 5, 1361 "Double Sunset" Eclipse Over Shanghai, China

The extant record of the May 5, 1361 "double sunset" eclipse over Shanghai (31°N, 121°E) states that: "The Sun was about to set. Then it dimmed suddenly, looking rather like a banana leaf. The sky darkened as if it were at night, and the stars shone gloriously. Totality lasted for but an instant. The sky brightened again, and the stars faded. The Sun set a little later (Songjiang District Gazette, 1663)." Songjiang is a district of Shanghai City. Tao Zongyi (1360–1368), a farmer in Tiantai, 200 km due south, also saw the same sequence of events.

Since totality was said to have lasted "for only an instant" (the maximum possible was 5 min) it appears that Songjiang and/or Tiantai were close to the edge(s) of the path of totality, and very near the evening terminator. Our computer simulation of the events gave a delta T of 8 ± 1 min. See the terminus of the path of totality, bounded by the two locations, in Fig. 2. In the next section we will merge our data with those of Zhang (1994), and Zhang and Han (1995) to combine the best features of these three up-to-date analyses of ancient and medieval Chinese solar eclipse records.

4. Merging of Three Datasets from Analyses of Ancient and Medieval Chinese Eclipse Records

We have plotted our rising/setting solar eclipse data as dotted circles in Fig. 3, which is based on Fig. 1 of Zhang and Han (1995). Included with the three delta T's from the previous sections, is an additional delta T value of 1.4 ± 0.1 hr from our analysis of the sunrise eclipse record of November 13, A.D. 532 (Pang *et al.*, 1988). All four points are dated and labeled. The delta T values from the analysis of timed solar eclipse records by Zhang and Han (1995) are plotted as filled circles. Results from analysis of central solar eclipse records by Zhang (1994) are plotted as unfilled circles.

The three datasets overlapped and blended well without any normalization. Zhang and Han (1995) have drawn a curve of bestfit through the filled circles. It fits the data of Zhang (1994) and ours as well, when extended to earlier and later times with the same degrees of local curvature. Our four datum points are especially close to it, separated by no more than the size of a circle. The merged datasets constitute a reliable source not only for studying the Earth's rotation rate in the Antiquity and Middle Ages, but can also serve as a stepping stone for extending the time series trend of delta T back into the High Antiquity. This extension requires that we analyze Shang dynasty oracle bone eclipse records, to be discussed in the next five sections.

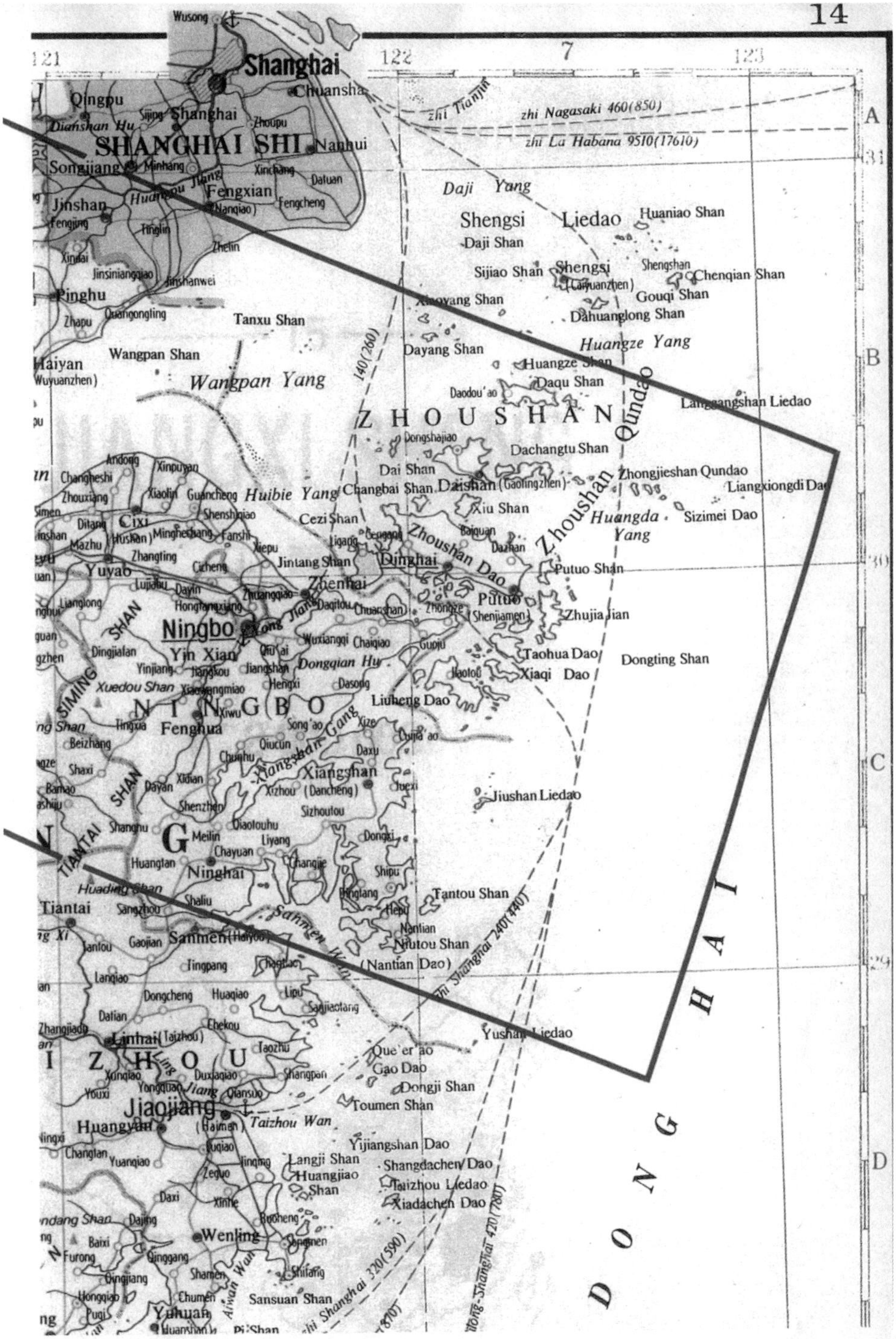

Figure 2 The path of totality of the solar eclipse of May 5, 1361. Note that Songjiang, Shanghai (31°N, 121°E) and Tiantai (200 km due south) astride the respective northern and southern limits of the path. "Double sunsets" had been observed from both sites on that day. Reproduced from *A Provincial Atlas of China* (1977), with the permission of Ditu Chubanshe.

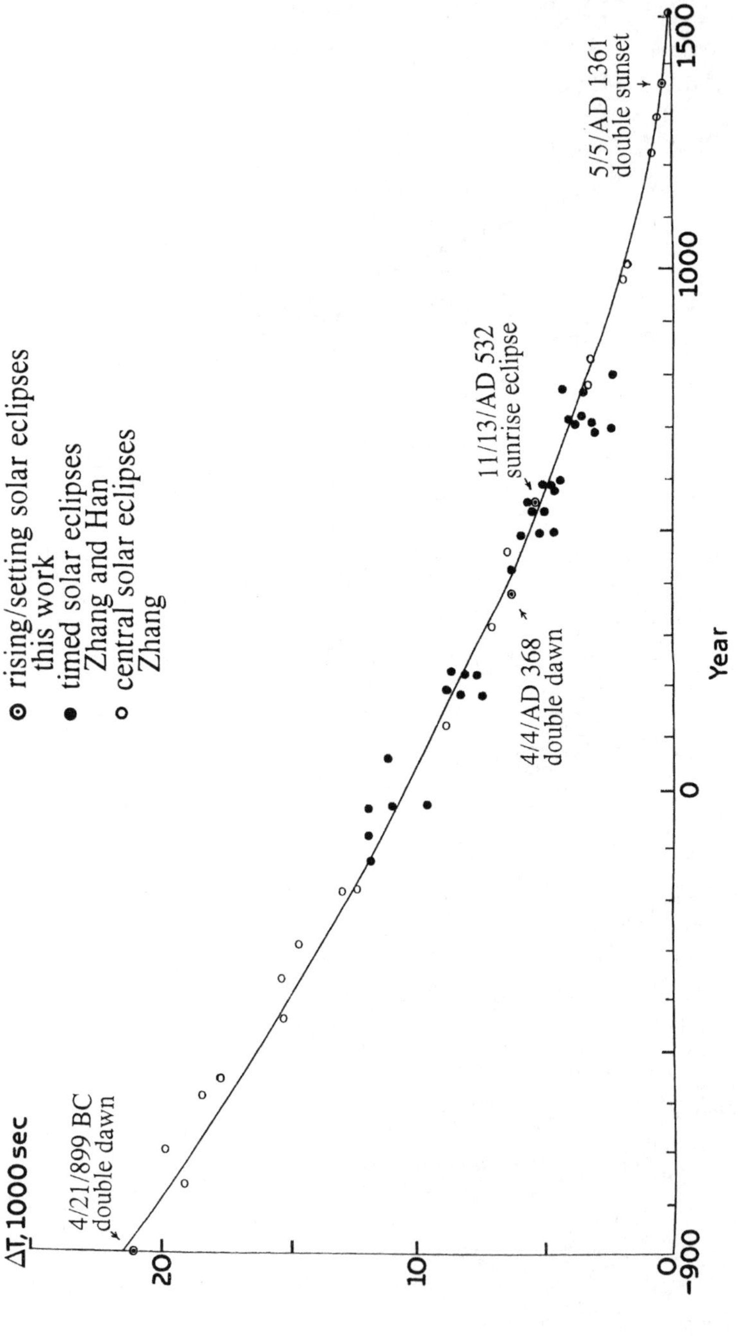

Figure 3 The clock error delta T *vs.* year. Plus years are A.D. and minus years – B.C. A curve has been fitted through the filled circles by Zhang and Han (1995), and we extended it to the edges of the diagram with the same degrees of local curvature.

5. Shang Dynasty Oracle Bone Eclipse Records and Royal Genealogy

The earliest Chinese writing that has been deciphered so far are the inscriptions on Shang (also called Yin) dynasty oracle bones. Earlier inscriptions, *e.g.*, from the antecedent Xia dynasty (*ca*. 1914 to 1600 B.C.), have been unearthed recently. They have not yet been deciphered, although some of their characters are very similar to Shang writing (*China Pictorial*, 1993). In contrast to the Xia artifacts, which have been systematically looked for by archaeological teams fanned out through China, the Shang dynasty ox scapulas and tortoise shells were accidentally discovered in a Beijing apothecary in 1899, sold as medicinal "dragon bones." The source was later traced to the ruins of the Shang capital Anyang (36.1°N, 114.3°E) in Henan Province. Subsequent excavations there have unearthed 160,000 oracle bones.

Among the astronomical data (Xu *et al.*, 1989) are six solar and seven lunar eclipse records that have a cyclic date (and for two, the lunar month as well), so as to be datable (Chang, 1980; Xi, 1984). The Chinese cyclic date is similar to our week in design but runs in 60-day cycles. It has been in continuous use from time immemorial. Additional advantages of using these inscriptions are not only their great antiquity, but also because they are written in an ancient form of the Chinese language that has been deciphered. One of us (H.H.C.) has reviewed the oracle bone inscriptions and their evolution into modern Chinese (Chou, 1979). All 13 Shang eclipse inscriptions have been successfully dated, and found to be almost perfectly correlated with the archaeologically verified Shang royal genealogy (Pang *et al.*, 1997). This discovery can be used to determine the absolute chronology of early China. We will discuss this briefly here, and then in greater detail in Section 13.

Before the Shang oracle bones were analyzed some scholars considered the dynasty to be legendary. It has now been proven that records had been accurately passed down to historian Sima Qian (91 B.C.) and the authors of the *Bamboo Annals* (299 B.C.). The kinglist, compiled from the Shang inscriptions, is identical to that of Sima Qian and the *Bamboo Annals*. The genealogy of the Xia kings has not yet been directly verified by archaeological finds, but the absolute chronology of early Xia from King Yu to Zhong Kang, as recorded in the *Bamboo Annals*, has been confirmed by astronomical evidence (Nivison and Pang, 1990; Pang and Yau, 1996).

We can combine the strength of royal genealogy – continuity – with that of astronomical dating – accuracy – by correlating them. Our process is similar to stretching and contracting an approximately ruled and slightly flexible (genealogical) scale here and there to line up selected ticks with independent benchmarks (eclipse dates). The dating of Shang dynasty oracle bone eclipse records gives us such dates, which is explained in the next four sections. The dating of the Xia eclipse records will be discussed in Sections 10 and 11.

6. Shang Dynasty Period One Oracle Bone Lunar Eclipse Records Dated

Five Period I (King Wu Ding) lunar eclipse records are listed in Table 1 in the order of decreasing relative ages, ranked according to evolutionary advances in the calligraphic style of their characters (Takashima, 1988):

Table 1 Circumstances of the Period I Lunar Eclipses, Recorded on Shang Dynasty Oracle Bones

	Cyclic day (Modulo 60)	Lunar month mentioned	Diviner name	Matching eclipse date, B.C.	Local time
31	*jia-wu*	Not stated	Bin	Dec. 25, 1322	3:30
57	*geng-shen*	13th (in Nov/Dec/Jan)	Zheng	Nov. 24, 1311	4:28
20	*gui-wei*	Not stated	Zheng	Feb. 27, 1278	3:54
22	*yi-you*	8th (in Jul/Aug/Sept)	Zheng	Sept. 2, 1279	3:05
9	*ren-shen*	Not stated	Anon.	Nov. 4, 1282	6:45

All five records use the phrase "yue you *shi* (Moon has been eaten)." The local time shown is a.m., and for mid-eclipse. Matching eclipse dates should range <59 years, as Diviners Bin and Zheng worked for King Wu Ding, who reigned for 59 years. Their names appear frequently in Period I inscriptions. The lunisolar Shang calendar is known to have initiated the year with the first new Moon after winter solstice, and added a leap (13th) month, when needed, to keep the ordinal number of the lunar months in step with the seasons of the solar year (Tung, 1960). Assuming that the Shang day was either "Egyptian" (dawn to dawn) or "Roman" (midnight to midnight) previous researchers could not find five computed eclipses that can match all the stated conditions (Chang, 1980). However the ancient Chinese day, unlike either, began at ~3 a.m. instead. Events that occurred between midnight and dawn had a 85% probability of being recorded with the old date, and 15% with the new (Kiang, 1981).

Armed with this new knowledge and a delta T, derived from the Period I (June 5, 1302 B.C.) Shang oracle bone solar eclipse record, that states "three flames ate the Sun, big stars were seen," Pang *et al.* (1989, 1995) successfully matched all five records. As they all occurred between midnight and dawn the odds are that one would carry the new date, and the rest–the old. This turns out to be exactly the case: The 1322 B.C. eclipse is the odd one. Our Julian calendar dates span only 44 years, and are in the same order as Takashima's (1988), except two of the latest inscriptions are transposed. Separated by less than four years they are simply too close together to be ranked by calligraphic style changes. Having successfully matched *all* Period I lunar eclipse records we now go on to date the contemporary Period I solar eclipse record, mentioned above.

7. The June 5, 1302 B.C. Total Solar Eclipse, Recorded on a Shang Dynasty Oracle Bone

Of the six solar eclipses recorded on Shang oracle bones only one was reported to have been total: Inscription No. 11506, written on both sides of a plastron. Unless otherwise noted we will use the *He-ji* index system, devised by Guo Moruo (1977). See Fig. 4 for an ink rubbings of the inscribed surfaces (Tung, 1953). One of us (H.H.C.) has transcribed the characters on the right side onto Fig. 5, and the left onto Fig. 6 to make them easier

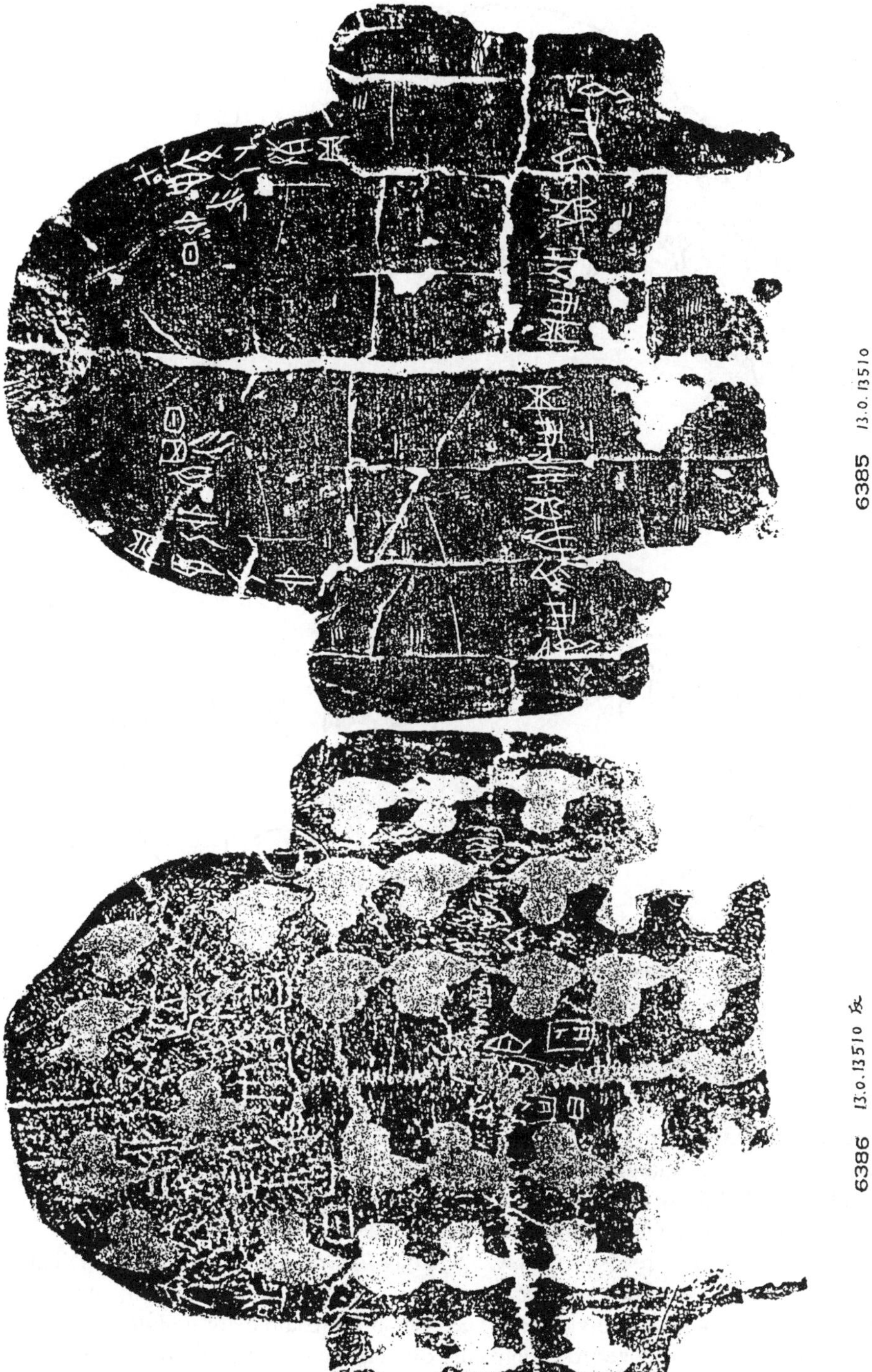

Figure 4 Ink rubbing of a Shang dynasty oracle bone (both sides of a plastron), describing a total solar eclipse. Transliterated in Figs. 5 and 6. Reproduced from Tung (1953) with permission.

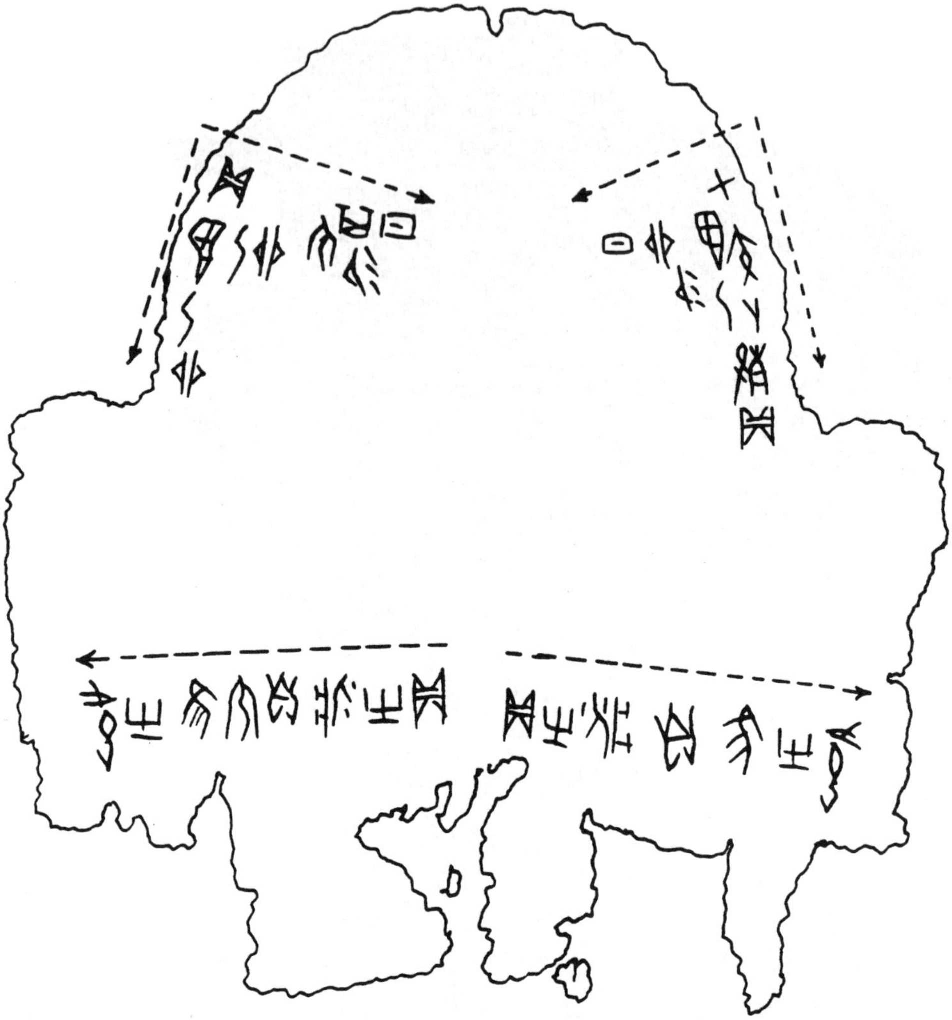

Figure 5 Transcription of the right panel in Fig. 4. Inquiry about the appearance of the Sun on the 52nd day of the Chinese 60-day cycle (*cf*. our week) was made on the 51st day. See the complete translation in the text.

for our readers to see. We have discussed in great details the provenance, etymology, previous scholarship, interpretation and dating of this inscription elsewhere (Chou and Pang, 1990), and will only briefly summarize our findings here. We have generally followed the decipherment of Yao and Xiao (1988) and the translation of Xu *et al*. (1989).

The inscription in Fig. 5 reads as follows: "On day *jia-yin* (51st day of cycle), Diviner Gu inquired: 'Will the next day, *yi-mao*, be a sunny day?' 'Inquiry about the next day *yi-mao*, will *yi-mao* not be a sunny day?' " The divination would be of no interest to us if it were not for the *post facto* verification of the divination, inscribed on the reverse side (Fig. 6).

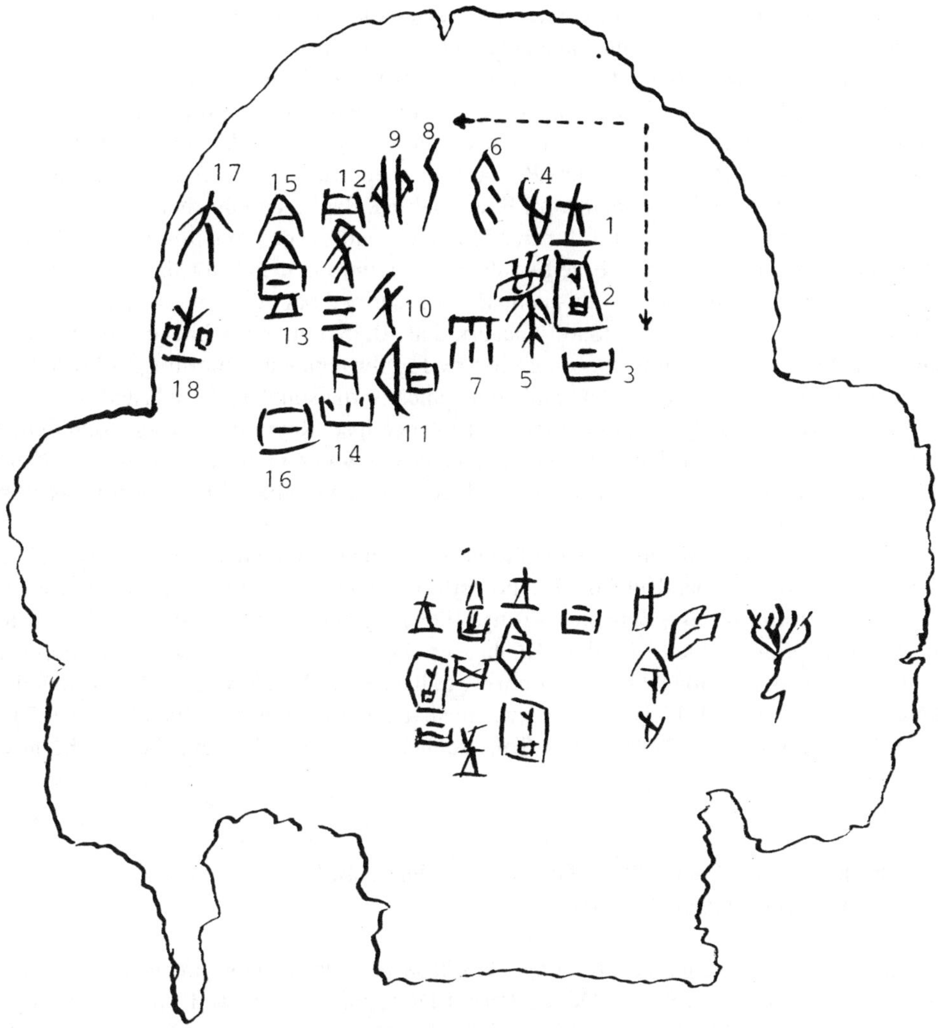

Figure 6 Transcription of the verification of the divination (left panel of Fig. 4): "... fog from the 52nd day till early (next) morning (then) three flames ate the Sun, big stars (were seen)."

Fig. 6, the verification, reads as follows: "The king made the prognostication and said that there was no misfortune and no rain. From day *yi-mao* (52nd day) to early (next) morning there was fog, (then) three flames ate the Sun, big stars (were seen)." The verification describes a total solar eclipse, as numerous oracle bone experts and historians of science have concluded, *e.g*., Liu (1945), Needham (1959), Liu (1978), Wang and Siscoe (1980), Chen (1987), Xu *et al*. (1989), and Chou and Pang (1990). Our reasons are quite straightforward: Only during totality can the spectacle of the Sun having been "eaten by flames" (a primitive people's description of a vanished solar disk, surrounded by coronal streamers and/or prominences) be seen. And only during totality can the stars and planets be observed with the naked eye in the daytime.

It is well known that Diviner Gu, with Bin and Zheng (see Table 1), worked for King Wu Ding (Chou, 1958), whose lunar eclipse inscriptions have been dated in the last section. So we look for a total solar eclipse that occurred on cyclic day 53 *bing-chen* between December 25, 1322 B.C. (the earliest Period I lunar eclipse) and February 27, 1278 B.C. (the latest), which could be seen from Shang China. To be certain that nothing is left out we have also done the search for the 51st and 52nd cyclic days *jia-yin* and *yi-mao*. The search gave no match for these two days, but a unique match for the 53rd: June 5, 1302 B.C. The circumstances of this eclipse are described below.

Our computations show that totality occurred at 10:45 a.m. local time, consistent with the oracle bone's description of obscuration by fog through early morning, followed by clearing, disappearance of the Sun, and appearance of the stars. Totality lasted for 6 min 20 sec, an extraordinarily long time for a total solar eclipse. Regrettably where totality had been seen was not noted. However, since Diviner Gu is known to have worked in King Wu Ding's court (Chou, 1958), it is reasonable to assume that the eclipse was total as seen from Anyang.

For a delta T of 7 hr 20 min the path of totality would be centered at Anyang, also called Yin. See the stippled band in Fig. 7. The path of totality can be moved to the east (or west) to cover almost all Shang sites (Tung, 1964), by adding (or subtracting) 17 min to (or from) the preferred delta T value. The broken lines in Fig. 7 indicate the new limits of totality if such an addition or subtraction were to be made. Therefore we conclude that the delta T value for 1302 B.C. is 7.3 ± 0.3 hr, consistent with the mean value of 7.2 ± 0.2 hr from the analysis of the 12th-century B.C. solar eclipse records, described in the next section.

8. Shang Dynasty Period Four Oracle Bone Solar and Lunar Eclipse Records Dated

Having successfully matched *all* Period I eclipse records we now proceed to date the Period IV eclipse inscriptions. Of the Period IV (Kings Wu Yi and Wen Ding) solar eclipse records listed by Zhang (1975) and Xi (1984), inscription Nos. 33694 and 33703 are only inquiries about whether an eclipse will occur. Since these are not observational records they are not useful to us (Hu, 1986).

The relevant contents of the valid records are: No. 33696 – "Day *yi-si* (42nd day) . . . Sun has *zhi* . . . reported to Shang Jia that evening-night, nine oxen sacrificed." No. 33698 – "Day *geng-chen* (17th day) . . . Sun has *zhi* . . . reported to He and Father Ding . . . nine oxen sacrificed . . ." No. 33699 –"Day *wu-zi* (25th day) . . . Sun has *zhi* . . . reported to He." This had been misdeciphered as day *wu-shen* (45th day), since corrected by Yao and Xiao (1988). No. 33700 –"Inquiry on day *yi-chou* (2nd day): 'Will the Sun have *zhi*?' (*verification*) 'It indeed had *zhi*.' " No. 33710–"Day *xin-si* (18th day) . . . Sun (has) *zhi* . . . reported to Father Ding." The large number of oxen sacrificed, and the prompt report to ancestors/deity, betray the Shang's great concern about the changes that had occurred on the Sun.

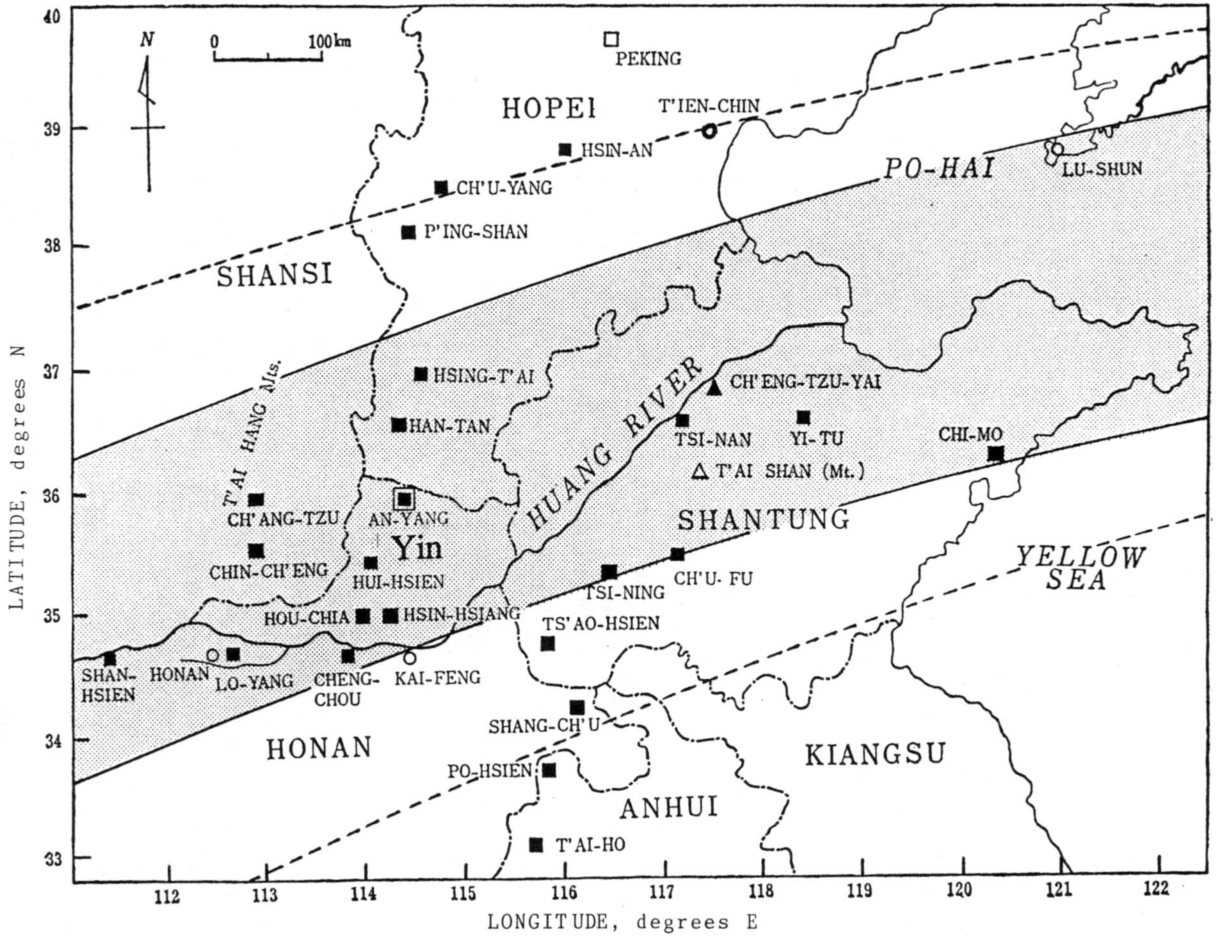

Figure 7 The path of totality of the June 5, 1302 B.C. solar eclipse. The stippled band was computed for a delta T of 7 hr 20 min, and can be moved to the east (or west) by adding (or subtracting) 17 min to (or from) this value to cover almost all Shang sites (filled squares). Modern cities are denoted by unfilled circles and squares. Reproduced from Tung (1966) with permission.

These inscriptions use the phrase "has *zhi*" rather than the "has *shi* (been eaten)," typical in Period I. The best explanation for this is that primitive Chinese characters that sound alike were used interchangeably (Xi, 1984). The "borrowed usage" of words with a similar sound is very common in Shang and Zhou inscriptions. Of the 15 Shang eclipse inscriptions with a cyclic date, eight use the term "*shi*" and seven use "*zhi*." The even split can be considered as a proof of dual usage until at least Period IV. We have found further proof of the interchangeability of the two terms by considering the context where *zhi* appears in various sentences.

First, *zhi* seems to refer to a *foreseeable* change on the Sun, as No. 33700 had shown. The anticipatable time could be two days, as No. 33703 also states – "Divination on day *ren-ji* (49th day): 'Will the Sun have *zhi* on day *jia-yin* (51st day)?' " Once some believed that *zhi* refers to sunspots. However in 1972 an ox scapula (*Tun-nan* No. 726) was unearthed from Anyang, inscribed with "Inquiry on day *ren-yin* (39th day): 'The Moon has *zhi* ...' " (Xi, 1984). Since the Moon does not have transient spots we are forced to conclude that *zhi* refers to a *foreseeable* change that can happen to *both* the Sun and Moon, and by analogy with *shi*, clearly also means "been eclipsed." Having established that these are *bona fide* eclipse records Pang *et al.* (1995) successfully dated them all.

The task of dating the five inscriptions is quite straightforward: We search eclipse canons for a solar eclipse that occurred on the stated cyclic day, and could be seen from Shang China in the late 13th or early 12th century B.C. Our dates agree with those of Zhang (1975), with the exception of inscription No. 33699. Zhang used the incorrectly deciphered cyclic day 45 *wu-shen*, and matched the record with the solar eclipse on May 7, 1161 B.C., which was not visible from China (Mucke and Meeus, 1983). Once this error was corrected the cyclic day 25 record has a good match with the solar eclipse on June 27, 1163 B.C. The *Tun-nan* No. 726 "The Moon has *zhi* ..." eclipse record, mentioned above, can be matched with the lunar eclipse of July 2–3, 1173 B.C.

The solar eclipses, all partial as seen from Anyang, were probably accidentally noticed during the performance of sunrise (and sunset) rituals at new Moon when "the king receives as guest the rising Sun" (Xu, 1990). The extinction by large airmasses had apparently made it easier to see the partial eclipses. Although the eclipse magnitude and solar elevation had not been noted, useful upper or lower limits on delta T can still be deduced from visibility constraints on the rising or setting eclipsed Sun from Anyang. The circumstances of the five eclipses are listed in Table 2.

No useful upper or lower limit on delta T can be deduced from the first record, because the eclipse apparently occurred at midday. However three of the four derived limits cluster around 7 hr 10 min, consistent with the 7 hr 20 min we got from analysis of the June 5, 1302 B.C. total solar eclipse record, discussed in the last section.

We have plotted our own results from Sections 2, 4, 7 and 8 in Fig. 8. Upper or lower limits are denoted by down and up arrows, respectively, and discrete values, by asterisks. The data of Liu and Yau (1990) and Stephenson and Yau (1992) are also plotted as filled and unfilled squares, respectively. The analysis of our data gave an equation of bestfit of delta $T = (30 \pm 2.5)T^2$, which is consistent with the bestfit curve in Fig. 3. We will now return to the oracle bone eclipse records.

Table 2 Circumstances of the Five Period IV Solar Eclipses, Recorded on Shang Dynasty Oracle Bones (in chronological order)

Inscription no.	Day no.	Cyclic day	Eclipse date, B.C.	Magnitude	Delta T	
					Hr	Min
33700	2	*yi-chou*	5/6/1226	0.78		
33698	17	*geng-chen*	10/21/1198	0.73	>7	26
33710	18	*xin-si*	6/7/1172	0.87	<8	52
33699	25	*wu-zi*	6/27/1163	0.45	>7	11
33696	42	*yi-si*	10/31/1161	0.86	<7	00

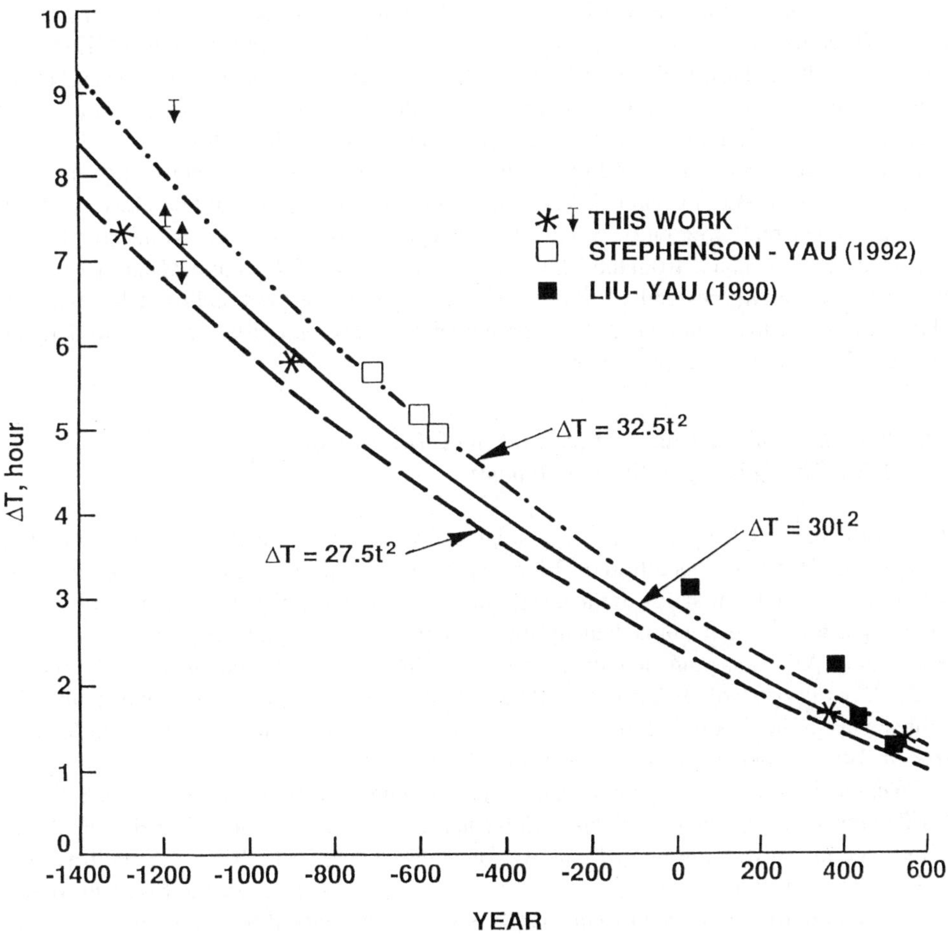

Figure 8 The clock error delta T *vs*. year. Plus years are A.D. and minus – B.C. The equation of bestfit to our data is delta T = $(30 \pm 2.5)T^2$, where T is the number of centuries before 1800.

Some have suggested that the phrase "has *zhi*" refers to red Sun or Moon. However it is highly unlikely that the atmospheric reddening processes can randomly generate 20 such sightings in 65 years (1226 to 1161 B.C.), with six coinciding with an eclipse day or night in a Saros cycle, a unique and highly systematic time series trend. Actuary tables put the odds of six such coincidences out of 20 at 1000 to 1 (Nolan, 1975)! The frequency of red Sun and Moon sightings has been estimated from a count of 145 such events, reported between A.D. 1030 and 1500 (*Gujin Tushu Jicheng*, a 1726 compendium of earlier events), *i.e.*, $(145/470) \times 65 = 20$. In the next section we will get a good feel for the degree of difficulty in matching just *one* solar eclipse record in 40 years.

9. Shang Dynasty Period Three Oracle Bone Lunar Eclipse Record Dated

Having successfully matched *all* Periods I and IV eclipse records we now go on the date a Period III eclipse inscription. The only oracle bone eclipse record from Period III (Kings Lin Xin and Kang Ding), No. 22067, states–"Divination on day *jia-yin* (51st day): 'There was an eclipse (*shi*). Should it be reported?' " It does not say solar or lunar, so it would seem easy to match. However the choices proved to be surprisingly limited.

First, we must choose in the 52 years between the latest (1278 B.C.) Period I eclipse and the earliest (1226 B.C.) Period IV eclipse. Second, our choice should be closer to 1226 B.C., as the Period II king Zu Jia reigned for 33 years. No solar eclipse, visible from China, occurred on a day *jia-yin* from the 1250's through the 1220's B.C., but the March 20, 1242 B.C. lunar eclipse fits perfectly. Viewed from Anyang the Moon rose eclipsed that evening. Having successfully matched *all* Shang oracle bone eclipse records, we go on to date the two Xia dynasty solar eclipse records.

10. The *Book of Xia* Solar Eclipse and the Fifth Year of King Zhong Kang of the Xia Dynasty

The world famous King Zhong Kang solar eclipse is based on three sources: (1) The "Yin cheng (Punitive Expedition of Prince Yin)" chapter of *Shujing (Book of Documents)* states that "... on the first day of the last month of autumn, the *chen* was not harmonious in (lunar mansion) Fang, the blind beat drums, junior officers galloped, people ran around ..." Astronomer Xihe was punished for failing to note the eclipse. (2) The *Zuo zhuan*'s citation of the *Xia shu* or *Book of Xia* (Duke Zhao, Year 17) is identical, except that it says the eclipse was in the first month of summer. (3) The *Bamboo Annals* sides with the *Book of Documents*, and puts it in King Zhong Kang's fifth year.

An extensive analysis of the *Zuo zhuan* passage (Pang, 1990) is summarized as follows: (1) The meaning of "*chen*" is given in *Zuo zhuan*, Duke Zhao, Year 7 (535 B.C.) – "*chen* is the meeting of the Sun and Moon." Before they understood what causes eclipses the Chinese thought that they were due to "inharmonious" meetings of the Sun and Moon. (2) The month of the eclipse in *Zuo zhuan* is wrong, because if a solar eclipse occurred in Fang it could not possibly have been the first month of summer. During this month the Sun was in Bi, a third of the sky away! Only during the last month of autumn could the Sun have been in Fang, which is only 5° wide (Chen, 1988).

The *Bamboo Annals* states that "a solar eclipse occurred on the first day of the ninth lunar month, day *geng-xu*, autumn, in the fifth year of King Zhong Kang, and Prince Yin was ordered to mount a punitive expedition against Xihe." The cyclic date in the *Bamboo Annals* is known to be an interpolation. Nivison and Pang (1990) have successfully reproduced the erroneous cyclic date by using mathematical formulas known to pre-Qin-dynasty astronomers. With the *Zuo zhuan*'s citation of the *Xia shu*, and the *Bamboo Annals*' cyclic date corrected, all three eclipse records are now completely consistent with one another. We will now discuss the dating of this eclipse.

Ancient Chinese astronomers could not even forecast contemporary solar eclipses with any degree of consistency. So the only way by which they could have learned about the Zhong Kang solar eclipse is that an observational record was accurately passed down to them through the ages. Pang (1985) has successfully dated the eclipse to October 16, 1876 B.C. On that day the eclipsed Sun, visible from China, was exactly in the middle of lunar mansion Fang. Nivison and Pang (1990) have found that the *Bamboo Annals*' chronology of the first four Xia kings – Yu to Zhong Kang – had been systematically displaced by 72 years. They have successfully reproduced this error using mathematical formulas known to pre-Qin astronomers, as discussed in the last paragraph. The corrected *Bamboo Annals* Xia dynasty chronology does put the fifth year of King Zhong Kang in 1876 B.C. Dating of the second Xia solar eclipse record reaffirms the accuracy of the astronomically verified *Bamboo Annals* chronology, which is discussed next.

11. The "Double Sunset" Eclipse of 1912 B.C. and the Third Year of King Yu of the Xia Dynasty

The recently discovered second Xia solar eclipse record is also based on three sources: (1) Philosopher Mozi (*ca*. 468–382 B.C.) wrote – "In ancient times the *san* (three) Miao tribes were in disarray. Heaven ordered their destruction. The Sun rose at night..." King Yu led a punitive expedition against the *san* Miao, and defeated them (*Condemnation of Offensive Wars* III). (2) The *Bamboo Annals* states "When the *san* Miao were about to perish... the Sun disappeared by day and reappeared at night..." (3) The *Sui Chaozi* account is same as the *Bamboo Annals*'.

As noted in Sections 2 and 3, a central solar eclipse occurring just before sunrise produces an optical illusion of "double dawn." Likewise such an eclipse right after sunset would be perceived as "double sunset." Primitive people, not understanding why the sky would brighten after sunset, described the optical illusion as "sunrise at night." In ancient times night began right after sunset.

The space and time intervals for matching this eclipse are extremely tight. First, historical atlases put the *san* Miao domain just south of the Yangzi River, west of Lake Pengli (now Poyang) and east of Mount Heng (Herrmann, 1966). Second, the astronomically verified *Bamboo Annals* Xia chronology puts Yu's reign in 1914–1907 B.C. (Nivison and Pang, 1990). There was indeed such an eclipse within the eight-year and 100-mile constraints!

The September 24, 1912 B.C. annular eclipse (in the third year of King Yu's reign), magnitude 0.97–0.99, produced a "double sunset" at 28°N, 114°E – exactly over the *san* Miao domain, as required! See the path of annularity in Fig. 9, computed for a delta T

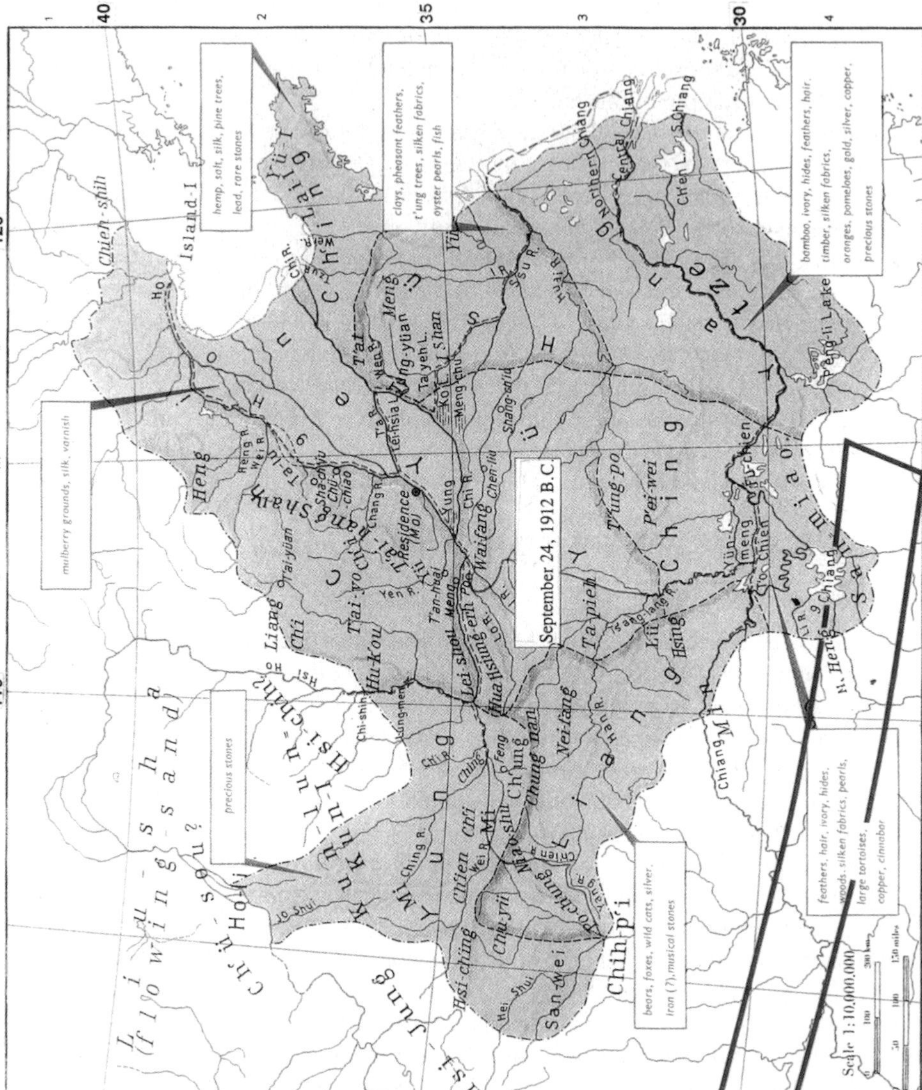

Figure 9 The path of annularity of the September 24, 1912 B.C. solar eclipse. The "double sunset" eclipse had been seen in the land of *san* Miao (28°N, 114°E) during the reign of King Yu, founder of the first dynasty Xia. Reproduced from Albert Herrmann's *Historical Atlas of China* (1966) with the permission of Edinburgh-Aldine Press.

of 12.2 hr, and Pang and Yau (1996) for more details. The corresponding coefficient of T^2, $c = 32\,\text{sec/cen}^2$, is also consistent with a c of $32.5\,\text{sec/cen}^2$, from the analysis of a late third millennium B.C. Mesopotamian lunar eclipse record by Huber (1987). We will analyze the five lunar eclipse records he dated in the next section to obtain useful upper or lower limits on delta T in the third millennium B.C.

12. The Earth's Rotation Rate in the Third Millennium B.C., From Analysis of Ancient Mesopotamian Lunar Eclipse Records

Middle Eastern astronomical tradition emphasizes observation of the Moon, as the month there begins only when the first crescent is sighted. This tradition is carried on in the observation of Ramadan, commemorating the flight of Prophet Mohammed from Medina to Mecca, under the light of the first crescent Moon. A rising eclipsed Moon has been regarded as an evil omen there. The May 22, 1453 dark lunar eclipse that "foretold" the fall of Constantinople a week later, for example, reinforced such phobia (Pang, 1993; Simarski, 1996). The prophecy is probably based on a long tradition of correlation between eclipsed moons, observed to be rising or setting, and disasters noted afterwards, like those cited below.

Humphreys and Waddington (1983) consider Joel's prophecy (*Acts* 2, pp. 14–21) that "the Sun will be turned to darkness and Moon to blood (at Christ's crucifixion)" to have been fulfilled by the rising eclipsed Moon of April 3, A.D. 33 (the 14th or 15th day of the first lunar month *Nisan* in the Jewish calendar). The A.D. 33 and 1453 events probably, in turn, reinforced much earlier correlation compiled in the Babylonian omina *Enuma Anu Enlil* (a set of cuneiform tablets inscribed in early second millennium B.C.). Among these are the three nearly identical *Nisannu* (first lunar month) lunar eclipses that "foretold" the deaths of successive Akkadian kings Manishtushu, Narasim and Sharkalisharri (Huber, 1987).

Huber (1987) used a delta T *vs*. time formula, based on much later data, to identify the lunar eclipses, noting that "we are extrapolating the astronomical theories 2,000 years beyond safe grounds." Inputting the 1912-B.C. delta T value from the last section as an initial trial value to our computer program we have, after successive iterations, derived useful upper or lower limits on delta T from the *Enuma Anu Enlil* lunar eclipses, using constraints on their visibility from Akkad or Ur at rise, set or change of watch. The night was divided into three watches there. Our results and those of Huber (1987) are combined and summarized in Table 3.

Note that the circumstances of the three lunar eclipses, that "foretold" the deaths of successive Akkadian kings, were indeed almost identical, as they all occurred in the first lunar month *Nisannu*, third watch, with the Moon transiting the Earth's shadow in an approximately upward direction. They also satisfy the requirement that "the Moon set while still eclipsed," as recorded in the *Enuma Anu Enlil*; and are consistent with the High Chronology for Mesopotamia.

The last two omina presaged King Shulgi's death by patricide "wronged by his son," and the destruction of Ur. An unearthed Sumerian king list (Weld-Blundell clay prism, 1923), inscribed in 1827–1817 B.C., states that these events were 42 years apart. The matching eclipse dates thus have to be, and are, similarly spaced. The matching lunar

Table 3 Circumstances of the five third millennium B.C. lunar eclipses, recorded in the Babylonian cuneiform Omina *Enuma Anu Enlil*, Tablets 20 and 21 (in chronological order)

Dynasty/King related	Lunar month stated	Eclipse began in watch no.	Enter-exit angle	Matching eclipse date, B.C.	Limit on delta T	
					Hr	Min
Akkad/Manishtushu	1	3	S to N	3/9/2302	> 13	38
Narasim	1	3	S to N	3/19/2265	> 13	26
Sharkalisharri	1	3	S to N	3/10/2237	> 12	25
Ur/Shulgi	3	1	E to W	7/25/2095	< 14	16
Ibbi-Sin	12	1	S to N	4/13/2053	> 12	49

eclipses occurred in the third lunar month *Simanu* and twelfth month *Adaru*, respectively, and during the first watch, as recorded in the *Enuma Anu Enlil*. The limits on delta T in Table 3 will now be merged with values from previous sections.

We have plotted the five upper and lower limits on delta T from Table 3, plus the discrete value for 1912 B.C. from the last section, in Fig. 10. The delta T values from Fig. 8 are also plotted here for comparison. All the historical eclipse data are consistent with the approximation, delta T $= (30 \pm 2.5)T^2$. The theoretical tidal delta T *vs.* time curve is plotted as a thin line above the experimental values for comparison. The difference between the theoretical and experimental curves is due to the effect of postglacial rebound.

To explain, during the Ice Age the Earth was flattened by mile-thick ice sheets pressing down at the high latitudes. Since the Ice Age ended the Earth, being an elastic body, has been gradually returning to its rounder former shape. The bounceback to its less oblate interglacial shape makes it spin faster, just as a spinning ice skater would, when outstretched arms are brought in closer to the body. The effect of postglacial rebound negates a third of the tidal braking. The net effect has been lengthening the day by 1.64 ± 0.14 millisec/century. For example a day in 2300 B.C. was 0.07 sec shorter than today. See Pang *et al.* (1998) for a more detailed discussion. As stated in Sections 1 and 5 an important goal of astronomical dating is the improvement of historical dates. We will now return to improving the Xia, Shang and Zhou absolute chronology.

13. Calibrating the Verified Xia-Shang-Zhou Royal Genealogies with 17 Eclipse Dates

As we had alluded to in Section 5 we can combine the strengths of verified royal genealogies and eclipse dating by correlating them. The 16 eclipse dates we have determined in the previous sections, along with the corresponding kings' names, and the number of generations before 841 B.C. (the earliest accurate historical date) they reigned, are listed in Table 4, and plotted in Fig. 11. We have also added the *Yi Zhou shu or Book of Zhou* cyclic day *bing-zi* lunar eclipse, recorded in the 35th year of the last predynastic Zhou King Wen, and dated to January 29, 1137 B.C. (Tung, 1960).

The bestfit curve in Fig. 11 has both the strengths of verified royal genealogies – continuity – and eclipse dating – accuracy. The overall slope is ~30 years/generation, as

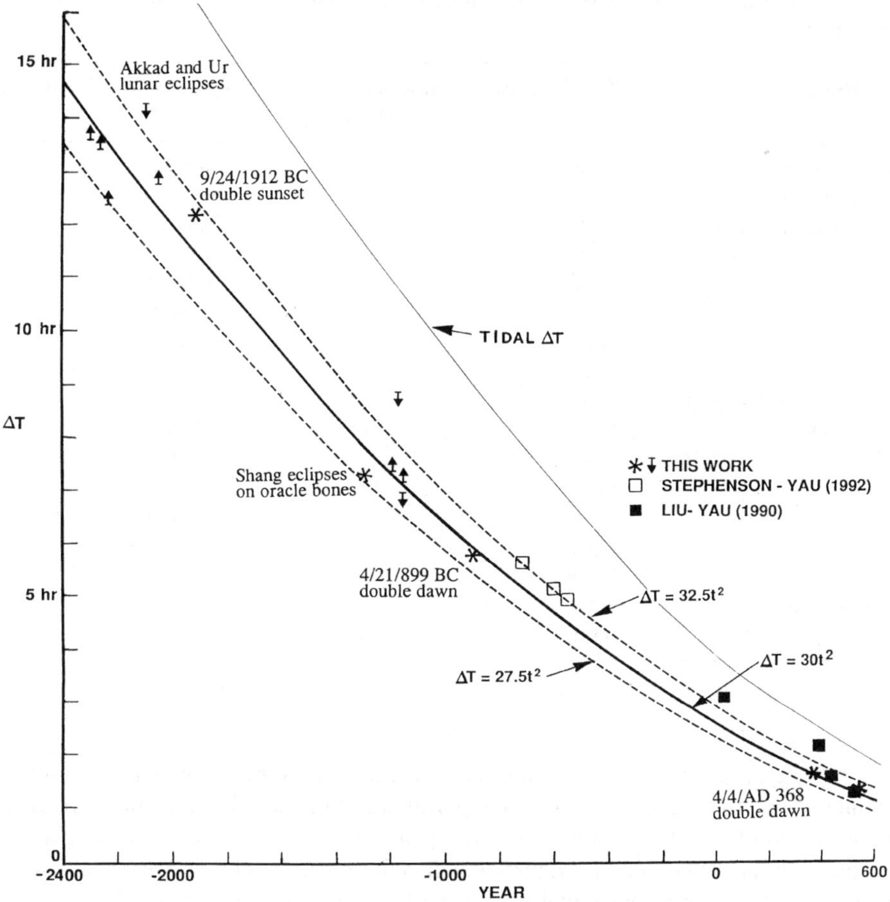

Figure 10 The clock error delta T *vs*. year. Plus years are A.D. and minus years – B.C.

the Chinese have known all along. The slightly less steep slope for Xia is due to earlier marriage and childbearing (Mozi, *Moderation in Expenditure* I; Pang, 1985). The curve is 99% accurate in times with eclipse data (Western Zhou, late Shang and early Xia), but not quite as accurate elsewhere (dashed portion). We believe that our data can be confidently used as a foundation for building a detailed absolute chronology for the Three Dynasties, an important project in China's current Five-Year Plan (Song Jian, *Sci. Tech. Daily*, May 17, 1996; Wehrfritz, *Newsweek*, July 7, 1997, 44–45). In the concluding section we will briefly outline the prospects for future research.

14. Epilogue

Many more historical eclipse records will likely come from a thorough search of China's extensive archives and ancient inscriptions now underway to support the State Council's

Table 4 The 17 eclipse dates of the Xia, Shang, and Zhou dynasties *vs*. their verified royal genealogies (in chronological order)

Dynasty/King	Type of eclipse	Generations from 841 B.C.	Eclipse date, B.C.	Ref. sec.
Western Zhou/Yi	Solar	−2	4/21/899	2
Western Zhou/Wen	Lunar	−9	1/29/1137	13
Shang/Wu Yi and Wen Ding	Solar	−10 to −11	10/31/1161	8
	Solar		6/27/1163	
	Solar		6/7/1172	
	Lunar		7/2-3/1173	
	Solar		10/21/1198	
	Solar		5/6/1226	
Shang/Lin Xin and Kang Ding	Lunar	−12	3/20/1242	9
Shang/Wu Ding	Lunar	−14	2/27/1278	6,7
	Lunar		9/2/1279	
	Lunar		11/4/1282	
	Solar		6/5/1302	
	Lunar		11/24/1311	
	Lunar		12/25/1322	
Xia/Zhong Kang	Solar	−35	10/16/1876	10
Xia/Yu	Solar	−37	9/24/1912	11

project of a scientific revision of China's ancient chronology, mentioned above. Of particular interest this time around are the local gazettes and historical government documents, numbering over ten million for the last two dynasties. For various reasons these were not all scanned in the 1974–1984 literature search. With much greater budgetary and technological support available now, hopefully this will be done. Spin-offs from this monumental effort would benefit research on the Earth's past rotation rate as well.

We have discussed the analysis of two Xia dynasty eclipse records in Sections 10 and 11. Extant historical texts mention other Xia astronomical records, although the passage of time seems to have rendered some of them difficult to interpret due to transmission errors. However just as the discovery of a Shang dynasty oracle bone a century ago led to 160,000 more, the recent discovery of a Xia era inscription gives us hope that many more will be forthcoming.

There is good reason for our optimism, because decades of intensive archaeological excavations have amply shown that Xia was not only a literate civilization, but a polity well-organized to construct kilometer-sized walled cities, forged bronze objects and undertake extensive flood control projects along the Yellow and Yangzi Rivers. It is reasonable to expect such an advanced civilization to have left many more astronomical (including eclipse) records, in addition to those that have barely survived the passage of time in extant historical texts. As we are about to begin a new millennium we look forward to working with more eclipse data, recorded four millennia ago, to further our knowledge of the Earth's past rotation rate, and improve historical dates.

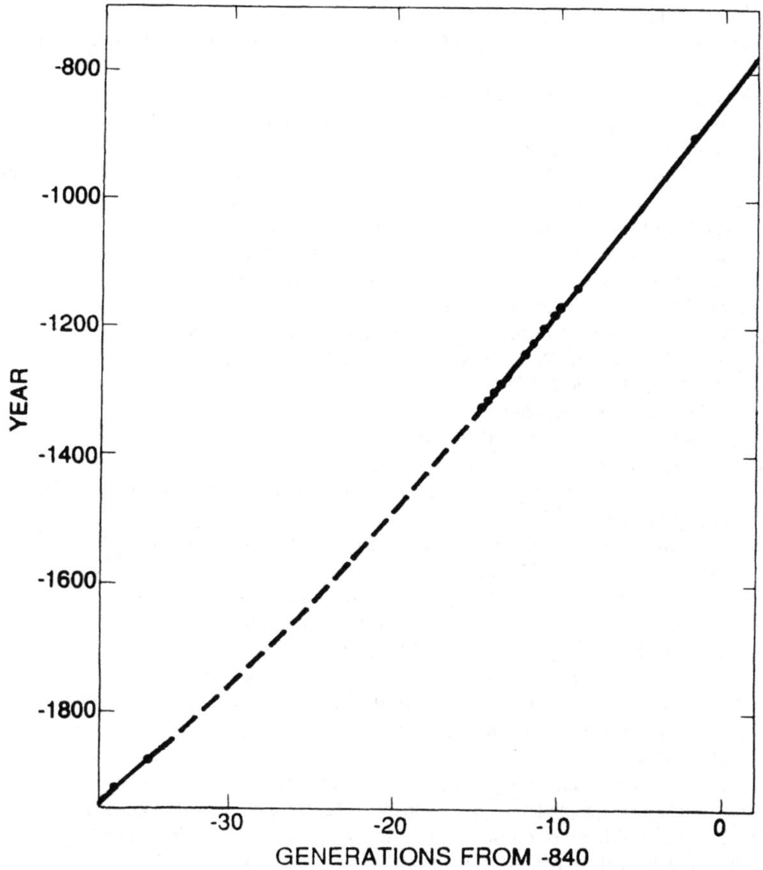

Figure 11 Eclipse dates *vs*. the number of generations the respective kings ruled before 841 B.C. (the earliest accurate historical date). The curve of bestfit has an overall slope of ~ 30 years per generation.

Acknowledgement

We thank John A. Bangert of the U.S. Naval Observatory. The Dudley Observatory provided financial support to K.D.P. The National Aeronautics and Space Administration's Planetary Astronomy, and Planetary Geology and Geophysics Programs supported K.K.Y. through a contract to the Jet Propulsion Laboratory.

References

Chang, Kwang-Chih, 1980, *Shang Civilization*, Yale University Press, New Haven, pp. 322–329, and 352–355.

Chen, Cheng-Yih and Xi, Zezong, 1993, in *Astronomies and Cultures*, C. L. N. Ruggles and N. J. Saunders, eds., University of Colorado Press, Boulder, pp. 32–66.

Chen, Zungui, 1987, *History of Chinese Astronomy*, rev. ed., Ming Wen, Taibei, Vol. 3, pp. 177–178, and Vol. 4, p. 42 (in Chinese).

China Pictorial, April 1993, No. 538, 33.

Chou, Hung-hsiang, 1958, *Imperial Records of the Shang-Yin Dynasty*, privately printed in Hong Kong (in Chinese).

Chou, Hung-hsiang, 1979, *Sci. Amer.* **240**, 135–149.

Chou, Hung-hsiang and Pang, K. D., 1990, paper presented at the International Symposium on Xia Culture, University of California, Los Angeles.

Guo, Moruo, 1977, *Combined Compilation of Oracle Bone Inscriptions*, Zhonghua, Beijing (in Chinese).

Han, Yanben, Li, Zhisen, Lin, Baisen and Wang, Xihong, 1984, *Chin. Astron. Astrophys.* **8**, 348–353.

Herrmann, A., 1966, *An Historical Atlas of China*, Edinburgh-Aldine.

Hilton, J. L., Seidelmann, P. K. and Liu, C. Y., 1988, *Astron. J.* **96**, 1482–1493.

Hilton, J. L., Seidelmann, P. K. and Liu, C. Y., 1992, *Astron. J.* **104**, 2250–2252.

Hu, Houxuan, 1986, *Early China*, Supplement No. 1, 35–38.

Huber, P. J., 1987, *Acta Historica Scientiarum Naturalium et Medicinalium* **39**, 3–13.

Humphreys, C. J. and Waddington, W. G., 1983, *Nature* **306**, 743–746.

Kiang, Tao, 1981, *Chin. Astron.* **5**, 111–121.

Levy, D. H., 1992, *Sky Teles.* **83**, 694–695.

Liu, Chao Yang, 1945, *Yuzhou* (Universe) **15**, 15–16 (in Chinese).

Liu, Ciyuan, 1988, *Chin. Astron. Astrophys.* **12**, 169–178.

Liu, Ciyuan and Yau, K. K. C., 1990, in *Earth Rotation from Eons to Days*, P. Brosche and J. Sundermann, eds., Springer-Verlag, Berlin, pp. 33–39.

Liu, Yuanlin, 1978, *Dalu Zazhi* (Continental Magazine) **57**, 197–206 (in Chinese).

Morrison, L. V., 1992, *Observatory* **112**, 289.

Mucke, H. and Meeus, J., 1983, *Canon of Solar Eclipses*, Astronomisches Buro, Vienna.

Needham, J., 1959, *Science and Civilization in China*, Vol. 3, Cambridge.

Nivison, D. S. and Pang, K. D., 1990, *Early China* **15**, 87–95.

Nolan, W. I., 1975, *The Facts of Keno*, Gambler's Book Club, Las Vegas, p. 17.

Pang, K. D., 1985, paper presented at the *U.S.–China Bilateral Symposium on the Analysis of Extraordinary Flood Events*, Nanjing (also *J. Hydrol.* **96**, 139–155).

Pang, K. D., 1990, *Early China* **15**, 173–195.

Pang, K. D., 1993, *Eos: Trans. Amer. Geophys. Un.* **74**, No. 43, 106.

Pang, K. D. and Yau, K. K., 1996, in IAU Symposium No. **172**, "*Dynamics, Ephemerides and Astrometry of the Solar System*," S. Ferraz-Mello, B. Morando and J.-E. Arlot, eds., Kluwer, Netherlands, pp. 113–116.

Pang, K. D., Yau, K., Chou, H. H. and Wolf, R., 1988, *Vistas Astron.* **31**, 833–847.

Pang, K. D. *et al.*, 1989, *Bull. Amer. Astron. Soc.* **21**, 753.

Pang, K. D., Yau, K. and Chou, H. H., 1995, *Pure Appl. Geophys.* **145**, 459–485.

Pang, K. D., Yau, K. K. and Chou, H. H., 1997, in *21st Century Chinese Astronomy Conference*, K. S. Chen and K. L. Chan, eds., World Scientific, Singapore, pp. 523–526.

Pang, K. D., Yau, K. K. and Chou, H. H., 1998, in *Dynamics of the Ice Age Earth: A Modern Perspective*, P. Wu, ed., Trans Tech, Switzerland, pp. 459–488.

Provincial Atlas of the People's Republic of China, 1977, Ditu Chubanshe, Shanghai.

Simarski, L. T., 1996, *Aramco World* **47**, 8–13.

Songjiang District Gazette, 1663 ed., **47**, 19 (in Chinese).

Stephenson, F. R. and Yau, K. K. C., 1992, *J. Hist. Astron.* **23**, 31–51.

Takashima, Ken-ichi, 1988, paper presented at 40th Annual Meeting of Association for Asian Studies, San Francisco.

Tao, Zongyi, 1360–1368, *Nancun Chuokeng Lu*, **19**, 14 (in Chinese).

Tung, Tso Pin, 1953, *Xiaotun II: Yinxu wenzi yipien* (Second Collection of Writings from Yin Ruin at Xiaotun), Part 3, Historical and Philological Research Institute, Academia Sinica, Taibei (in Chinese).

Tung, Tso Pin, 1960, *Chronological Tables of Chinese History*, Hong Kong University Press.

Tung, Tso Pin, 1964, *Fifty Years of Studies in Oracle Inscriptions*, Centre for East Asian Cultural Studies, Tokyo.

Wang, P. K. and Siscoe, G. L., 1980, *Solar Phys.* **66**, 187–193.

Weld-Blundell clay prism, 1923, Oxford Ashmolean Museum Serial No. 1923.444.

Xi, Zezong, 1984, *Archaeoastron.* **7**, 34–45.

Xu, Zhentao, 1990, *Phil. Trans. Roy. Soc. Lond.* **A330**, 513–515.
Xu, Zhentao, Yau, K. K. C. and Stephenson, F. R., 1989, *J. Hist. Astron.* **20**, S61–S72.
Yao, Xiaosui, and Xiao, Ding, 1988, *A General Compilation of Copies and Explanations of Oracle Bone Inscriptions from Yinxu*, Zhonghua, Beijing, p. 758.
Yau, K., Yeomans, D. K. and Weissman, P., 1994, *Mon. Not. Roy. Astron. Soc.* **266**, 305–316.
Yeomans, D. K. and Kiang, T., 1981, *Mon. Not. Roy. Astron. Soc.* **197**, 633–646.
Zhang, Peiyu, 1975, *Acta Astron. Sinica* **16**, 210–224 (in Chinese with English abstract).
Zhang, Peiyu, 1994, *Publ. Purple Mount. Obs.* **13**, 23–30 (in Chinese).
Zhang, Peiyu and Han, Yanben, 1995, *Acta Astron. Sinica* **36**, 314–320 (in Chinese with English abstract).
Zhuang, Weifeng and Wang, Lixing 1988, *Union Compilation of Ancient Chinese Records of Celestial Phenomena*, Jiangsu Science and Technology Press (in Chinese).

Part 2

Modern Astronomy in the Orient

2.1. Philippe de La Hire at the Court of Jayasiṃha

David Pingree
Brown University

In 1727 Jayasiṃha, the Mahārāja of a Rajput state which lies about 150 miles South Southwest of Delhi, took two important steps in his eventful career.[1] He founded his new capital a few miles from Amber, its predecessor, naming it Jayanagara or Jayapura, and he sent a Jesuit missionary, Father Manuel de Figueredo, along with several others to Portugal to obtain for him astronomical books and instruments, and a skilled astronomer.[2] In November of 1730 Father Manuel returned, to the new city of Jayapura, with the items Jayasiṃha had requested. Though we are nowhere informed of what instruments were brought back from Europe, the books included the 1727 Paris reprint of Philippe de La Hire's *Tabulae astronomicae*;[3] and the "skilled" astronomer selected by the Portuguese was the young Pedro da Silva, who had studied astronomy in Portugal under J. B. Carbone, who was using de La Hire's tables in the 1720's.[4]

The notice concerning the "puer a patre educatus India oriundus maximo ingenio praeditus nomine Petrus de Silva," who "etiam Lusitaniae apud reverendissimum et clarissimum patrem Ioannem Baptistam Carbone<m> astronomiae operam dedit" is found in a manuscript that formerly belonged to the Mahārāja, but is now part of the City Palace Museum in Jayapura where I and a half-dozen colleagues, with the kind permission of the Mahārāja and financial support from the Dharam Hinduja Center at Columbia University, were able to catalogue the approximately 300 astronomical manuscripts in January of 1998. The manuscript in question, Khasmohor 7832, is a handwritten copy in 275 pages of the printed exemplar of de La Hire's work, which is no longer to be found. The copying was done by a Frenchman, Joseph du Bois who, after 15 years of wandering in Asia, had ended up in Delhi whence, after several years of pretending to be a physician, he had been called to Amber in the service of Jayasiṃha. He completed his task of copying on 10 September 1732. The need for a second copy of the *Tabulae astronomicae* may have arisen because the tables were in fact to be used by two groups of astronomers working at Jayasiṃha's court, the Muslims who were preparing the Persian *Zīj-i-Muḥammad Shāhī*, in which the mean motion tables of the planets are adopted from de La Hire's to fit the Muslim calendar,[5] and the equation tables were simply borrowed from the same source, and the Hindus who tried to explain de La Hire's astronomy in Sanskrit. It is this latter effort that I shall primarily pay attention to in this article.

It is, of course, most probable that Jayasiṃha had a Sanskrit version of some sort made soon after he received the book in 1730. The evidence for this lies in a most interesting manuscript, Khasmohor 5183, in which are recorded the discrepancies, measured in minutes of arc, between the Jayapura observations of the Moon and, first, the lunar longitudes computed by means of de La Hire's tables, secondly those computed by means of the Navīna or New Tables; the latter are undoubtedly from the *Zīj-i-jadīd* or *New Tables* composed for Ulugh Beg of Samarqand by a group of astronomers including himself in ca. 1440. Jayasiṃha acquired a copy of the *Zīj-i-jadīd* in 1727;[6] and there is a Nāgarī version of its lunar tables at Jayapura, Khasmohor 5484, on the first leaf of which Ulugh Beg's

S.M.R. Ansari (ed.), History of Oriental Astronomy, 123–131.

tables are compared with those found in the Sanskrit version of the *Zīj-i-Shāh Jahān*, the *Siddhāntasindhu* of Nityānanda, which was the object of some serious study at Jayapura in 1727.[7]

The years for which the comparisons of lunar longitudes contained in Khasmohor 5183 were made extend from 1727/8 to 1737; of course, no computations based on de La Hire's tables could be made before it had been translated so that the rules for manipulating the tables were thoroughly understood.

The evidence is that the first translation did not successfully convey de La Hire's meaning. A compilation of all the differences between de La Hire and observation above 0;30° in Khasmohor 5183 reveals two between 0;50° and 1° – one of 0;51° in 1728/9 and another of 0;56° in 1731/2; eighteen between 0;40° and 0;49° – 3 in 1727/8, 2 in 1728/9, 10 in 1730/1, 3 in 1731/2, and 1 each in 1732/3 and 1733/4; and only two above 0;30° after 1733/4-in fact, in 1735/6. Either the observations improved dramatically in 1733/4 or the understanding of how to use de La Hire's tables got better; the latter seems to be the correct explanation.

We can be certain that the understanding of the *Tabulae astronomicae* was not easy for the Jayapura astronomers, to whom neither Pedro da Silva nor Joseph du Bois seems to have been able to offer sufficient assistance. The main difficulties lay in de La Hire's mathematics, which employed logarithms, including logs of trigonometrical functions, and which did not clearly present the geometrical bases for some of his rules. One of the Jayapura astronomers whose failure to grasp de La Hire's meaning is egregious was Jayasiṃha's *jyotiṣarāja*, an astronomer from Gujarāt named Kevalarāma, who entered the Mahārāja's service in 1725. He was a very traditional Indian; he wrote, for instance, probably in the late 1720's, a *Bhāgavatapurāṇajyotiṣayor khagolabhūgolavirodhaparihāra* in which he tried to reconcile the flat earth of the popular Indian *purāṇas* with the spherical earth and heavens of the scientific *siddhāntas* by proposing the existence of two earths, one large and flat and one small and spherical;[8] and a *Brāhmapakṣanirāsa* in which he asserted the superiority of the *Sūryasiddhānta*, which pretends to be the revelation of the Sun-god, over the Brāhmapakṣa, which he claims to be inferior because it was authored by men.[9] But suddenly we find him composing – presumably in 1731 or 1732 – a Sanskrit poem, the *Dṛkpakṣasāriṇī*, expounding de La Hire's computations of the longitudes (and latitudes where appropriate) of the Sun, the Moon, and the planets, and lunar and solar eclipses.[10] There must have existed a prose version on which he could base this versification. And Jayasiṃha, in a passage we shall examine shortly, referred in 1732 to an unsatisfactory translation.[11]

I know of six manuscripts of the *Dṛkpakṣasāriṇī*, of which I have been able to collate five.[12] None contains the same text as any of the others. It would appear that there were originally three *adhyāyas* and most of de La Hire's tables; the first *adhyāya*, on the true longitudes of the Sun, the Moon, and the five planets, is found in the manuscripts at Benares, Jodhpur, and Kota, and in one of those at Poona (BORI 537 of 1895/1902); the second, on lunar eclipses, is found in these same four manuscripts, but the Benares and Kota manuscripts have each added a different, variant version of *adhyāya* two; and the third, on solar eclipses, is found in the Benares manuscript and in both Poona manuscripts. The tables are found in Baroda manuscript 3162; a facsimile of the first page of the mean motion tables of the Moon in the Baroda copy has been published,[13] and it agrees with the later Sanskrit prose version to which we will turn shortly. There is also found in the later Sanskrit prose version Kevalarāma's revised Indian calendar with its epoch date, which I will discuss in more detail in describing the later Sanskrit prose version, and many of his new technical terms and new meanings of old technical terms. That Kevalarāma precedes the later Sanskrit prose version is indicated by his failure to understand de La Hire's mathematics and the utter confusion of his presentation of the Frenchman's rules, which he puts in the wrong order, thereby breaking the mathematical logic, and which he causes to suffer from omissions, repetitions, unwarranted supplements, and gross errors. Kevalarāma had serious difficulty in recognizing and comprehending de La Hire's use of logarithms and of logarithms of sines, cosines, and tangents in his rules for computing the longitudes of the planets: these functions, of course, must be added to and subtracted from each other, as is done in de La Hire's examples, rather than multiplied and divided by each other as Kevalarāma instructs his poor reader to do. In chapter 2, on computing lunar eclipses, he at first did not even recognize that de La Hire was using logarithms; the two variant versions of this *adhyāya* correct this mistake, but one is still instructed to multiply and divide them by each other.

For use when de La Hire instructs one to employ these functions Kevalarāma provided a table of the logarithms (*vibhāgas*) of all integer numbers from 1 to 10,000[14] and a table of the logarithms of sines (*jīvas*) and tangents (*chāyās*) for every minute of arc from 0° to 90°;[15] both are carried to seven decimal places as was noted by Hunter of the tables he saw at Ujjayinī in 1792/3 in the hands of Kevalarāma's grandson.[16] Hunter also saw "the treatises of plane and *spherical trigonometry*, and on the construction and use of logarithms, which are annexed to Cunn's or Commandini's edition. In this translation, the inventor is called Don Juan Napier, an additional presumption that Jayasinha's European astronomers were of the Portuguese nation."[17] Though I have found one manuscript of the table of logarithms and two of the table of the logarithms of sines and tangents, I have as yet come across none of the treatise on the use and construction of logarithms.

As noted above, it is not yet clear for whom du Bois' transcription of the *Tabulae astronomicae* was intended. Some technical terms in the tables have their equivalents in Sanskrit (written in Nāgarī or Roman script) above or beside them, others their equivalents in Persian (written in Persian or Roman script). On p. 265 of du Bois' manuscript, several folia after the end of the tables, someone has written in a crude Roman script: "kisna discipulus magistro suo se commendat die 11 octobris 1734". This indicates that a Hindu named Kisna—that is, Kṛṣṇa—was learning Latin in 1734, perhaps from du Bois and perhaps in order to be able to translate de La Hire into Sanskrit. However, a very incomplete comparison of the Sanskrit copies of the tables in the later Sanskrit prose version with those found in the printed text (I have used the first edition, printed at Paris in 1702) and those found in du Bois' copy shows the Sanskrit copyist agreeing at times with one, at times with the other, and more often than not introducing errors of his own. Further study may clarify this confusing situation.

Now we need to say something about this later Sanskrit prose version. We found at Jayapura two copies of it; this version was not a translation, but an adaptation. The two manuscripts are Khasmohor 5295, which contains the entire prose version on ff. 1–39v, and Khasmohor 5609, which is a fragment whose beginning is lost. It now starts with f. 33, which contains a table relating to Jupiter; the prose version ends on f. 39v. There are strong indications, even though a detailed collation has not yet been made, that 5609 is a copy of 5295. A third manuscript is Baroda 3162.

The first problems faced by the astronomers who adapted de La Hire in Sanskrit are that the Frenchman used the Julian calendar, or the Gregorian after the Reform, and situated his Prime Meridian so that it passed through Paris, whereas the Indians naturally chose to utilize a variant of their local calendar and to have their Prime Meridian run through Jayapura. They avoided using the calendar as it normally existed in Rājasthān, where the beginnings of the months vary radically from year to year with respect to solar longitudes because they coincide with syzygies of the Sun and the Moon, but rather they adopted an East Indian style wherein the months vary only gradually over time since their beginnings are determined by the Sun's entry into the zodiacal signs. Moreover, in order to avoid extensive recomputation to adapt de La Hire's anomalistic year- length to the sidereal year of the Indian calendar, they simply copied de La Hire's tables in which in each quadrennium the first three years contain exactly 365 days, the fourth 366. In this they showed themselves to be less assiduous than their Muslim counterparts, who made the necessary adjustments to fit de La Hire's parameters into the Muslim calendar, which is strictly lunar.

The longitudinal difference between Paris and Goa according to de La Hire's Table IV is 4 hours and 46 minutes, while that between Paris and Bologna is 38 minutes. Since du Bois, on p. 142 of his manuscript, at the bottom, notes that Jaipur is 4 hours and 11 minutes from Bologna, he must have reckoned it to be 3 minutes East of Goa and therefore 4 hours and 49 minutes East of Paris.[18] An analysis of the Sanskrit tables of mean lunar motions shows that 4 hours and 49 minutes is precisely the longitudinal difference assumed in their computation. On p. 141 of his manuscript in the right margin, du Bois wrote: "Sauayaipur differentia a penatibus 4′40″". This indicates that du Bois' hometown lay 9 minutes of time (or 2;15°) East of Paris.

Because the calendar-date in Jayapura would be given in terms of the current tithi (a thirtieth of a synodic month) in the current synodic month, the Sanskrit prose version of the *Tabulae astronomicae* begins with a table of the *abdapa* (weekday on which the year begins) and of the *śuddhi* (accumulated epact expresssed in tithis) for the time intervals used in de La Hire's annual tables of mean motion, namely: 1 to 20 Julian years; 40, 60, 80, and 100 Julian years; 200, 300, 400, and 500 Julian years; and 1000 and 2000 Julian years. Since the *śuddhi* for 2000 years is rounded off to exactly 10 tithis, that for 100 years is 26 tithis, that for 20 years is 11;12 tithis, and that for 4 years (i.e., for 1461 days) is 14;14,24 tithis or, since there was in this interval one intercalary month, 44;14,24 tithis. The annual epact, then, is 11;3,36 tithis, and each Julian year contains 12;22,7,12 mean synodic months. This is a pretty good parameter to come out of 10 tithis in 2000 years. There follows a table of the *abdapa* and *śuddhi* for every 100 Julian years since Saṃvat 1786, Śaka 1651 (A.D. 1729), and of the accumulated days in the year at the end of each solar month. Since the entry of the mean Sun into tropical Aries occurred, according to our computations based on the mean motion tables of the Moon, on Saturday 19 March 1729 (which date falls in Saṃvat 1786 and Śaka 1651), that is the epoch of the Sanskrit tables;[19] the *abdapa* for that epoch year is given as Monday, which was 21 March, wrongly supposed to be the date of the *true* Sun's entry into Aries. This confusion arises because the *mean* Sun on 21 March had the same longitude as the *true* Sun had on 19 March. The accumulated days at the ends of the solar months indicate that the true motion of 30° of the Sun in each of the solar months in order took place in 31, 31, 32, 31, 31, 31, 30, 30, 29, 29, 30, and 30 days respectively, to give a year of exactly 365 days.

From these data, in the normal fashion of an Indian *siddhānta* or *karaṇa*, the user of the prose version is asked to compute the *ahargaṇa* – that is, the days that have elapsed since the epoch, which in this case is 21 March 1729 noon at Jayapura. The instructions, following de La Hire, then tell us to correct the hour by the equation of time and the longitudinal difference of one's locality from the meridian of Jayapura. While the mean motion tables are computed on the basis that the time-difference between Paris and Jayapura is 4:49 hours, the longitude of Jayapura (for longitude in this case the prose version uses the Arabic/Persian word *ṭūl*) is given as 112;17° instead of the 72;15° implied by a time-difference of 4:49 hours. The reason for this discrepancy is that the user of the Sanskrit text was expected to use a Muslim geographical table, in Arabic or in Persian; for in such a table Jayapura would be indeed about 112° from al-Jazā'ir al-Khālidāt (the "Fortunate Islands").

The author of the later Sanskrit prose version then proceeds to compute, as examples, the mean longitudes of the Sun, the Moon, and the five star-planets with their apogees and nodes at 6:34 PM of Thursday 14 śuklapakṣa of Vaiśākha in Saṃvat 1789, Śaka 1654, which was 27 April Julian or 8 May Gregorian in the year 1732; this date must have been only shortly before he wrote this work. The order in which he treats the planets is that of the weekdays – Sun, Moon, Mars, Mercury, Jupiter, Venus, and Saturn – as is normal in Indian astronomy; the same order of the weekdays is used in his copies of de La Hire's planetary tables though the latter employs a European order: Sun, Moon, Saturn, Jupiter, Mars, Venus, and Mercury. Moreover, instead of calling the mean motions of the two inferior planets in their orbits about the Sun by the names of the planets themselves, he uses the traditional Sanskrit terms – *budhocca* and *śukrocca* – which refer to the *śīghra* apogees of Mercury and Venus, that is, to the sums of their proper mean motions and the apparent mean motion of the Sun even though only the proper mean motions are given in his tables.

It is only on f. 3 that the Sanskrit prose version begins to coincide with the Latin text of de La Hire, beginning with his praeceptum II, on finding the true longitude of the Sun. The Sanskrit, however, is not a translation, but a summary of the Latin. The order in which steps are prescribed in computing the true longitudes of the Moon and the planets by de La Hire is not strictly observed, but the changes are not so great as to alter the results significantly if the instructions are actually understood, which would have been no mean feat if one could not read the Latin original. The Sanskrit completely omits praeceptum VII, on the computation of the Moon's horizontal parallax, since this little treatise completely ignores both lunar and solar eclipses; and it fails to reproduce the important diagram of a heliocentric model for computing the true longitude and latitude of a superior planet that follows de La Hire's praeceptum VII; this diagram is essential for understanding the geometry of de La Hire's rules, and it is the only place in the *Tabulae astronomicae* where the heliocentric theory, which is never mentioned, is at least illustrated and shown to be the basis for de La Hire's work. This diagram, however, was copied and fully explained in the *Phiraṅgicandracchedyakopayogika* which we will discuss later.

The Sanskrit author restates, with some minor omissions, the procedure for computing the true longitude of a planet set forth by de La Hire in praeceptum VIII, but misleadingly employs the technical vocabulary employed in Indian geocentric astronomy to identify elements of the heliocentric theory.

At the bottom of f. 8 of Khasmohor 5295 the anonymous author ends his discussion of the method of finding the longitudes of the planets; he left the verso of this leaf blank. Another scribe, presumably of a later time, has written at the top the equivalent in Sanskrit of: "in the year 1735 precession amounted to 19;52°; its annual motion is 0;0,51°." Actually, the rate of precession used is 1° in 70 years or a bit more than 0;0,51,25,42° per year. The value for precession in 1735 is consistent with one given by Jagannātha Samrāṭ, Jayasiṃha's guru, namely, 19;47° in 1730,[20] but the year from which these values are computed works out to be, improbably, in the middle of the fourth century A.D. instead of the early sixth. Regardless of this curiosity, however, the note indicates that the later Sanskrit prose version was composed before 1735; and 1732 remains the most probable date.

One of the difficulties in understanding de La Hire that the Jayapura astronomers faced was the difficulty of understanding his model of the Moon; for he provided no diagrams to illustrate the geometric rationale (he indicates in his preface that he had none[21]) for his second equation of the Moon, that is, the annual equation, which depends on the Moon's distance from both the Sun and the solar apogee, and whose maximum in his tables is 0;12,33°, and the third equation, the evection, whose maximum is 3;5°; we have in Khasmohor 5182 two virtually identical diagrams showing what the Jayapura astronomers must have thought this lunar model looked like. The earth rotates about the Sun, with the Sun's apsidal line going through the Sun from the lower right to the upper left. About the earth and the Sun are sets of two orbits absolutely parallel to each other; the concentric orbit is that of the mean Moon, the eccentric one that of the true Moon; the true Moon is also on an eccentric orbit about the Sun. It is unclear to me how this model works, but it embodies all the elements involved in de la Hire's three equations of the Moon.

We know independently that Jayasiṃha was worried about de La Hire's lunar model from a letter written at the beginning of 1733 – it is dated 24 January of that year – by the French Jesuit, Father Calmette, from Veṅkaṭagiri in the Karṇāṭaka to M. de Cartigny, the intendant-général of the French naval forces.[22]

"We are six missionaries in the land of the infidels; two others are prepared to come, while in the Kingdom of Bengal a vast field for establishing a new mission opens up: that is all of North India. The Prince of Orissa calls us; another Prince, even grander than he in Hindustan, Raja (perhaps he means Rajput) by caste and able in astronomy, invites and urgently begs the missionaries of Bengal to come into his territory, where he hopes to establish them. He loves the sciences, and one can judge the extent of his understanding by the questions which he has already posed to them. They are:

1. Whence comes the difference that he (that is, Jayasiṃha) finds between the observed longitude of the Moon, and the calculation made with the tables of M. de La Hire, which he (Jayasiṃha) has caused to be translated? This difference is close to one degree; however, the instruments with which he has made his observations are large

and exact, and the observations were made with all necessary care. Is this difference also found for the meridian of Paris?

2. Are there tables which give the movements of the Moon in perfect agreement with observations? If there are, who is their author and what astronomical hypothesis does he follow?
3. What is the hypothesis that M. de La Hire follows, and on what geometric basis has he made his tables of the movements of the Moon?
4. In what way do they observe in Europe the longitude of the Moon when it is not on the meridian, and with what instruments?
5. On what foundation has M. de La Hire established his third equation of the movements of the Moon, and how could one reduce it to an hypothesis and calculate it geometrically?

In response to Jayasiṃha's letter of 1732, Fathers Boudier and Pons set out for Jayapura from Candranagara in Bengal on 6 January 1734.[23] All along their thousand-mile journey they made observations, with a quadrant, of meridian transits of the Sun and of some fixed stars; at Delhi, between 17 and 28 May, they used a gnomon to observe the distance of the Sun from the zenith at noon; and throughout the trip they observed the immersions of the first satellite of Jupiter with a 17-foot telescope in order to determine local longitudes. Their first observation at Jayapura, of a meridian transit of Lyra, occurred on the night of 7 August 1734; and they observed there, with a telescope, the immersion of the first satellite of Jupiter on 13 August. From this observation they determined that Jayapura was 4:55 hours east of Paris; from the observation of the total lunar eclipse of 1 December 1732 made at Jayapura by the Brāhmaṇa astronomers they determined the longitudinal difference between Paris and Jayapura more precisely to be 4:55:38 hours. Their new value for the longitudinal difference between Jayapura and Paris differs by about 6-2/3 minutes from the one associated with Joseph du Bois and the later Sanskrit prose version; the latter, therefore, clearly antedates the summer of 1734.

Thus Fathers Boudier and Pons arrived in Jayapura in the summer of 1734, and began to make observations there in August. They presumably answered the Mahārāja's questions concerning de La Hire's lunar theory. The answer to his fifth and last question is found in the *Phiraṅgicandracchedyakopayogika*; it is a diagram allegedly illustrating the geometry of the third equation surrounded, on the preceding and following folia, by an explanation of how it works – an explanation I have not yet fathomed. The presumption must be that this, and the rest of this little text, was provided by the two French Fathers.

The *Phiraṅgicandracchedyakopayogika* is preserved on ff. 40–52 of Khasmohor 5295, on ff. 40–48 of Khasmohor 5609, and on all 15 folia of manuscript 727 at the Rajasthan Oriental Research Institute at Kota. The text discusses ten diagrams; accompanying the ninth, which depicts a lunar eclipse, is an example dated Sunday 28 May 1732 Julian; there was to have been an example for the tenth diagram, of a solar eclipse, but all three manuscripts break off in its heading.

While it is of interest to note that the first figure involves a conic section because it suggests that this is the "treatise on conic sections" that William Hunter said that he saw in the hands of Kevalarāma's grandson in Ujjayinī in 1792/3, it is more useful for our purposes simply to recall that this is the text that contains as figure 6 the diagram of de La

Hire's heliocentric model of a superior planet and the illustration of de La Hire's third lunar equation. The Sanskrit explanation of the first of these two diagrams is correct, and begins with the simple statement: "in this figure it is assumed that the earth moves, and the Sun is assumed to be fixed." This is the only straightforward assertion of the heliocentric system that I have found in any of the Sanskrit or Persian works written at Jayasiṃha's court.

At the beginning of this paper I referred to the *Zīj-i-Muḥammad Shāhī*, the Persian astronomical tables that a group of Muslim astronomers on Jayasiṃha's staff had been working on since the 1720's. This *Zīj's* chronological tables, tables of trigonometry and spherical astronomy, geographical table, and star-catalogue were mostly derived from Ulugh Beg's *Zīj-i-jadīd*, though a few relevant to the latitudes of Indian cities were the products of the Jayapura astronomers.[24] In recent years, however, it has been demonstrated, as noted above, by Mercier, and now even more convincingly by van Dalen, that the mean motion tables of the Sun, the Moon, and the planets are based on those in de La Hire's *Tabulae astronomicae*, suitably modified to fit the time-units of the Muslim calendar: it also appears, on inspection, that the equation tables are directly copied from de La Hire. This adaptation of the Frenchman's tables, though not explicitly of the heliocentric theory on which they were based, must be due to the influence and advice of Fathers Boudier and Pons; for this and other reasons the *Zīj-i-Muḥammad Shāhī* was not finished before 1735 at the earliest. The role of the two French priests in the final shaping of the *Zīj* was alluded to by Father Pons in a letter he wrote at Karikal near Tanjore on 23 November 1740:[25] "The Rāja Jayasiṃha will be regarded in the coming centuries as the restorer of Indian astronomy. The Tables of de La Hire, under the name of this prince, will have spread everywhere within a few years".

Notes

1. Of fundamental importance for the life of Jayasiṃha is Bhatnagar (1974).
2. Much has been written concerning this mission, including the reference to it by Jayasiṃha himself in the introduction to his *Zīj-i-Muḥammad Shāhī*; see Hunter (1797), esp. pp. 187–189. See as well Kaye (1973), pp. 5–7; Forbes (1982), esp. p. 237; Sharma (1982), esp. p. 335; and Sharma (1995), pp. 294–298. Unfortunately, none of these accounts is entirely accurate. See Ansari (1987) for the use of telescope by Jayasiṃha and the connected *problématique*.
3. The *Tabulae astronomicae* were originally published in 1702, a fact to which Jayasiṃha refers in the introduction to his zīj; see Hunter (1797), p. 187. However, it was the reprint of 1727 that was in Jayapura for it was copied by Joseph du Bois in that city in 1732. Du Bois' preface to that transcription was published in Mercier (1993).
4. See Forbes (1982), p. 236, and Mercier (1993), p. 164.
5. Mercier (1984), and van Dalen (2000).
6. Manuscript 11 in the collection of Arabic and Persian manuscripts in the Maharaja of Jaipur Museum.
7. There are four enormous copies of the *Siddhāntasindhu* at Jayapura; one of these, Maharaja of Jaipur Museum 23, was copied at Jayasiṃha's command by Gaṅgārāma in 1727.
8. An edition is being prepared by C. Minkowski and D. Pingree.
9. An edition by D. Pingree will be published soon.
10. An edition is being prepared by S. Ikeyama and D. Pingree.
11. This reference by Jayasiṃha is quoted in the letter of Père Calmette translated below.
12. The five collated manuscripts are: Benares 35726; Rajasthan Oriental Research Institute (RORI) (Jodhpur) 37733; RORI (Kota) 728; Poona, Bhandarkar Oriental Research Institute (BORI) 926 of 1886/92 and BORI 537 of 1895/1902. As yet unavailable is Calcutta Sanskrit College jyotiṣa 55.

13. Sharma (1995), p. 246.
14. Calcutta Sanskrit College jyotiṣa 93.
15. RORI (Jodhpur) 23958 and RORI (Kota) 726.
16. Hunter (1797), p. 210.
17. *Ibid.*, p. 209.
18. In the *Zīj-i-Muḥammad Shāhī* the time-difference between Paris and Delhi is $4 : 54^{h}$, so that the longitudinal difference between Jayapura and Delhi is $73; 30° - 72; 15° = 1; 15°$, a value also found in the *Zīj-i-Muḥammad Shāhī*; see B. van Dalen (2000), p. 63, Note 7.
19. The epoch of the tables in the *Zīj-i-Muḥammad Shāhī* is 20 February 1719.
20. R. S. Sharma (1967), p. 1164.
21. P. de La Hire, *Tabulae astronomicae*, Paris 1702, 2nd page of the Epistola.
22. LEC (1819), vol. 7, pp. 503–509.
23. *Ibid.*, vol. 8, pp. 439–454. On Boudier see also Sharma-Huberly (1984), esp. pp. 101–105.
24. Pingree (1987), esp. pp. 323–324.
25. LEC (1819), vol. 8, pp. 37–53, esp. pp. 44–45.

References

Ansari, S. M. R. (1987): "Sawai Jai Singh and European Astronomy," *Istoriko-Astronomitcheskie Issledovania*, Moscow, pp. 151–170.

Bhatnagar, V. S. (1974): *Life and Times of Sawai Jai Singh*, Delhi.

van Dalen, B. (2000): "The Origin of the Mean Motion Tables of Jai Singh," *Indian Journal of History of Science*, **35**(1), pp. 41–66.

Forbes, E. G. (1982): "The European Astronomical Tradition: its Transmission into India, and its Reception by Sawai Jai Singh II," *Indian Journal of History of Science* **17**, pp. 234–243.

Hunter, W. (1797): "Some Account of the Astronomical Labours of Jayasimha, Raja of Ambhere, or Jayanagar," *Asiatick Researches* **5**, pp. 177–211.

Kaye , G. R. (1973): *The Astronomical Observatories of Jai Singh*, repr. Delhi.

LEC (1819): *Lettres édifiantes et curieuses*, Lyon.

Mercier, R. (1984): "The Astronomical Tables of Rajah Jai Singh Sawai," *Indian Journal of History of Science* **19**, pp. 143–171.

Mercier, R. (1993): "Account by Joseph Dubois of Astronomical Work under Jai Singh Sawā'ī," *Indian Journal of History of Science* **28**, pp. 157–166.

Pingree, D. (1987): "Indian and Islamic Astronomy at Jayasiṃha's Court," *From Deferent to Equant*, New York, pp. 313–328.

Pingree, D. (1999): "An Astronomer's Progress," *Proceedings of the American Philosophical Society*, **143**, pp. 73–85.

Pingree, D. (?): "Paurāṇika versus Siddhāntic Astronomy," to be published in the Proceedings of the Tenth World Sanskrit Conference, held 3–9 January 1997 in Bangalore, India.

Sharma, Ram Swarup (1967): *Samrāṭjagannāthaviracitaḥ samrāṭsiddhāntaḥ*, Vol. 2, New Delhi.

Sharma, V. N. (1982): "Jai Singh, his European Astronomers and the Copernican Revolution," *Indian Journal of History of Science* **17**, pp. 333–344.

Sharma, V. N. (1995): *Sawai Jai Singh and His Astronomy*, Delhi, pp. 294–298.

Sharma, V. N. and L. Huberly (1984): "Jesuit Astronomers in Eighteenth-Century India," *Archives intern. d'histoire des sciences* **34**, pp. 99–107.

2.2. European Astronomy in Indo-Persian Writings

S. M. Razaullah Ansari
Formerly at Aligarh Muslim University

1. Introduction

After the fall of the Mughal empire, when the first war of independence against British colonizers failed in 1857, the East India Company's management was transferred to the British Crown in 1858. The period from 1858 to 1947 is known as the British Period of Indian History. It should be noted that only in 1910, a Department of Education was established by the British Government of India and in the following decades modern universities were established in various important Indian towns, wherein the West European type of the education and training with *English* as the medium of instruction were imparted. However, more than a century before, Indian scholars came into contact with the scholar–administrators of the East India Company (EICo), either through employment or social interaction. Thereby, Indians became acquainted with the scientific (and also technological) advances in Europe. A few of them had already visited England and other European countries, Portugal, France etc. in the last quarter of 18th century, in order to experience and to learn *first hand* about European sciences in general.

It is an established fact that modern science, as we understand it today, is actually a product of the so-called *Scientific Revolution* of 17th century Europe. Its most important milestones are considered to be the Heliocentric System of Copernicus[1], Galileo's work in mechanics and astronomy[2], and Kepler's hypothesis of elliptical orbits and his laws for planetary motion[3]; all these culminated in the Newtonian treatise: *Principia Mathematica*[4].

We also know that with the arrival of EICo and its British officers comprising administrators, medical doctors, engineers and the like, Persian-speaking Indian scholars came into direct contact with the British scholar–administrators. This contact resulted in a very fruitful academic interaction and exchange of scientific ideas. Consequently, we present in this paper a survey of selected Indo-Persian writings dealing with Modern European Astronomy, particularly theories of Copernicus (1473–1543), Galileo (1564–1642), Kepler (1571–1630) and Newton (1642–1726). We present in this paper first the works of those Indian scholars who travelled to Europe and had direct experience of the European sciences, for instance Mīr Muḥammad Ḥussain Landanī (travelled 1774–77), Mirzā Abū Ṭālib Landnī (travelled 1799–1803), but also those of other experts on European sciences, e.g. ^{c}Abdul Qādir Jaunpūrī (b. 1728), Ghulām Ḥussain Jaunpūrī (fl. 1850), and Raja Ratan Singh Zakhmī.

2. Mīr Muḥammad Ḥussain

Mīr Muḥammad Ḥussain ibn ^{c}Abdul ^{c}Aẓīm Iṣfahānī (alias Landānī, since he travelled to London) was a famous *Ḥakīm* (Unani physician) and poet of Murshidābād.[5] He died in 1790. He is reported to have been an expert on natural (rational) sciences (*Ma^{c}qūlāt*).

S.M.R. Ansari (ed.), History of Oriental Astronomy, 133–144.

Since he was very much interested in modern European Sciences due to his acquaintance with British officers, particularly with Henry Elliot, he had the opportunity to travel to England and Europe during 1774–77. On his return he wrote a tract in Arabic that is known by several titles: "A Tract on the Description of Europe and the Modern World, with some Problems of Modern Mathematical Sciences according to European Scholars", manuscript (Ms.) in M. A. Library (Aligarh), No. 33 Arabiya ᶜUlūm; "A Tract by Mīr Muḥammad Ḥussain on the Theories of European Scholars, concerning Planets and Heavens", Ms. in Mullā Fīroz Collection (Bombay); other Ms. copies in Sālār Jang Museum and Osmania University (Hyderabad), in Raza Library (Rampur), and in the private collection of Mr. Muslim Ziayi (Karachi). This tract was written at the request of S͟haik͟h ᶜAbdul Qādir Jaunpūrī, who was himself an expert on modern Sciences.[6] This monograph[7] consists of an introductory historical-geographical chapter on Europe, followed by an account of modern European astronomical development. We summarize his important observations.[8]

Having studied the magnet and its use in the mariner's compass, Mīr Muḥammad Ḥussain described first the discovery of America by *Columbus*. He stated that the telescope was invented by the English philosopher Sir Isaac Newton and that it brought about a revolution in astronomy leading to wonderful observations by European scholars; for instance, observations of thousands of fixed stars, 70 stars in the constellation of *Pleiades*, elliptical bright rings around the planet *Saturn* and its five satellites, but four of *Jupiter*, three dark parallel lines on the surface of *Jupiter*, the changing shape of *Venus* like the Moon, sunspots[9] with their period of 25 days (from which fact Mīr Muḥammad Ḥussain concluded correctly the axial motion of the Sun). Further, he emphasized the *planetary* nature of the Earth and also the modern heliocentric planetary system. He explained that the philosopher *Copernicus* named it the solar system (*waḍaᶜ –i shamsī*). He also mentioned that with the aid of the telescope the inferior conjunctions of the Sun with *Venus* and *Mercury* (i.e., their transit over the solar disc) could be observed as dark specks. Using this fact Mīr Muḥammad Ḥussain argued that those two planets revolved around the Sun. He also listed the periods of revolution of planets and their sizes in units of the Earth's size. Further, he mentioned the *great comet* of 1680 A. D., and tried to explain its largest elliptical orbit.[10] Finally he remarks that each of the fixed stars (*T͟hawābit*), along with its planets and satellites, like the Sun, forms a universe by itself. He therefore concludes that there exist an infinite (*g͟hair mutanāhiyah*) number of Universes (*ᶜAwālim*).[11] The text is as follows:

> "Each of the fixed stars along with its planets and satellites is a universe by itself. That is why, the Sun with its Planets and Satellites is a Universe (*ᶜĀlam*) by itself. It is therefore proved that infinite universes exist . . ., that is, the signs of the Nature (God) are not confined to this Universe. On the contrary, the existence of the infinite Universes manifests the omnipotence (of God)."

This is exactly the modern view of the universe comprising billions of galaxies, each galaxy consisting of billions of stars, and statistically a million stars or celestial bodies may have a system of their own just like Mīr Muḥammad's "infinite Universes" with an extremely large number of solar-like systems. The connotation of the word *Universe* (of galaxies) is evidently different today. We are not able to identify Mīr Muḥammad's source.

3. Mirzā Abū Ṭālib

Mirzā Abū Ṭālib bin Mirzā Muḥammad Bég Iṣfahānī (1752–1805/6) was a well-known personality in 18th century India. He was in the services of Nawāb Āṣaf ad-Dawlah of Avadh, of Col. Alexander Hannay (the Collector of Gorakhpur) and of Mr. Middleton (the Resident at Lucknow).[12] However, he became quite well known both in Europe and India with the publication of his travelogue *Masīr-i Tālibī fī Bilād-i Afranjī* (in Persian), an account of his journey to Europe during 1799–1808.[13]

Here we are naturally interested in his scientific writings, particularly on modern astronomy, on which he wrote before embarking on his European journey and also after his return. In the former category three manuscripts are extant:

(1) "A Tract on the Proof (*Athbāt*) of Modern Astronomy", Ms. in Kutub Khāna-i Gharb (Hamadān), No. 1639, written ca 1772 A. D.[14]

(2) "A Tract on Modern Astronomy" (*Risālah Hay'at–i Jadīdah*), in Raza Library (Rampur), Ms. No. 1227, dated 1797 A. D.

(3) "Astronomy in Persian" (*Fārsiyah Hay' at*), in Quṭbuddīn Collection at M. A. Library (Aligarh Muslim University), Ms. No. 74/2, written ca 1798 but scribed in 1818, with 54 folios.

In the last category we have found only one manuscript (extant in two copies), with the titles

(4) "The Culmination of the Divine Unity" (*Ma^c rāj al-Tawḥīd*), University Persian Collection in M. A. Library (Aligarh Muslim University), Ms. No. 1812, scribed in 1806 by ^cĀshiq ^cAlī at the instigation of Khwājah Farīduddīn Aḥmad[15], comprising 10 folios (f. 20a – f. 29a).

This manuscript is anonymous, but we have identified its author as Abū Ṭālib for the following reason. Its beginning coincides with another manuscript with the same title but attributed to Abū Ṭālib Khān Iṣfahānī of Lucknow. It is extant in the New College Collection (Edinburgh), Ms. No. 93, consists of 22 folios and is dated 1807.[16] Storey mentions that this Ms. was composed in 1804 A. D. (1219 A. H.) and was dedicated to Abū al-Fatḥ Sulṭān Muḥammad Ṣafawī. The beginning of this Ms. seems to coincide with that of the above-mentioned Aligarh Ms., hence its identification. The internal evidence, which is given in the sequel, shows that this tract was composed in about 1802, just *before* his return from Europe.

This tract of Abū Ṭālib is in fact a prose commentary on his own poem of 65 couplets.[17] In his preface (fol. 20a) he explains that modern knowledge[18] is presented in this tract, namely regarding fixed stars, planets, satellites (moons), comets, spheres, the pure air, the nature of colours, seas, mountains, the motion of the Earth, the reasons for solar and lunar eclipses, ebb and flow etc. In particular, he gives the following information about the solar system: planetary heliocentric distances, their axial and orbital periods, diameters and speeds; for instance, he states that Jupiter and Saturn are 49,490,976 and 907,956,130 miles from the Sun respectively. These heliocentric distances are quite comparable with the modern values.

He mentions also another planet beyond *Saturn* by the name of *Jāj* or *Jārj*. That is, *Georgium Sidus*, named after the king of Britain George III and which the astronomer J. E. Bode named *Uranus*. We know that the British astronomer F. W. Herschel discovered it in 1781.

In the eighth couplet and its commentary, Abū Ṭālib enumerates 18 satellites or moons in the solar system, namely, for *Earth*, 1, for *Jupiter*, 4, for *Saturn*, 7, and for *Uranus*, 6.

To understand these numbers we first note the following.[19] Galileo had already discovered the four satellites of *Jupiter* with a telescope in 1609/1610, and the fifth one was spotted by C. G. Perrine in 1904. Today fourteen Jovian Satellites are known. In the case of *Saturn*, the seven satellites were discovered by Christiaan Huygens (in 1655), three by G. D. Cassinni (in 1671, 1672, two in 1684) and two by F. W. Herschel (in 1789). Today we know that *Saturn* possesses ten satellites. As for Uranus, its discoverer, Herschel, thought that he had found six of its satellites in 1787, although only two were real; the rest turned out to be stars of low luminosity. The third and fourth satellites were found by W. Lassell in 1851, and the fifth by G. P. Kuiper in 1948. All this information actually proves that Abū Ṭālib's data were *up-to date, since* he incorporated in it the discoveries of 1787 and 1789. Further, he clearly states in his tract that "without the use of a telescope the satellites of planets [beyond Earth, called today outer planets] could not be seen".[20] In connection with *Uranus*, he again adds the same remark:[21]

> "*Jaj* is a star [i.e., planet] which was not known to ancient astronomers. It was seen after the invention of the reflecting telescope. Within the last 20 years, it has been observed astronomically. Within 2–3 years two other planets by the side of *Jāj* have been discovered; their speed is still unknown. One of them is called *Sīras* and the other *Palas*."

We use here the Persian orthography for the names of the "new" planets, as given by Abū Ṭālib. From these lines we may deduce the date of writing of this tract as follows. Herschel discovered Uranus in 1781. Now 1781 + 20 years = 1801, i.e. Abū Ṭālib is writing these lines in 1801 A. D. Further, he talks about two more planets. These are actually the *minor* planets as we call them today. *Ceres* was discovered by the Italian astronomer G. Piazzi on January 1, 1801 and *Pallas* by the German astronomer W. Olbers in 1802. The last one indicates that Abū Ṭālib was writing his tract, No. 4 , in about 1802 just before his return to India in 1803.

In fact, Abū Ṭālib stresses at various places in his tract the importance of the telescope. For instance, with its help "3000 stars have been observed in addition to one thousand by the naked eye observation".[22] In passing we may mention that in one of his earlier tracts, No. 3 above, written in 1798, before his journey to Europe, he mentions the length of a telescope as 40 yards, with which "the English nation observed mountains on the moon and now [1213 A. H./1798 A. D.] they have taken to manufacturing more powerful telescopes for observations".[23]

In his discussion of the comets, Abū Ṭālib writes: "The comets are numerous and three of them, possessing elliptical orbits around the Sun, have been observed more or less better [than others]. One of them has a period of 75 years and the period of the second and third are 229 and 575 years".[24] For the latter he gives also the aphelion and perihelion distances of the comet, and even the speed of transit through the perihelion as 880,000 miles/hour.

The terminology used by Abū Ṭālib is the following: For aphelion *Muntahāyi Bu^{c}d*, for perihelion *Muntahāyi Qurb*, and for the heliocentric system *Markazīyat–i Shams*. In fact Abū Ṭālib argues in favour of the heliocentric system by giving the famous example of the transits of *Venus* and *Mercury* as dark spots on the solar disc, i. e., their conjunctions (*Muqārnah*) with the Sun, and states clearly that due to such observations the "hypothesis of geo-centric system (*Markazīyat–i Arḍ*) is impossible".[25] All these details are an indication of Abū Ṭālib's quite detailed knowledge of the European astronomy of the early 19th century.

It is interesting to note that in connection with the transit of *Venus* he mentions the name of Quṭbuddīn Shīrāzī (1236–1311), to whom it was known. We know today that Ibn Sīnā (d. 1073), Ibn Bājja (d. 1138) and others had observed "the two planets as black spots on the face of the Sun",[26] evidently due to the transit(s) of *Venus* and/or *Mercury*. In fact, Quṭbuddīn has referred to the observation of the *Venus* transit by Ibn Bājja.[27]

4. ^{c}Abdul Qādir Jaunpūrī

^{c}Abdul Qādir Jaunpūrī bin Khairuddīn ^{c}Imādpūrī (1727–1787) was also a scholar of European sciences and philosophy. He was employed as a Munshi in the office of the East India Company (Calcutta). He was quite a well-known scholar among the *^{c}Ulamā'* (Muslim academics) of the time, particularly through his correspondence with them, for instance, with the famous Indian theologian Shāh Walīullāh. His intense interest in the European sciences can be gauged by the fact that Mīr Muḥammad Ḥussain (see section 2 above) wrote his tract with the encouragement of ^{c}Abdul Qādir. He has been reported as the author of the following books on European science/philosophy:[28]

(1) A tract on modern chemistry;
(2) A comparative study of oriental and occidental sciences;
(3) A critical study of Francis Bacon's writings.

The last work seems to be very significant. No further details are found in the literature known to us to-date. Most probably it refers to the treatise *Novum Organon*, written in 1620 by the philosopher of science, Francis Bacon (1561–1626). In this book Bacon analyzed the scientific method, which concerns mainly the union of theoretical reasoning not only with experimentation but also with the mechanical arts, or what we term today technology.[29]

It is not known whether these writings, originally written in Arabic, were translated or commented upon in Indo-Persian. We are listing them (especially No. 2–3) here due to their great importance for the modern scientific method.

5. Ghulām Ḥussain Jaunpūrī

Ghulām Ḥussain Jaunpūrī *Karbalā'ī bin Fatḥ Muḥammad* (1790–1862) was a court astronomer of the Raja of Tikārī, Raja Khān Bahādur Khān Nuṣrat Jang, Muslim son of Maharaja Mitrajīt Singh. Tikari is an Indian town in South Bihar, where Ghulām Ḥussain carried out his astronomical observations. Later he left for Benaras where he was appointed Chief Justice at the Court of the Rajah of Benaras. After a few years he moved to Murshidābād to serve its Nawab. He died at Dawoodpur, on the way to his home town of Jaunpūr, in 1862.[30]

The following extant works of Ghulām Ḥussain are of interest to us here:[31]

(1) On the construction of the astrolabe and its use in solving some astronomical problems, the latter written in 1818;
(2) On the terminology of almanacs;
(3) Key to mathematical astronomical sciences. This encyclopaedia was completed on October 29, 1833, and printed lithographically in Calcutta in 1835. The Persian title is *Miftāḥ al-Raṣad*, or *Jāmi*ᶜ *Bahādurkhānī.*
(4) Astronomical-mathematical tables (*Zīj*) for calculating the position of planets, the timing of lunar and solar eclipses, and solving calendric problems. It was written in 1838, and printed in 1855 at the Cadre Press, Benaras. Ghulām Ḥussain dedicated it to his patron Bahādur Khān, hence its title, *Zīj-i Bahādurkhānī.*[32]

Ghulām Ḥussain's *Encyclopaedia of Sciences* is in Persian and comprises six books ("Treasures"). Each book is divided into chapters (called "castles"). The six books deal with the following topics: geometry, optics, arithmetic, mensuration (with sections on trigonometrical functions and solutions of both plane and spherical triangles), the science of astronomy, of astronomical tables and almanacs. In the sequel, we are interested in his fifth book, particularly those sections that concern modern astronomy.[33]

Ghulām Ḥussain describes the heliocentric system of the planets by a diagram (chapter 1, p. 476). He includes therein the minor planets *Pallas* and *Ceres* between *Mars* and *Jupiter*, also *Uranus*[34] beyond the orbit of *Saturn*. His diagrammatic representation consists of satellites around Jupiter, Saturn and Uranus. He is fully aware that "contrary to all ancient scholars, who believed that planets could not be more than seven, Christian scholars had found the following" (p. 472):

(a) "On March 10, 1781 A. D. Piyarlas [actually Piazzi] found a planet moving above [the orbit of] *Saturn*. It was faster than the fixed star with its speed 42.294 seconds/day. It was named after the King Jārjīs", i.e., George.
(b) "On January 1, 1806 A. D. a planet was observed, moving below [the orbit of] *Jupiter* but above [that of] *Mars*, with the speed 12.853 minutes/day. It was called *Sarīsh*" [actually *Ceres*].
(c) "A scholar named 'Baktar Albars' [actually Walter Olbers] discovered on March 28, 1802 A. D. another planet between *Mars* and *Ceres* with the speed 12.6835 minutes/day and which was named *Pallas*." It is worthy of note that Ghulām Ḥussain records his own observations of *Pallas* with the aid of a telescope (*Dūrbīn*), carried out in Mirzapur in 1242 A. H./1826 A. D. under the guidance of Captain Darānīs, who is not identifiable.

We may note here that whereas the dates and the names of the discoverer for the items (b) and (c) above are correct and exact, Ghulām Ḥussain's information for (a) is incorrect. Evidently, *Uranus* was discovered by the English astronomer F. W. Herschel on March 13, 1781 and not by the Italian G. Piazzi, who found the planet *Ceres*. However, his information on speeds seems to be correct.

Furthermore, G͟hulām Ḥussain knows quite well that "European scholars (*Ḥukamā'–i Farang*) do not consider the Moon as a planet. On the other hand they say that it is a satellite (*Tawābi*[c]) of the Earth around which the Moon revolves. Actually they have found with the aid of a telescope (*Minẓār*) four satellites of Jupiter, seven of Saturn and six of *Jārjīs* [i.e., *Uranus*] . . . they grant [a force of] attraction (*Anjad͟hāb*) existing between the Sun and all planets . . . the planets being suspended without any connection" (p. 473).

Here G͟hulām Hussain is referring to the force of gravitation as theorized by Isaac Newton, whom he calls Ḥakīm Ḥād͟hiq Niyūtan. He attributes to Newton the concept of axial rotation of the Earth and also its revolution around the Sun on an elliptical orbit (p. 462). In this connection he also argues that "the European scholars do not recognize the theory of ancients that the Sun moves around the Earth, since a bigger body can not revolve around a smaller body". In the section on comets, G͟hulām Ḥussain confronts the readers with the opinion of the ancients – that comets belong to the terrestrial (sublunar) world – and that of the European scholars, according to whom they belong to the family of superior planets, moving on an oval-like orbit.[35]

6. Rāja Ratan Singh Zak͟hmī

Rāja Ratan Singh (1782–1851), with *nom de plume* Zak͟hmī (the wounded), belonged to an illustrious family in the service of the rulers of Awadh. He began as a servant of the East India Co. at Calcutta (ca 1803), but returned to his birthplace, Lucknow, in 1815, when he joined the court of the Awadh ruler, G͟hāzīuddīn Ḥaidar (reigned 1814–27), and later worked at the court of the succeeding Nawābs, Naṣīruddīn Ḥaidar (reigned 1827–37) and Naṣīruddawlāh. He was a scholar of many languages, namely Arabic , Persian, Turkish, Sanskrit and English. He is known to have been a prolific writer, *littérateur*, historian and particularly a famous scholar of astronomy.[36] In astronomy he wrote two works:

(1) *Ma*[c]*yār al–Zamān*, a treatise on calendar and chronology in two chapters, completed in 1819. In this tract he discusses timings of day, night, month, year and the eras: Christian, Hindu, Greek, and Egyptian. The extant manuscripts are: 1 in Karachi, 1 in Calcutta, and 1 in Patna, the latter being scribed by G͟hulām Ḥussain and dated 1822.[37]
(2) *Ḥadā'iq al–Nujūm* (Gardens of Astronomy), written in 1837 on the orders of Naṣīruddawlāh, alias Muḥammad [c]Alī Shāh. A number of its manuscripts are extant in Aligarh.[38] It was published in a few editions, in 1837, 1841 and 1843, by Muḥammad Ḥussain, and printed at Muḥmmadī Press at Lucknow.

In the following, a brief sketch of the second work (abbreviated hereafter as HN) is presented.

Ḥadā'iq al–Nujūm is a treatise of 1158 pages, consisting of nine Chapters (*Ḥadīqah*). A chapter is divided into several sections (*Chaman*), and a section into several subsections (*Gulban*). In the following we do not discuss this chapterisation. Furthermore, we refer to the text only by page numbers in parentheses, using the printed edition of 1841, available in the M. A. Library (Aligarh Muslim University). The author states in the beginning that he "describes in this book the celestial bodies, solution of their astronomical problems according to the past and present European scholars and also some researches of his own in this respect" (p. 8).

In this excellent and systematic treatise, he mentions the astronomical work and discoveries of Copernicus, Tycho Brahe, Kepler, Galileo and Newton, and the then recent work of Hevelius, Flamsteed, John Herschel, Cassini and Laland, to name but a few. He is well aware of the works of his European contemporaries. The instruments mentioned are telescopes (*Dūrbīn, Sitārahbīn*), and even micrometers (*Raizah-i Paimā*) with cross-wires (p. 154). He informs us about the establishment of the Greenwich and Paris observatories. Besides he notes that "although the Peking observatory does not match with the (modern) European ones, yet its (old) instruments are quite large" (p. 156).[39] He attaches a map of the Old Peking observatory. Strangely enough, there is no mention of Raja Swai Jai Singh's observatories in India, despite the massive stone instruments which were also extant even in those times.

HN is, in fact, a very comprehensive treatise dealing with modern European astronomy. It is a unique book of its kind in Indo–Persian. It comprises 129 figures (*Shakl*), including geometrical diagrams and drawings. There are 180 tables (*Lawḥ*) of various kinds. Some of them we list in the sequel. HN contains 19 maps, mostly geographical, but also of celestial constellations. On one map (fig. 56), there are diagrams of *Mars*, crescent of *Venus, Jupiter* with bands, *Saturn* with two rings, and also *Uranus* (called by him Jārjīs, i.e., *Georgium*, original name of the *Uranus*). In another map (fig. 57), the author has represented diagrammatically the solar disc with sunspots, and three comets of 1835, 1819 and 1832, in that order.[40] About the first one he writes that he observed it himself for several days, its period being 76 years and 248 days. He devotes to this (*Halley's*) *comet* a whole section (pp. 805–810), in which he discusses its earlier sightings and his own first observation on Oct. 15, 1835 near *Corona Borelis* (in Persian, *Iklīl Shimālī*) with the length of its tail between 29°–30° (p. 805). Of note is that his section on comets is a very detailed one. He explains their physical nature and elliptical orbits. There is also a unique table (No. 112) of 106 comets with all relevant information, spanning the period of their appearances: the first in 835 A. D. and the rest from 1231 to 1835 A. D. (pp. 776–799).

In the section dealing with arithmetical and geometrical preliminaries, Ratan Singh deals with the geometry of ellipses in detail (p. 56 *et seq.*). He clearly states that "the orbits of all planets and of their satellites, together with those of the comets, are elliptical", *Baiḍī* in Persian (p. 135), and he gives a heliocentric diagram of the solar system (p. 179), in which the order of the planets is as follows: *Mercury, Venus, Earth* (with 1 moon), *Mars, Vesta, Juno, Ceres, Pallas, Jupiter* (with 4 moons), *Saturn* (with 7 moons), and *Uranus* (*Jārjīs*) with 6 moons. He then devotes a full section to each one of them. His knowledge of the planetary system is up-to-date, i.e., for his times. Note that the title of one of his sections is "*Proof that the Sun is stationary*". For instance, as a proof of the heliocentric system, he deals with the *transits of Venus* in particular. In fact in one table, 51, all information about the transits of *Venus* is listed for the period 900 A. D. to 3000 A. D. (pp. 310–315), his source being Laland. In this connection, he also mentions that "Ibn Sīnā, Abū ᶜImrān in Baghdad, Muḥammad abī Bakr . . . observed *Venus* on the face of the Sun, [while] Ibn Sīnā, Ibn al-Māja and Ibn al-Haytham observed *Venus* and *Mercury* each as a black dot on the disc of the Sun" (p. 309).[41]

Finally, we may mention that he presents some important tabulated material. For instance, Table 58 for obliquity according to Indian, Greek, Arab, Iranian and European astronomers (pp. 363–367); and Table 109 for lunar and solar eclipses, spanning the years 1837–1900 (pp. 709–722). Besides his knowledge of Islamic Astronomy, which is quite apparent,[42] Ratan Singh also indicates his good knowledge of Ancient Indian (Hindu) astronomy, since he gives in his Table 133 (pp. 1000–1004) the planetary revolutions in a *Kalpa*, according to *Brahāmasiddhānta, Sūrya–* and *Panca–siddhānta*, Āryabhaṭa and Pārāshar, and lists relevant information about this Indian *Yuga* astronomy.

7. Concluding Remarks

Due to the limited time available we cannot go into other writings on modern European astronomy. Worthy of note is that in all Asian civilizations astronomy was considered to be the most important science; even mathematics was subservient to it. In fact, on our sub-continent astronomy was taken particularly seriously throughout the Ancient and Medieval periods. The works which have been summarized above are just a small selection from scores of Persian tracts concerning *modern* European astronomy, written particularly in the first half of the 19th century. They are clear indicators that the Indian mind was quite receptive to astronomy – the most modern European science during the 18th and 19th centuries. In fact, the initiator or predecessor of all the above-mentioned Indian scholars was Raja Swai Jai Singh II (1686–1743), who is particularly famous for the construction of observatories in a number of Indian towns (Delhi, Benares, Mathura, Ujjain and Jaipur) and who sponsored the compilation of an astronomical table, *Zīj-i Muḥammad Shāhī* (ZMS) which in turn is mainly based on Phillipe de La Hire's astronomical tables.[43] We intend to publish a comprehensive paper on this *problématique* elsewhere. However, in passing we would add here that Jai Singh was quite aware of the use of the telescope for astronomical purpose. And he clearly states in his *Zīj*, that "telescopes were constructed in my kingdom and using them a number of observations were carried out".[44]

We conclude this paper on an *enigmatic* and disappointing note! Despite quite substantial writings on Modern European Science – including not only astronomy but also mathematics – all those efforts remained *outside* the circuit of indigenous *madrasah* education in India. Although the writers were quite important personalities, yet they could not influence the *madrasah* syllabus, wherein the ancient out-dated textbooks on astronomy and mathematics were used quite seriously. For instance, the compendium on astronomy by Maḥmūd bin Muḥammad ᶜUmar al-Chaghmīnī (d. 1344) and its commentary by Qāḍī Zādeh Rūmī, or treatises on astronomy and mathematics ("Description of Celestial Spheres", "The Essence of Arithmetic"), by Bahā'uddīn al-ᶜĀmilī (d. 1622). This lack of interaction between the *madrasah* teachers and the new team of scholars, who were extremely receptive to European knowledge, requires a detailed study. Its results will be published elsewhere.

The author acknowledges gratefully the generous financial assistance offered by Dr. J. Andersen, General Secretary of IAU for his participation in the 23rd General Assembly of IAU at Kyoto.

Notes

1. This is in contra-distinction to the *geocentric* system of antiquity particularly prevalent since Claudios Ptolemy (ca 85–165 A. D.). The geocentric system of the then known universe was adopted by astronomers in India as well as those of Islamic countries. For the influence of Islamic Astronomy on Copernicus and particularly its Ptolemaic critique, see Swerdlow and Neugebauer (1984), "Arabic Astronomy and the Marāgha School", pp. 41–48, use of Ibn Shāṭir's Model, p. 61 *et seq.* For the transfer of Islamic astronomy to Medieval Europe in general, see Sayili (1958), esp. pp. 358 *et seq.*
2. Galileo Galilee (1564–1642) is famous for his theory of *free fall* of bodies and for using a telescope for astronomical purposes, particularly for his observation of the motion of satellites around Jupiter. We shall refer to this discovery in the sequel. He is also known as the *first* (European) scholar to use the so-called *scientific method*, i.e., the mathematical-experimental method as the basis of the 17th century scientific revolution. Incidentally this assertion of being the *first ever* scientist to propound the above-mentioned scientific method has been debated, for details see Schramm (1963) and Ansari (1994).
3. The discoveries of Johannes Kepler (1571–1630) were quite unique to 17th century Europe. His predecessors, Greek, Indian and Islamic astronomers, stuck only to circular orbits. In India, Raja Jai Singh (1686–1743) was the *first* to use the elliptical orbits, see section 7 in this paper. For the introduction of Kepler's laws into China, see the contribution of Hashimoto in this Volume.
4. We may mention here that the Indian scholar Tafaḍḍul Ḥussain Khān (d. 1801) had presumably translated *Principia* into Arabic. He is famous for his knowledge of Latin and English besides Arabic and Persian. He also wrote on modern European algebra and astronomy, Raḥmān ^{c}Alī's *Ta<u>dh</u>kirah* (Urdu translation) pp. 139–40, and Salīm, p. 65.
5. For details of his life, see Barkatī's article on Muḥammad Ḥussain Landanī (in Urdu).
6. <u>Sh</u>ai<u>kh</u> ^{c}Abdul Qādir bin <u>Kh</u>airuddīn ^{c}Amādī Jaunpūrī (1727–1787) studied modern Sciences in Calcutta, where he was an employee of the East India Co, see Section 4 above.
7. Besides this monograph, he also wrote a tract each on the body, the heart, and Persian grammar for Europeans, and collections of poems, etc. cf. Barkātī, p. 74.
8. We use Ms. No. MFR IV. 51, Mullā Fīroz Collection at K. R. Cama Oriental Research Library (Bombay), containing 20 folios. It is the Persian translation of his Arabic Version. For the first report on this manuscript, see Ansari (1987/1992), pp. 123–124.
9. He uses the terms *Shāmāt Sawdā* in Arabic and *Lakkāhāyi Siyāh* in Persian.
10. The great comet of 1680 A. D. which appeared in late autumn was discovered by Kalendermacher Kirch, the founder of the Berlin observatory. This comet was particularly conspicuous by its tremendous tail. It is said to have been one of the brightest comets in the history of mankind and therefore was the subject of many paintings and wood-carvings. Its importance can be gauged by the fact that Isaac Newton had discussed it in his *Principia*. The astronomer E. Halley calculated its period as 575 years.
11. Cf. Salar Jang Museum Collection, Persian Ms. *Hay'at* (astronomy) No. 2, fol. 45b.
12. For detailed life-sketch see Beale, p. 32, see also Abū Ṭālib/ Sarvat Ali's introduction in Urdu.
13. It was translated and published in English (1810, 1814), in French (1811, 1819), in German (1813) and Urdu (1904, 1984). The Persian text was edited by his son Mirzā Ḥussain ^{c}Alī and Mīr Qudrat ^{c}Alī and published from Calcutta in 1812, see Storey (1972), Vol. 1, Part 2, pp. 878–879.
14. Cf. a detailed study of this Ms. by Hamdānī (1984). Hamdānī has wrongly identified Abū Ṭālib as an Iranian.
15. Khwājah Farīduddīn Aḥmad was the maternal grandfather of Sir Sayyid Aḥmad Khān (1817–98), who is famous as the founder of the Aligarh Muslim University, Aligarh. Farīduddīn is well known as a mathematician and astronomer.
16. See Storey Vol. II, Part 1, p. 97.
17. For a detailed comparative study of Abū Ṭālib's four Mss., see Ansari (1987/92), pp. 125–129.
18. He uses in fact the phrase "knowledge according to modern scholars" *Aqwāl–i Ḥukamā'–i Jadīd.*
19. Herrmann (1975), p. 89, where a table for the satellites and their discoveries is given.
20. Ms. 4, *op. cit.* fol. 21b, line 17.
21. *Ibid*, fol. 21a.

22. He is referring here to the catalogue of stars by Sulṭān Ulugh Beg (1393–1449), wherein only 1022 stars are listed and which were observed by the naked eye at the Ulugh Beg's observatory at Samarqand, built in 1420.
23. It should be noted that Herschel became quite famous not only as an astronomer but also as an instrument maker. He built literally hundreds of reflecting mirrors for telescopes. His best and most famous reflecting telescope had a mirror of 48.8 inches diameter, and the length of the telescope-tube was 40 ft., see Herrmann, p. 208 *et seq.*, the section on the development of astronomical instruments. So Abū Ṭālib was quite aware that observations of planetary satellites and new planets (like *Uranus* and the minor planets) were only possible by (perfecting the technology of) the reflecting telescope.
24. Ms. No. 4, fol. 23b.
25. *Ibid*, fol.25a.
26. Cf. Sayili (1958), p. 360, esp. f.n. 24.
27. Goldstein (1969), esp. p. 55.
28. See Salim pp. 62–63, [c]Abdul Ḥayyī, p. 366.
29. Cf. Mason (1961), pp. 170–172.
30. See details of his life in Ansari (1995/96).
31. Ibid, see also Storey, Vol. II, Part 1, No. 43, pp. 19–20, No. 161, p. 99.
32. See Ansari (1995/96) for some details.
33. We use the printed copy as available in Maulana Azad Library (Aligarh Muslim University), with the stamp of the Lytton Library. We refer to pages of this copy and give them in parentheses.
34. He calls Uranus by *Jārjīs.*
35. The cometary orbit as a conic-section was quite well known at the end of 18th /beginning of 19th century. We have published a diagram, given in a Persian Ms., showing the parabolic path of two comets, see Ansari (1987/92), photograph 1, p. 131.
36. Cf. Beale (1971), p. 295, Storey (1972), Vol. II, Part 1, p. 99, No. 163.
37. The Libraries are: Anjuman Taraqqī-i Urdu (Karachi), Asiatic Society of Bengal (Calcutta) and Khuda Bakhsh Oriental Public Library (Patna).
38. Ḥabīb Ganj and Subḥānullāh Collections in M. Azad Library, Aligarh Muslim University (Aligarh).
39. By this he seems to mean that they are quite accurate, the main reason for constructing large brass and stone instruments.
40. For the comet of 1819, he names Herschel as the discoverer, and for the comet of 1832 Dolhos/Wolhos, not identifiable. Neither of these comets are listed in his Table 112, see above.
41. Cf. Sec. 3, last paragraph on this point, p 137. Note that Ibn Bājja is wrongly scribed as Ibn Māja.
42. Cf. HN, p. 309, also his Table 120, pp. 889–893, for a catalogue of stars according to [c]Abdurraḥmān al-Ṣūfī
43. Cf. David Pingree's contribution in these *Proceedings.*
44. See Ansari (1985a) p. 381, wherein we have published photographs of the marginal diagrams showing surface features of Saturn, Jupiter with four moons, Mars, crescents of Venus and Mercury, found in one of the manuscripts of ZMS. On the *problématique* of the use of telescope during his time, see the detailed discussion in Ansari (1985b), Sec. 1.3.1, pp. 2–6.

Bibliography

[c]Abdul Ḥayyī, Sayyid, *Islāmī [c]Ulūm wa Funūn Hindustān Mein*, Urdu translation of *Al-Thaqāfatul Islāmiya fī al-Hind*, by [c]Abul [c]Irfān Nadwī, Azamgarh, 1969.

Abū Ṭālib, Mirzā, *Masīr-i Ṭālibī fī Bilād-i Afranjī*, (An account of a Journey to Europe), Urdu translation of the Persian text by Sarvat Ali (Urdu Promotion Board), New Delhi, 1984.

Ansari, S. M. Razaullah (1985a), The Observatories Movement in India during the 17th–18th Centuries, *Vistas in Astronomy*, Vol. 28, pp. 379–385.

Ansari, S. M. Razaullah (1985b), *Introduction of Modern Astronomy in India during 18th–19th Centuries*, IHMMR, Hamdard Nagar, New Delhi, 81pp.

Ansari, S. M. Razaullah (1987/92), Modern Astronomy in Indo-Persian Sources, in *Transfer of Modern Science & Technology to the Muslim World* (Proceedings of the International Symposium, held on Sept. 2–4, 1987), Editor: Ekmeleddin İhsanoğlu, Istanbul 1992, pp. 121–144.

Ansari, S. M. Razaullah (1994), Modern Scientific Method in the Works of Ibn al-Haytham, in *Fakhruddin Ali Ahmad Memorial Volume*, edited by Nazir Ahmad and Asloob Ahmed Ansari (Ghalib Institute, New Delhi), pp. 248–271.

Ansari, S. M. Razaullah (1995/96), The Indian Astronomer Ghulām Ḥussain Jaunpūrī and his Zīj-i Bahādurkhānī, *Studies in History of Med. & Science* (New Delhi), Vol. XIV, No. 1–2, New Series, pp. 181–188.

Barkātī, Maḥmud Aḥmad (1975), Mīr Muḥammad Ḥussain Landanī (in Urdu), *Al–ᶜIlm* (Karachi), No. 1 (Jan.–March,), pp. 64–78.

Beale, T. W. (1971), *An Oriental Biographical Dictionary*, new edition by H. G. Kenne, reprint of the last edition (London 1894) by Manohar Reprints (Delhi).

Goldstein, B. R. (1969), Some Medieval Reports of Venus and Mercury Transits, *Centaurus*, Vol. 14, pp. 49–59.

Hamdānī, Ḥussain Maᶜṣūmī (1984), Risālah dar Athbāt-i Hay'at-i Jadīdah by Abu Ṭālib, *Majallah Maᶜārif* (Tehran), Series 1, No. 2, pp. 117–185.

Herrmann, D. B. (1975), *Geschichte der Astronomie, von Herschel bis Hertzsprung*, VEB Deutscher Verlag der Wissenschaften, Berlin.

Mason, S. F. (1961), *History of Science* (London 1953), German translation by K. M. Meyer-Abich (Alfred Kroener Verlag), Stuttgart.

Raḥmān ᶜAlī, *Tadhkirah ᶜUlamā'-i Hind*, Urdu translation by Muḥammad Ayyūb Qādrī, Pakistan Historical Society, Karachi, 1961.

Salim, Syed Mohammad (1993), *Maghrabī Zubānon ke Māhir ᶜUlamā'* (before the establishment of Aligarh College), in Urdu, Lahore, 1993.

Sayili, Aydin (1958), Islam and the Rise of Seventeenth Century Science, *Belleten* (Ankara), Vol. 22, No. 87, July issue, pp. 353–368.

Schramm, Mathias (1963), *Ibn al-Haythams Weg zur Physik*, Franz Steiner Verlag, Wiesbaden.

Storey, C. A. (1972), *Persian Literature: A Bio-bibliographical Survey*, Vol. I, Part 2; Vol. II, Part 1 (on Mathematics, Weights and Measures, Astronomy and Astrology, Geography), London, Luzac & Co. Reprint, first printed in 1958.

Swerdlow, N. M. and Neugebauer, O. (1984), *Mathematical Astronomy in Copernicus's De Revolutionibus*, Part 1, Springer Verlag, New York.

2.3. Toshio Takamine's Contact with Western Astrophysics

David DeVorkin
Smithsonian Institution

Introduction

In the fall of 1920 two young scholars happened to meet in London and found they had many interests in common in laboratory spectroscopy and astrophysics. One was an Indian and the other Japanese, and their paths crossed as they visited American and European centers to acquire the tools and techniques of the new physics and to hone their skills. Their chance meeting bears witness to the growing excitement over how the new physics made spectroscopy a central ingredient not only in the exploration of the atom, but of the structure and compositions of the sun and stars.

The Japanese scholar was Toshio Takamine from Tokyo. The Indian he met was Meghnad Saha, the young Calcutta physicist who was then just putting the finishing touches on the fourth paper in his famous quartet on ionization equilibrium in the atmospheres of the sun and stars. Their single meeting produced a trickle of correspondence over the next twenty years that sheds light not only on their personal interests and capabilities in their home institutions, but reveals how they viewed the practice of western physics and astrophysics and where they fit into the picture. Historians like Lawrence Badash, Helge Kragh and Bruno Carazza have used the correspondence of the Japanese physicist Hantaro Nagaoka to illuminate the state of physics in Europe in 1910 when Nagaoka visited the West (Badash, 1967; Carazza and Kragh, 1991). Here we take a different tactic, looking at how Takamine viewed his work in the context of the growth of problems in the specific area of laboratory astrophysics, a hybrid enterprise that gained a few crossover workers in the 1920s and 1930s in America and Europe, but very few in the Far East. Takamine was virtually alone in the field in Japan, save for the sympathies of his powerful mentor, Nagaoka, and his growing circle of students in the 1930s.

Much has been written on Saha's life and work (Kochhar, 1992; DeVorkin, 1993; 1994) but very little on Takamine, aside from an extensive Nachlass. The two were very different scientists, from very different circumstances. Saha was to become famous for his work on thermal ionization equilibrium processes but he was continually frustrated by his inability to establish a major school or institute where he could pursue these and other problems in his native land, and thus establish physics as an autonomous activity in India. Takamine, however, enjoyed the fruits of a robust, growing economy and culture that had embraced western science for its own nationalist ends after the Meiji Restoration. A result of the Meiji Restoration was that astronomy was transformed from being a court-oriented calendrical science into a wholly modern enterprise. The change was abrupt and discontinuous. Modern astronomy began from scratch, as did modern physics in Japan (Nakayama, 1969; Koizumi, 1974; Nakayama, Swain and Yagi, 1963).

S.M.R. Ansari (ed.), History of Oriental Astronomy, 145–157.

Takamine's Life and Training

Takamine was born in 1885 in Tokyo, the second son of a samurai who had served a local fief before the Meiji restoration. The family's high samurai status insured that Toshio was well educated, and his father had official contact with Americans in the late 19th century (Takamine, 10 January 1925). Takamine acquired an interest in science early in life, graduating from the University of Tokyo in 1909 with a Bachelor of Science in physics. Upon graduation he joined the Army for the requisite one year of active duty, and when he was discharged in 1911 he married and returned to graduate study (Takamine, 1964a). He was much influenced by the optical physicist Taro Aqaba and entered graduate school to study optics and spectroscopy, receiving his doctorate in 1915. There he came under Nagaoka's wing, and it is largely through Nagaoka's mentoring that Takamine enjoyed a successful career in spectroscopy.

Takamine came of age when strong nationalistic role models in science were available to him. The first generation of Japanese physicists, such as Yamagawa and Tanakadate, had pursued physics in service to the nation, but those who followed, particularly Nagaoka, chose physics more for its intellectual motivations (Koizumi, 1974, p. 96). Earlier in his career, Nagaoka had suggested a Saturnian model for the atom, in the spirit of Maxwell's gravitational study of the stability of Saturn's rings, but was frustrated when it was never accepted either in Japan, or in the West, especially because it could not account for improved line and band spectra then being gathered. This led him to concentrate on improving the observational database for atomic structure studies. Therefore his main activities when Takamine encountered him centered on the laboratory study of spectra.

Working under Nagaoka, in 1912 Takamine studied the spectrum of mercury and found many of the lines resolvable into sharp components, using the Zeeman Effect to separate out lines of elements according to how their energy states were influenced by magnetic fields. He also explored interferometric methods of determining wavelength. On one memorable occasion Nagaoka led Takamine to a small observatory, trying to interest him in exploring the solar spectrum with a Rowland diffraction grating. With Torashiro Tanakadate, Takamine studied the solar spectrum and became fascinated with the subject at a time when no one in Japan worked in the field (Takamine, 1964a).

In 1917 Takamine was assigned to teach physics at Kyoto University, where he taught spectroscopy with Usaburo Yoshida and they collaborated in the study of the Stark Effect. By late 1918, in the wake of the First World War, which re-kindled Japanese conviction to build up scientific, technical and industrial strength in order to remain competitive with the West, funding from private foundations and from the Ministry of Education began to flow and a new cohort of Japanese scholars began visiting the West, not only for post-graduate study, but, as in Takamine's case, for research (Bartholomew, 1989, 83–84; Kiyonobu and Yagi, 1974). Takamine had just been invited to join the staff of the newly formed Institute of Physical and Chemical Research in Tokyo where Nagaoka was head of the physics section, when he was granted a leave of absence to work in the West. He obtained funding through awards and scholarships, and toured spectroscopic laboratories spending some eight months at George Ellery Hale's Mount Wilson Solar Observatory, where he worked in the Pasadena laboratories of J. A. Anderson and A. S. King, continuing to explore the Stark Effect for metals. Hale's observatory was then the premier establishment for laboratory astrophysics: not only did the observatory possess the largest telescopes in the world, but also there were few places that could match its experimental expertise and technical infrastructure.

First World Tour

Takamine arrived in Pasadena in September 1918, two months before the Armistice and when a good fraction of the shops and staff of the Mount Wilson Observatory were engaged in the production of optical devices for the Ordnance Department, or were absent entirely, like Hale, on warwork. With so many staff diverted to wartime activities, W. S. Adams, Hale's deputy, had a manpower problem keeping astronomy alive. Especially languishing was new apparatus developed by J. A. Anderson, including a short focus plane-grating spectrograph, small electric furnaces for use between the poles of large magnets, large dynamos to produce strong electric fields, a Langmuir high vacuum pump and an interferometric spectrograph (*Carnegie Yearbook* #18, 1920, pp. 250–251). Adams was unwilling to hire assistants to allow Anderson to continue his work on the Zeeman and Stark phenomena, but Takamine was welcomed as free labor.

Takamine's talents and interests paralleled Anderson's, and he fit right in, learning quickly to use Anderson's large grating spectrograph. Takamine examined metallic spectra using field strengths sometimes as high as 60,000 volts per centimeter. By December 1918 Takamine had already produced some valuable results working on the spectrum of iron searching for correlation between the Stark effect and a curious "pole effect" that had been observed by two Mount Wilson staff members, Charles St. John and H. Babcock. The pole effect was a slight shift in the positions of spectral lines, when viewed near the pole-tips of large electromagnets, that was thought to be a complex function of pressure, length of the arc, velocity of the electrons and ions, the potential gradient, and other causes – a function of the apparatus, in other words. Nagaoka had been fascinated with the effect, and was studying the phenomenon in his own laboratory in Tokyo. It engaged many spectroscopists because it could produce systematic error (Hentschel, 1997).

Anderson had been unable to find any correlation before he became totally involved in warwork, but Takamine, using new powerful dynamos in series to create stronger electric fields, found the evidence Anderson sought, leaving little doubt that somehow the pole effect and the Stark effect were related. Not only had Takamine thus made himself extremely useful, but also his temperament fit nicely with Anderson's: "Takamine is a delightful fellow, and we are very fond of him." (Anderson, 22 December 1918; *Carnegie Yearbook* #18, 1920, pp. 257–258) Adams was similarly impressed, as was Hale, who hoped that his own staff could soon return to astrophysics following Takamine's lead on the Stark effect (Hale, 28 January 1919).

Takamine continued his observations through mid-March, wrote it all up for the *Astrophysical Journal*, and then planned out the rest of his tour, which included touring East Coast centers and then Europe. Takamine knew he had to return to Tokyo, but wanted to extend his stay another year, hoping that Nagaoka would grant the extension. Anderson endorsed Takamine's request, asking Hale to pay for it (Anderson, 8 April 1919) but the request was not granted. Nevertheless it demonstrates Takamine's interest to stay in the West as long as possible, rather than to return home to build up a domestic facility. During this first tour he worked at the Johns Hopkins University in Baltimore in R. W. Wood's laboratory (Takamine, 1964b).

What Takamine found at Hale's Pasadena laboratories were spectroscopic facilities far greater than any he had known. Further, Anderson had perfected a method for producing very small cathodes that could reside within strong electric fields, something Takamine and Yoshida at Kyoto had been working on in 1916–1917. Clearly much of this work was a continuation of Takamine's interests, and they dovetailed nicely with Anderson's. As Takamine reported back to Hale after his long sojourn, he was delighted to hear that they were continuing to work on the Stark effect: "With the wonderful equipment you have at Pasadena, quite unique in the world, I have no doubt that a great help should come out of your results" (Takamine, 16 November 1921).

Takamine made a lasting impression on the Mount Wilson staff. He was not the only Japanese national touring laboratories at this time. Shinzo Shinjo, from the Imperial University, Kyoto, was interested in variable stars and visited Mount Wilson in 1919 as well. Just before he returned home, Shinzo described his visit as a "program of pilgrimage," using equipment and taking observations he could never duplicate at home (Shinzo, 4 December 1919). Takamine, however, was not travelling to complete his own work, but to participate in the work at hand. Much of this was, of course, serendipitous. Sailing for Europe, he worked in Oxford with T. R. Merton, also visited Alfred Fowler's laboratory in South Kensington. Fowler, a highly respected astronomical spectroscopist, was just finishing his seminal *Report on Series in Line Spectra*, and his laboratory was a beehive of activity. He opened his doors to foreign physicists, and it was here that Takamine met Saha, who, like himself, worked far from the centers of the emerging new physics.

In January 1921 Takamine wrote Saha expressing his delight at their meeting. They discussed Sommerfeld's theories and pledged to keep in touch. In the spring Takamine moved to Bohr's new Institute in Copenhagen. This last opportunity came by accident: one of Takamine's reasons for visiting London was to look for new optical devices at Adam Hilger for Nagaoka's Institute and for his own laboratory. During his visit, he learned of Bohr's new Institute and of their interest in having experimentalists collaborate. Following Bohr's suggestion, Takamine studied the excitation of forbidden spectral lines in the mercury spectrum with H. M. Hensen and S. Werner, and continued to examine the Stark effect on the mercury lines, which, Bohr recalled, "brought much new information for the testing and elucidation of the theoretical view points." Also influenced by Bohr, Takamine studied the Balmer series of hydrogen using the Stark Effect as a probe. His researches, according to recollections in his *Biography*, earned him an Imperial prize in 1923. Indeed, in later years, largely because of Nagaoka's influence and Takamine's industry, Otto Laporte recalls that Sommerfeld believed that the Japanese had a lead on the investigation of the Stark effect in hydrogen lines – Takamine and Kokubu, Nuitta, Fukuda and Fujioka were "constantly being quoted" (Laporte, 1964; Takamine, 1964a; Fukuda, 1964; Bohr, 1964, pp. 384–386).

Return to Japan

Back in Japan in the fall of 1921, Takamine found that spectroscopy was no longer isolated. Several universities now had courses in the subject and new laboratories were being built.

Takamine returned to the Institute of Physical and Chemical Research, which was then housed in the University of Tokyo Physics Department before its own building was built, and much of his time was taken over preparing the facilities and equipment for the new Institute (Fukuda, 1964, pp. 319–323; Koizumi, 1974, p. 92). The Institute was another product of the lessons of war. Japanese scientists had long lobbied for a national facility that could compete with those in the West. At the time, even the Imperial universities were not thought of as research facilities, rather they were for training. Itakuru Kiyonobu and Yagi Eri, among other Japanese historians, view the establishment of the Institute in 1917 as a "major turning point in the history of Japanese science." Clearly, Takamine had returned to the one place in Japan where he could build a spectroscopic laboratory that could compete with the West. By the mid-1930s the Institute consisted of some 27 laboratories, though with each successive year more and more of them were engaged in military and industrial research (Kiyonobu and Yagi, 1974, pp. 158; *Minerva Jahrbuch Der Gelehrten Welt*, 1937, p. 1349).

Takamine refined his techniques for studying the Stark and Zeeman effects and their use in the analysis of spectra. Most of his new equipment (mainly from Adam Hilger in London) was installed and the building opened in the spring and summer of 1923, but on 1 September 1923 Tokyo suffered a huge earthquake and though the building was not terribly damaged much of the delicate equipment was smashed. Takamine also suffered personal loss when his wife died that fall and left him with four children. Rebuilding his life and laboratory occupied his mind and body for some time, though his laboratory issued its first reports in 1924 (Fukuda, 1964; N. Takamine, 1964a).

Later Tours

Takamine's misfortunes only heightened his passion to revisit the West. In May 1924, aided by the Imperial Prize he had won the previous year, he wrote to William Meggers of the National Bureau of Standards announcing how he was going to use the Prize: "There has been, in recent years, so great an advance made in spectroscopy in your country that I expect & rejoice in thinking I may learn a great deal in the United States ..." (Takamine, 3 May 1924). He hoped to travel once again in July to attend meetings, but wanted to spend most of his time in Pasadena and Washington. After meeting physicists in Toronto, however, he added Michigan to his itinerary, spending several months in Harrison Randall's laboratory learning infrared spectroscopic techniques and sitting in on special lectures by Friedrich Paschen. It seemed almost like there were too many laboratories to visit, and too little time. By January Takamine was in Washington at the Bureau of Standards, and did more networking than laboratory work. Anderson was in Washington then, and Takamine was delighted to hear him lecture.

As in 1918, Takamine was highly impressed with the capabilities of the Pasadena laboratories, especially Anderson's ability to achieve very high temperatures. Again, Takamine's interest centered on the design of experiments and being able to achieve means to search out phenomena in a variety of experimental set-ups with different sources of excitation. After spending eight months in Bohr's laboratory once again, where his new wife gave birth to a child, an event which endeared the happy couple to the members of the Institute, Takamine returned to the United States to work with Anderson. By May 1926 he was again in Pasadena working with King, Anderson and St. John (Schultz, 17 May 1963, p. 14). Writing to Harrison Randall, Takamine described what was most important to him about Mount Wilson. It was the scale of the enterprise, and the comparative freedom one had to pursue the latest lines of research with the most powerful equipment (Takamine, 6 January 1925).

Each time Takamine visited Mount Wilson, he hoped to perform his own research, but he was also sensitive to what the Mount Wilson people were doing and did his best to fit in. He well knew that the Pasadena laboratories of Mount Wilson were usually quite crowded, and more applicants were rejected than accepted. When he tried to secure a place for his friend M. Kiuchi in late 1921, Hale said the laboratory was full. Takamine apologized for bothering Hale with his request: "it appears as if Pasadena is attracting the world-famous physicists from every corner of the world, & I cannot help feeling envious of the wonderful opportunities you have there!" Takamine therefore knew his own chances would be better the closer his work dovetailed with the interests in Pasadena, and knew too that his expertise on the Stark Effect "is doubtless one of the most hopeful ways to get insight in the atomic structure." (Takamine, 16 November 1921).

Hale seconded Takamine for a coveted Carnegie Research Associateship for another longer stay at Mount Wilson, but this time it was for another purpose. Delayed a year by health problems, he finally arrived in November 1930 and worked there to try and secure absorption spectra of the higher order terms of the Balmer series, hoping to reproduce what was seen in the spectra of hot stars, just what King and Anderson were trying to do. "Through the kindness of [Adams, Anderson and King] I could obtain immense facilities at the observatory" he later told John C. Merriam of the Carnegie Institution. "I attempted to obtain Balmer spectral lines which more or less would resemble the manner in which they appear in some of the stellar spectra." He did not obtain definitive results and hoped this would leave the door open to return someday (Takamine, 3 May 1931).

Access to larger and more complete facilities was only one reason why Takamine repeatedly traveled to the West. Although laboratory directors like T. R. Merton advised him to stay at home "and develop my subject, the Stark Effect, more deeply," as Takamine recalled wistfully late in life, he knew he had been caught up like other Japanese scientists who were too anxious to "follow and to catch the western scientist's trendy research subjects." (Takamine, 1964a, p. 79). Takamine constantly felt the need of further stimulus from Western physics. Each time he visited the West, he was engaged on another research project that paralleled interests in the West. But Takamine had deeper motives for his travel, which come clear in his correspondence with Saha.

By the end of the 1920s Takamine had built a competent research laboratory at the Institute, concentrating on the power of the Stark Effect to unravel the structure of matter. His second and third visits to the West, in 1925 and 1930, however, were departures into astrophysics. In 1925 he told Hale that he wanted to build a small solar observatory near Tokyo, where he could retire to concentrate on spectra, and asked Hale for advice on how to secure American funds to build such a facility. Takamine knew that he should apply first to Japanese sources rather than American friends, but, as he admitted to Hale, "somehow I am rather disappointed with most of the [Japanese wealthy men] on account of their being too anxious to get immediate profit from the contributions." Hale's advice was to start small and search for local support (Takamine, 19 January; Hale, 19 March 1925). But

Yoshio Fujita and the observatory director Sotome built the first modern Japanese solar observatory, reminiscent of the Potsdam Tower, at the University of Tokyo's Astronomical Observatory in the early 1930s. Funding was wholly Japanese. Takamine had supported such projects, but limited himself to developing vacuum spectroscopy for astronomical studies. In 1938 he started a monthly colloquium on the subject and brought together physicists from the University and his Institute with astronomers from the observatory, including Hagihara, Fujita and Hatanaka (Tanaka, 1964, 356–360; Fujita, 1934; 1964, 351–354). At about this time, Takamine also renewed his contact with Saha when he asked about the action of ultraviolet sunlight on the upper atmosphere. Takamine had been interested in the extreme ultraviolet for the past several years and wanted Saha's advice on UV experiments he might make in his laboratory and, once again, in Pasadena (Takamine, 30 September 1937).

Takamine's Final Tour and View of the West

Takamine wanted to parlay suggestions from Saha into a research program attractive to American astronomers. Saha, happy to comply, advised Takamine to build a UV monochromater to search for the specific wavelengths that could stimulate the ionization of molecular nitrogen. Knowing that Anderson was interested in the fundamental problem of determining the nature of the resonance lines of the hydrogen series in absorption, a particularly difficult problem, Takamine also had a compact grazing incidence spectrometer built by his staff to be used in Anderson's laboratory in Pasadena. Knowing too that Anderson himself was heavily involved in the 200-inch telescope project, Takamine hoped that he might be welcomed as a helping hand. But he also worried that the growing war would make it difficult for him to travel, as he confided to Saha: "The great turmoil we are in just now seems really forbidding to such an attempt, at least for some time to come" (Takamine, 30 September 1937).

By this time, Takamine was feeling isolated and alienated from his own world. Inviting Saha to visit in 1937, and then learning that Saha would do so only if he could inspect Japanese nuclear physics laboratories, obviously interested in many applications, Takamine admitted that he was personally "of very retiring disposition, & have little influence in my own country." He would be happy to introduce Saha to those with influence, such as Nagaoka, Viscount Okochi, and especially to his colleague Y. Nishina, who was soon to complete another "powerful cyclotron" for the Institute. Takamine was not political, and seemed to harbor certain unstated tensions that, in late 1938 at least, affected his already failing health. He admitted to Saha that he "suffered from a rather serious nervous over-strain" and had to stay away from large gatherings of scientists, work on committees, or confrontations of any kind (Takamine, 6 September 1938). Such feelings intensified in 1939, as the military exerted more and more influence on Japanese scientific affairs. By then, military expansionists gained control of much of domestic life in Japan and the rights and freedoms of Japanese citizens were severely restricted, especially when the government abolished political parties. Saha continued to complain to Takamine of the pitiful conditions in India for experimental research, as well as the difficulties he had maintaining his scientific activity because of the demands of his political and administrative work. In turn, Takamine revealed his own frustrations, as well as his priorities:

> I think in many respects Japan is like India. When a scientist becomes famous & distinguished, he is burdened with many other duties which hamper his own research. (Personally I am rather avoiding to become famous in Japan, but of course this is not a very healthy attitude). I think that the ideal atmosphere of a civilized country ought to be that when a scientist does good work, he should get more & more facilities & *leisure* to push on his own research. But, evidently, the ideal conditions seems to be very far off! (Takamine, 12 August 1939).

Searching for ideal research conditions, in 1939 Takamine applied to the Carnegie Institution for another extended visit as he continued to work away in his laboratory on molecular nitrogen, as well as helium and neon spectra. Saha continued to provide him with new insights into molecular photoionization that his theoretical students were developing, and made various predictions for nitrogen's behavior that Takamine tried to confirm in his own laboratory. Throughout this time, Takamine expressed more and more concern that the growing war would eventually stifle all pure research. He was clearly not sympathetic with expansionist military priorities, but never said as much to Saha. Adams was happy to have Takamine visit as a Carnegie Research Associate, and endorsed his request. Anderson did too, claiming that Takamine's presence at Mount Wilson "would be exceedingly valuable." His continuing devotion to vacuum ultraviolet work in Japan was very important to them; any weakness he might have in theory and interpretation was of little concern to the Americans, who were less interested in theory than most Europeans (Adams, 17 January 1938). Takamine was granted the stipend, and was doubly grateful because, as he emphasized more than once, it demonstrated both his and the Carnegie's "exceptional sympathy toward the pure scientific research . . ." Evidently he did fear that this sympathy was rapidly disappearing in his own Institute, which was, as historians have shown, by then dominated by military and industrial contracts; the only area of physics still being fully funded was nuclear physics (Takamine, 17 June 1938; Hirosige, 1963). Eventually, Takamine secured funds to travel, mainly from the United States, and this time he brought along an assistant, Yoshio Tanaka.

One episode during his final visit in late 1939 highlights just how naive Takamine was concerning the expense of maintaining an infrastructure for research in the United States. When Takamine arrived, he asked the Mount Wilson shops to fabricate some complicated vacuum couplings, which, Adams later explained to the Carnegie Institution, "involved the employment of a glass blower at rather high expense." There were also considerable costs for liquid air. Takamine was deeply embarrassed when Adams confronted him with the bills. He assured Adams that under what he called "normal conditions" his Institute in Japan would have had no trouble paying, but he was also astounded at how expensive technical work like glass blowing was in America. Takamine admitted that he took for granted "the enormous difference in the cost of labor here, as compared with Japan." "In my laboratory in Tokyo," he explained to Adams and Vannevar Bush, the new CIW President, "I have two glass blowers each getting 2 [yen] ($0.50) per day, but here the professional glass blower demands 1$ 50ct per hour." Nowhere were the cultural differences between

Japan and America more starkly in contrast for Takamine (Adams, 15 February 1940; Takamine, 6 February 1940). The funds were soon found, however, and by March 1940, as Takamine reported, he and Tanaka were working seven days a week, literally all of their waking hours, and were beginning to get "some interesting results on the absorption of helium in the extreme UV . . ." (Takamine, 15 March 1940). In Pasadena, Takamine rekindled many friendships, and managed to have the venerable astrophysicist Henry Norris Russell read his paper before the National Academy in April 1940, the month his Carnegie fellowship expired.

One of Takamine's highest priorities was finding a way to extend his visit. He and Tanaka had work permits until the end of the year, and secured an invitation from Berkeley to spend a few months there. Their work was progressing slowly, however, Takamine reported to Russell, and there was much left to do. Therefore, "we are hoping to extend our stay, if possible, to 1941" (Takamine, 12 August 1940). Takamine wrote to many influential Americans trying to secure an extension of his work permit and additional funds. But as relations between Japan and the United States, though still officially neutral, were getting tenser every day, he was finding Americans less willing to help out. Adams was willing to endorse his extension in August, but after the Tripartite Plan of Mutual Assistance was signed between Germany, Japan and Italy in late September, Adams' sympathies were considerably lessened. Adams wrote to a Carnegie Institution official "Takamine's request is altogether too indefinite to justify a grant in any event." They would tolerate his presence, but not at their expense. The work was valuable, but Adams, King and Anderson felt that they had gone as far as possible without another large infusion of funds to upgrade the equipment and this was not possible. Besides, Adams felt there was also a moral question: "The truth of the matter is that Takamine does not want to go back to Japan. Living conditions are very bad and he is very much out of sympathy with the military party which is dominating Japan's foreign policy." (Adams, 2 October 1940)

Notwithstanding his support in good times, Adams was intensely xenophobic. By 1940, observatory men like Adams were feeling very pressed because of the increasing influx of war refugee scientists from Europe. Given the pressure, it is rather surprising that Adams tolerated Takamine and Tanaka at all. Quite likely the reason was that, unlike elite European scientists, who could be very demanding and even threatening to Americans, Japanese physicists like Takamine worked very hard to ingratiate themselves to their American hosts, and aligned their research interests closely with those programs in need of help. Most definitely Takamine was very good at this, but his unexpected requirements for glass blowing and liquid air came at a particularly unfortunate time. At the very least a staunch moralist, Adams felt that Takamine had a responsibility to his family: "personally I think it is his duty to return and do the best he can to assist in their support and comfort." (Adams, 2 October 1940).

Takamine hoped that the crisis would pass and that relations would improve between their two countries. He was disappointed that no additional Carnegie funds would be provided, but he was determined to stay in the United States. He assured Adams that he would not again ask for money and was embarrassed if this had put Adams in a difficult situation. Takamine and Tanaka returned briefly to Pasadena in the spring of 1941 to complete a paper "Absorption Spectra of Gases in the Extreme Ultraviolet." Adams had approved this plan, expecting they would soon leave for Japan. But once again Takamine got an extension, this time when Japanese Ambassador Nomura asked him to attend National Academy meetings in Washington "in the hope of maintaining, at least, the good intellectual intercourse between two countries." Takamine used this trip to visit George Harrison's vast laboratory at MIT where he found the apparatus for studying the Zeeman Effect huge in scale: "I even felt as if I was in a factory." This sojourn also helped to delay his return to Japan until June, which gave him time to visit Pasadena once more and "talk shop" with Adams and the others, and seek out theoretical insight into the photoionization processes he was observing (Takamine, 1964a; Takamine, 24 April 1941).

By October, Takamine and Tanaka were back in Japan. Takamine reported to Adams that his family was safe and that he and Tanaka were again working away on NO absorption in the extreme ultraviolet. He expressed great melancholy and fondness for his lost Pasadena life and hoped to stay in touch. But the war soon stopped all further contact. In January 1942, Adams reported to Joel Stebbins that the "last Japanese were supposed to have left yesterday, so that now any Mongolian we see on the streets may be assumed to be a Chinese. I think this will be a great help to most of us . . ." And in the Annual Report of the CIW for 1941, President Bush summed it all up: "After Pearl Harbor this country ceased to be an oasis in a world at war, and entered on a period of strife and sacrifice" (Adams, 63.1126; Bush, 1941, p. 3).

Takamine had lost his oasis to the terrors of war. He doggedly managed to maintain a low level of work, even though the majority of his staff was dispersed or drafted outright. On 13 April 1945 the Physical Institute was burned to the ground in a B-29 raid, and in August the entire neighborhood was fire bombed and destroyed. Takamine and his family escaped to Kamakura. Even so, in the face of chaos, Takamine tried to keep active. Without assistants, he did all the work himself, whatever he could – his interest in vacuum spectroscopy never wavered. Even before the nation was normalized after the war Takamine was commuting from Kamakura to Tokyo to continue work. He had rescued a 1-meter grazing incidence spectrometer and used it to study the absorption spectrum of nitrogen molecules, which turned out to be his last research (Tanaka, 1964, pp. 356–360).

Once exchanges were possible after the close of war, Russell arranged to have literature sent, and the Mount Wilson staff put together packages of clothing and living essentials. Takamine thanked them all, admitting that: "It is a great consolation to me that I began to hear from many of my old acquaintances in U.S." Over a dozen physicists and astronomers provided help, a testimony to the degree to which Takamine sought both intellectual guidance and personal solace in the West. In the wake of war, the Japanese NRC was disbanded and the Research Institute was "legally dismembered" to make way for new institutions, including a new Science Council of Japan, with Nagaoka as its president. When the Institute was reorganized as the Science Research Institute in the early 1950s, senior physicists over 65 were forced out, including Takamine. Takamine took this almost as a relief, telling a friend "Let me accept a first lead to the passage to heaven." He died in 1959 (Kinoshita, 1964, pp. 269–273; Bartholomew, 1989, 278).

Takamine sought out a "spiritually peaceful life" free from administrative responsibility. His health was never robust after the 1920s, and it is clear that he sought the solitude of research. His penchants for research visits to the West have to be appreciated in this light. While on travel, working as a guest, Takamine was free to pursue his muse unfettered by politics or the personal vicissitudes of the Institute, which was clearly moving in directions by the late 1930s that he found unacceptable. As he recalled late in life: "Every time when I come back to my school from abroad, I always feel that we lack the means for cooperative work" Takamine sensed that cooperation between Mount Wilson and Caltech and the University of California was far more open and productive than cooperation between the Institute and the University of Tokyo: "Japanese academia is not as cooperative . . . as it is in the United States . . . From a feudal society to a post-war democracy, Japanese society has now changed significantly . . . why can't we adopt the openness and cooperative attitude from the American academy?" (Takamine, 1964a, p. 90).

Takamine expressed similar concerns in his letters to Saha in the late 1930s. The insular nature of Japanese science Takamine deplored was a remnant of the traditional chair system that Japanese senior scientists, including Takamine, enjoyed. Thus his penchant for travel away from his home, to be completely free for research, takes on deeper meaning in his apparent sense of isolation from his own colleagues. He knew that Merton's advice had been right; if he had stayed home and concentrated on his work, and fought for his laboratory, or even for a solar observatory, he probably would have gotten much farther in the Japanese hierarchy, "but I didn't have the confidence and courage to seek such success" (Takamine, 1964a, p. 91). He had made a conscious choice, a choice, in fact, that was hardly unique. Americans like Russell also deserted any plans to build up their local facilities when faced with the prospect of competition from Mount Wilson. Russell, like Takamine, spent as much time in Pasadena as possible. Even Nagaoka, for all his prominence and political power, claimed to hate politics and administrative responsibility. But he accepted it and used it to build up Japanese physics, partly because, unlike Takamine, he possessed an "aggressive, almost agonizing sense of racial rivalry with the West." (Koizumi, 1974, pp. 88; 91).

Takamine did not share Nagaoka's competitive fire. The West always enraptured him. There he found not only the freedom for research, but inspiration as well. That he also sought out Saha's advice on occasion, and in the late 1930s based his own research on some of Saha's suggestions, reveals that it was not only his fascination with the West, but his search for physical insight, that propelled him to travel as often as he did. This seems to be much in keeping with Japanese scientists generally, and of Japanese culture, which, Bartholomew and others have claimed, had a tradition of "self-conscious borrowing" in the late 19th and early 20th centuries (Bartholomew, 1989, p. 83). But for Takamine, though the Mount Wilson Observatory was a world center for spectrum analysis as applied to the stars, it was also an "oasis" from administrative responsibility. Takamine was always mindful to adjust his choice of problems to maximize his welcome there. In so doing, he inserted yet another social factor into the formula that shapes the face of science.

Sources and Acknowledgements

General trends in Japanese astronomy were gathered from many sources, including Transactions of the IAU (1922–1938); Annals of the Tokyo Astronomical Observatory, 1–47 (1915–1935); Japanese Journal of Astronomy and Geophysics – Transactions, 1922–1935; Japanese Journal of Astronomy 1 (1949). A fundamental source of information for this paper came from Yoshio Fujioka, ed., 1964. *Biography* [*of Toshio Takamine*], cited herein as *Biography*. I want to express my deep appreciation to Seiko Green and Robin

Halter, who volunteer for the Smithsonian's translation program, as well as to Yasuko Fujiki, who provided additional translations. I am also indebted to the archives consulted: The American Institute of Physics Niels Bohr Library (NBL); the George Ellery Hale Papers microfilm edition, Caltech (GEH); the Mount Wilson Papers, Henry Huntington Library (HEH); the Carnegie Institution of Washington Archives (CIW); The Henry Norris Russell Papers, Princeton University (HNR); the Harrison Randall Papers, University of Michigan, Bentley Historical Library (MICH); Meghnad Saha Papers, Nehru Papers, Nehru Library, New Delhi (NEHRU) and Saha Institute of Nuclear Physics, Calcutta (SINP).

References

Adams, W. S. to Gilbert, 2 October 1940. CIW.

Adams, W. S. to Gilbert, 15 February 1940. CIW.

Adams, W. S. to Merriam, 17 January 1938. "Takamine" Personnel Files. CIW.

Adams, W. S. to Stebbins, n.d. Adams Papers, HEH.

Anderson, J. A., to Hale, 8 April 1919. Reel 82, GEH.

Anderson, J. A., to Hale, 22 December 1918. Reel 82, GEH.

Badash, Lawrence (1967), "Nagaoka to Rutherford 22 February 1911," *Physics Today (April 1967)*, pp. 55–60.

Bartholomew, James R. (1989), *The Formation of Science in Japan*. Yale.

Bohr, N. (1964), "Recollections of Professor Takamine," in *Biography*, pp. 384–386.

Bush, V. (1941), "Report of the President," *Annual Report of the Carnegie Institution of Washington*, No. 39.

Carazza, Bruno and Helge Kragh (1991), "Document: An Oriental in Europe: Nagaoka, Righi, and the State of Physics 1910," *Historia Scientiarum*, Vol. 41, pp. 37–44.

DeVorkin, D. H. (1993), "Saha's Influence in the West: A Preliminary Account," in S. B. Karmohapatro, ed., *Meghnad Saha Birth Centenary Commemoration Volume*. Saha Institute of Nuclear Physics, pp. 154–202.

DeVorkin, D. H. (1994), "Quantum Physics and the Stars IV: Meghnad Saha's Fate," *Journal for the History of Astronomy*, Vol. 25, pp. 155–188.

Fujioka, Yoshio, ed. (1964), *Biography* [*of Toshio Takamine*]. Tokyo: Institute of Applied Optics. Herein cited as *Biography*.

Fujita, Yoshio (1934). "On the Prism Spectrograph of the Tower Telescope in the Tokyo Astronomical Observatory," *Annals of the Tokyo Astronomical Observatory*, appendix 41.

Fujita, Yoshio (1964), "The Very Essence of a Scholar," in *Biography*, pp. 351–354.

Fukuda, Mitsuharu (1964), "The Origin of Takamine's Research Laboratory," in *Biography*, pp. 319–323.

Hale, G. E. to Takamine, 19 March 1925. Reel 34, GEH.

Hale, G. E. to Anderson, 28 January 1919. Reel 82, GEH.

Hentschel, Klaus (1997), "An Unwelcome Discovery: The Pole Effect in the Electric Arc, a Threat to Early 20th Century Precision Spectrometry," *Archive for History of Exact Sciences*, Vol. 51, pp. 199–271.

Hirosige, Tetu (1963), "Social Conditions for the researches of Nuclear Physics in Pre-War Japan," *Japanese Studies in the History of Science*, Vol. 2, pp. 85–87.

Kinoshita, Masao (1964), (Recollections of Takamine), in *Biography*, pp. 269–273.

Kiyonobu, Itakuru and Yagi Eri (1974), "The Japanese Research System and the Establishment of the Institute of Physical and Chemical Research," in Nakayama, Swain and Yagi, pp. 158–207.

Kochhar, R. K. (1993), "Science in British India. I. Colonial tool; II. Indian response," *Current Science*, Vol. 63, No. 11, 10 December 1992, pp. 689–694; Vol. 64, No. 1, 10 January 1993, pp. 55–62.

Koizumi, Kenkichiro. (1974), "The Emergence of Japan's First Physicists: 1868–1900," *Historical Studies in the Physical Sciences*, Vol. 6, pp. 3–105.

Kumar, Deepak. ed. (1991), *Science and Empire: Essay in Indian Context*. Anamika Prakashan, Delhi.

Laporte, Otto (1964), "Recollections of Professor Takamine," in *Biography*, pp. 386–389.

Nakayama, Shigeru (1969), *A History of Japanese Astronomy*. Harvard, p. 231.

Nakayama, Shigeru, David Swain and Yagi Eri (1963), *Science and Society in Modern Japan*. MIT.

Schultz, Betty, Oral History, 17 May 1963, NBL.
Shinjo, Shinzo to Adams, 4 December 1919. Reel 98, GEH.
Takamine, T. to Adams, 24 April 1941. CIW.
Takamine, T. to Bush, 6 February 1940. CIW.
Takamine, T. to Hale, 10 January 1925. Reel 34, GEH.
Takamine, T. to Hale, 16 November 1921. Reel 98, GEH.
Takamine, T. to Hale, 16 November 1921. Reel 98, GEH.
Takamine, T. to John C. Merrian, 17 June 1938. CIW.
Takamine, T. to Meggers, 3 May 1924. Meggers Collection, NBL.
Takamine, T. to Merriam, 3 May 1931. CIW.
Takamine, T. to Merriam, 15 March 1940. CIW.
Takamine, T. to Randall, 6 January 1925. MICH.
Takamine, T. to Russell, 12 August 1940. HNR.
Takamine, T. to Saha, 13 January 1921. Nehru/SINP.
Takamine, T. to Saha, 30 September 1937. SINP/NEHRU.
Takamine, T. to Saha, 6 September 1938. SINP.
Takamine, T. to Saha, 12 August 1939. SINP.
Takamine, T. (1964a), "Memory of Spectrum Study," in *Biography*, pp. 77–91. Written at his retirement in 1961.
Takamine, T. (1964b), "Recollections of Professor Wood," in *Biography*, pp. 91–99.
Tanaka, Yoshio (1964). "Professor Takamine and Spectroscopy," in *Biography*, pp. 356–360.
Venkataraman, G. (1988), *Journey Into Light: Life and Science of C. V. Raman*. Indian Academy of Sciences, New Delhi.

2.4. The Earliest Evidence of the Introduction of Kepler's Laws into China as is observed in the *Lifa wenda*

Keizo Hashimoto

We have recently found a manuscript, entitled the *Lifa wenda* (Dialogue on Astronomy) by Jean-François Foucquet (Fu Shengze, 1665–1741), in the British Library.[1] Together with the other partial but otherwise identical version, which Jami had located in the Vatican Apostolic Library,[2] the manuscript from the British Library, especially Book V, Part 1, gives us details of how Kepler's first and second Laws were introduced into China as early as the 1710s. This means that we can trace back the history of the introduction more than two decades earlier than was previously thought.

On the other hand, when the *Chongzhen lishu* had been compiled for several years starting from 1629, although Kepler's optical astronomy was extensively introduced, we cannot find in it the slightest evidence of a description of his Laws. Thus, the *Lifa wenda* can be regarded as the earliest work in which Kepler's Laws were openly discussed.

In connection with Kepler's Laws, Copernicanism was also discussed, particularly concerning the instrumental model of the solar system, that is the Orrery, manufactured by O.C. Roemer, which had been brought to China by the French Jesuit missionaries and had been presented to Kangxi Emperor on behalf of the French King, Louis XIV.[3] On several pages of the manuscript, Copernicanism is also referred to in connection with the explanation of planetary motions. The latter evidences show that Copernicanism had been introduced into China almost half a century before M. Benoist dealt with it in his *Treatise on the Earth (Kunyu quantushuo)* in 1767.[4] Both of them can drastically change our understanding of the history of astronomy in China. In the following we should like to discuss first the problem of the early introduction of Kepler's Laws.

1. Kepler's Laws

First of all, let us see how Kepler's Laws were accepted and developed in Europe after Kepler's introduction of elliptic orbits, replacing the combination of circular motions. According to C. Wilson, Ismael Boulliau objected to the magnetic mechanism hypothesised by Kepler to account for the eccentricity of the planetary orbits.[5] He composed the *Astronomia philolaïca* in 1645. Boulliau had imagined another way of deriving elliptical orbits from uniform circular motions, proposing for the first time, "from the general circumstances of planetary motion", that these orbits are elliptic. According to Boulliau, the circles lie in a cone and the mean motion takes place about the axis parallel to the base.[6] He fails to recognise that this implies an equivalent uniform angular motion about the non-solar focus; the hypothesis is thus equivalent to the empty-focus equant that Kepler's "Uranian friend" Albert Curz had proposed for the Moon, to which Kepler refers in the *Rudolphine Tables* (1627).

In the *Astronomia geometrica* of 1656, Seth Ward assumes that the "simple elliptical hypothesis" with superior focus as equant point is true. Both Kepler and Boulliau had

S.M.R. Ansari (ed.), History of Oriental Astronomy, 159–168.

failed to recognise its truth. Another proponent of the simple elliptical hypothesis was Emile-Francois Pagan, who in 1657 published *La théorie des planetes*.[7]

Now let us see how Foucquet describes and introduces the discovery and development of the theory of elliptic orbits in the *Lifa wenda* (Dialogue on Astronomy), which was probably prepared between 1712 and 1716.[8] In Book V, Part 1, he begins the explanation of the development of European astronomy with the development over the previous 60 years of astronomical instruments and observational achievements in Europe,[9] particularly after the establishment of the Paris Observatory, and alludes to the installation of the telescope mounted with micrometer (*liang wei ge*[10]) there.

First of all Foucquet discusses the necessity of introducing the elliptical orbits in place of the combination of circular motions. Before the discussion of the problem of Martian motion, he particularly emphasises the shift of the perihelion of the orbit of Mercury in order to demonstrate the inadequacy of circular motions. Then he describes how the observational results of his times show the discrepancy from the theoretical calculations.

Foucquet tries to show how recent telescopic observations had become precise by reporting Huygens's determination of Saturn's ring (making use of his long telescopes) from 25th March 1655, through 16th January 1656, to 12th 1659.[11] In the manuscript *Lifa wenda* he explains the result with the heliocentric model of the solar system as Huygens did. Foucquet also tries to emphasise the importance of Cassini's telescopic observations of the surface of Jupiter, including a dark spot which appeared between 1690 and 1691.[12]

Kepler had derived the elliptic orbit of Mars by making use of Tycho Brahe's observations of opposition. We can observe four observational data derived from Brahe's observations in the *Lifa wenda*.[13] We can find this data cited from the *Almagestum novum* in 1651 by the Bolognian Jesuit astronomer G. B. Riccioli, which had been used by Boulliau. Based on these data, together with other observational results by various astronomers in Europe, Foucquet tries to emphasise the inevitability of the introduction of the non-circular motions of the planets.

Then he discusses the mechanistic "necessity" of the introduction of elliptic orbits, alluding to the Cartesian physics, which we can observe immediately after his discussion of the non-circular motion of Mars.[14] We shall see shortly this problem below.

Although Foucquet fails to give any illustrations in the manuscript to explain the geometrical orbit of planets, we can reconstruct what he means to describe, that is, the elliptic orbits in terms of Kepler's method. Following the explanation of Kepler's Laws, he discusses Boulliau's so-called revised method, Pagan's (Bagan) simple method, and Riccioli's (Li-zhuo-li) spiral orbit, successively.

Let us examine his description of the so-called Kepler's method (*Ke-bo-er zhi fa*), in place of the areas rule, in the manuscript itself. He writes as follows:

> (After having studied the record of observations of Mars by Tycho Brahe) 'Kepler for the first time abandoned circular motions, and adopted the ellipse (*Dan-xing-xian*) for the orbit of Mars.'[15]

Here he does not use the term *Tuo-yuan*,[16] which became the standard representation after the compilation of the *Shuli jingyun* in 1723. In order to explain the new method (*Xinfa*), he describes the geometry of the ellipse, emphasising the importance of understanding its character.

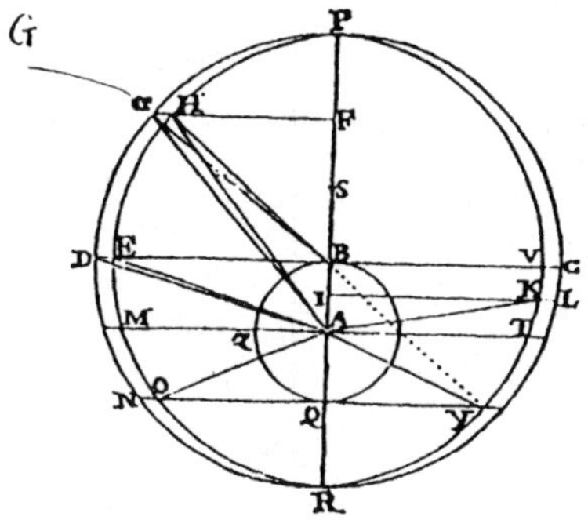

Figure 1 From Riccioli's *Almagestum Novum* (1651), p. 531.

Foucquet constructs the ellipse, on which a planet moves, and also its auxiliary circle, and discusses the properties of the auxiliary circle. Firstly, he draws the auxiliary line, which is perpendicular to the major axis, then he shows that the ratio HF:GF is proportional to the ratio of the minor axis of the ellipse to the diameter of the auxiliary circle. Here HF is named the *liexian* and GH the *chengxuan*. He goes on to show that the ellipse has two focuses (*juguangjuhuodian*, or abridged as *judian*), on the lower of which the sun is located (Fig. 1).

After that Foucquet explains the areas rule of Kepler.[17] He discusses how the planet moves on the ellipse about the sun. For the explanation, he makes use of Riccioli's representation, which we can find in the *Almagestum novum*. Riccioli tries to discuss the areas rule introduced by J. Kepler.[18] Obviously Foucquet must have been describing the method as a whole, relying on the contents of the explanation as well as the illustrations from Riccioli's book (Fig. 1).

He also alludes to the causes of the elliptical motion of planets, such as magnetic force, which Kepler used. But he declines to accept Kepler's analogy, and follows Descartes' physics of the cause of motion. By doing so, he starts to introduce Boulliau's method as well as Pagan's so-called simple method, which used the second focus as the equant.

It is interesting to observe Cassini's method explained as the fourth method in the *Lifa wenda* under discussion. He assumes the simple elliptic hypothesis: if S is the focus where the eye (or Sun or Earth) O is the centre of the ellipse, ARP a circle with radius equal to the semi-major axis of the ellipse, and if the true anomaly is $v_R =< ASR$ and the mean anomaly M_R, then $< AOR = 1/2\ (M_R + v_R)$, see Fig. 2, where $AB = v_A - v_B$, $DF = M_A - M_B$, $BC = v_B - v_C$, $DF = M_B - M_C$; only the differences in true and mean anomaly being observationally determinable.[19] The intersections G and H fix the line GH on which point O, the centre of the ellipse, is found by dropping a perpendicular from B (Fig. 3). As the text suggests,[20] he invented *cassinoid*, with the aim of obtaining a possible orbit for the planets in which the superior focus would serve as equant point, and introduced another kind of elliptic orbit.[21]

As the fifth method, he also explains Riccioli's spiral orbit of the sun (or planets). We can show this using the figure from Riccioli's original book (Fig. 4). In general, we can repeatedly say that he is rather faithfully following the *Almagestum novum* by G.B. Riccioli.

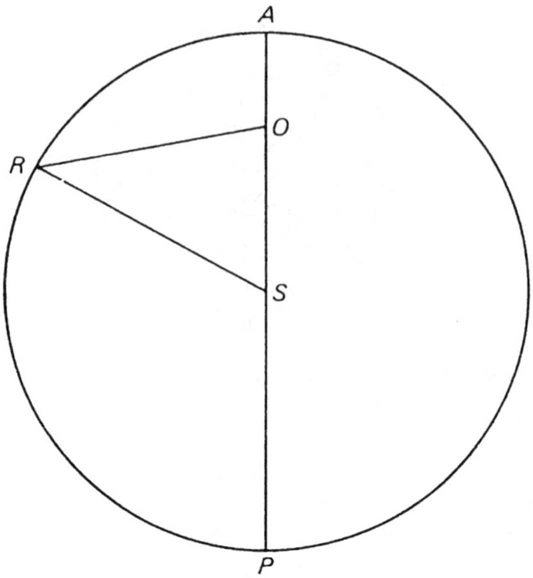

Figure 2 The Elliptic Hypothesis by Cassini I (from C. Wilson, 1989, p. 182).

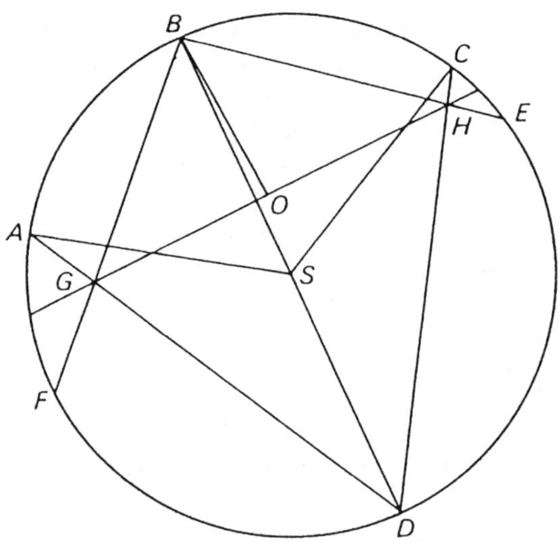

Figure 3 Cassini's procedure for determining eccentricity and aphelion from observation (C. Wilson, 1989, p. 183).

However, we must point out that, notwithstanding his long discussion of Kepler's Laws, Foucquet instead eventually transcribed as the astronomical tables La Hire's *Tables* of 1702. La Hire had produced the *Tables*, totally relying on his own observations made over a long period at the Paris Observatory after he succeeded Jean Picard. It was, indeed,

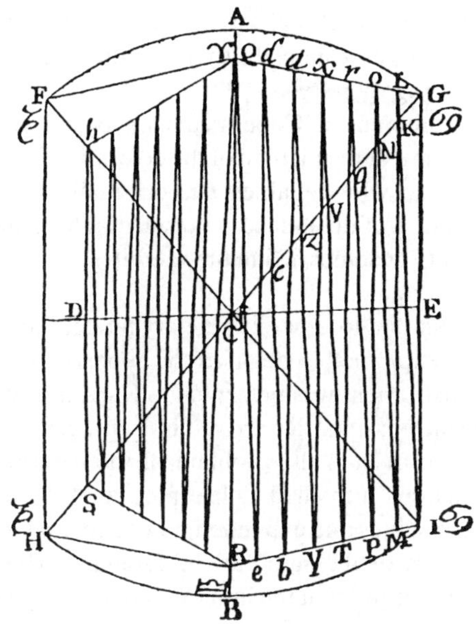

Figure 4 Riccioli's spiral orbit (G. B. Riccioli, *Astronomia Reformata*, 1665, p. 66).

practical astronomy that Foucquet was trying to introduce in the manuscript, although he tried to explain even the physics of planetary motions.

In this connection it is also worthwhile pointing out that he brought to China the achievements in determining the astronomical constants which were remarkably improved after the invention of the telescopes mounted with micrometer. A few examples are the obliquity of ecliptic, Cassini's determination of horizontal parallax of the Sun as well as the refraction up to the zenith. Chinese astronomers had only known Tycho Brahe's refraction, which had been introduced several decades ago.

2. Copernicanism

In the *Almagestum Novum* of 1651, G. Riccioli rejects the ellipse because he does not find its empirical evidence sufficiently convincing. Moreover, in his later book, the *Astronomia reformata* of 1665, Riccioli is as anti-Copernican as ever, claiming in fact to have proved the immobility of the earth. However, he had by then adopted the ellipse as a provisional basis for planetary theory.[22] Nevertheless, he discussed the theory of Copernicus (Ge-bai-ni) as the hypothesis of the universe, when introducing the semi-Tychonic world system.[23]

It seems likely that the Jesuit missionaries in China, who were in charge of the Imperial Astronomical Observatory in Beijing, eventually followed Riccioli's choice for the adoption of Kepler's Laws. It was crucial for them in order to compile the *Lixiang kaocheng houbian* (Sequel to Astronomical Compendium) in 1742. This was until now believed to be the first, formal introduction of Kepler's theory of elliptic orbits, for which the

German Jesuit missionary Ignatius Koegler (Dai Jinxian), together with the Portuguese missionary Andreas Pereira, took responsibility for compiling, as the Astronomer Royal (*Qintianjianzheng*) of the Qing dynasty.

In connection with the problem of Copernicanism, or heliocentrism, we should not overlook Foucquet's discussion of the method of the determination of longitude, by making use of the satellites of Jupiter, which Galileo had originally suggested. Note that in the discussion of the determination of differences in terrestrial longitude, the discovery of the finite velocity of light by O.C. Roemer, using the satellites of Jupiter, played an important role.[24]

But here we must point out that Roemer's planetary model (i.e. an Orrery), based on the heliocentric idea, has been explained in detail in the appendix to the Treatise on Eclipses in the *Lifa wenda*.[25] The instrument was presented to Kangxi Emperor on behalf of Louis XIV, when the French Jesuit missionaries arrived in Beijing.[26]

We shall now continue to discuss the problem of the determination of geographical positions. Cassini's instructions furnished a clear picture of the best seventeenth-century research methods and at the same time explained how terrestrial longitude could be determined by timing the eclipses of the satellites of Jupiter.[27] The most satisfactory time observations of Jupiter could be made at the immersions and emersions of the first satellite.

When Jean de Fontaney, a Jesuit professor of mathematics at the College of Louis le Grand, was preparing to leave for China as the head of the first French Jesuit mission, G. D. Cassini trained him. On his way to China he collected data on the longitudes of localities in the Orient.[28] Later the French missionaries worked for the imperial enterprise of the cartographic survey of China in the late 1710s.

Foucquet, referring to the *Traité de la Lumière* of 1690 by C. Huygens, discussed this method of geodesy. We can see the explanation of the method, which faithfully reflects Huygens's discourse (cf. Fig. 5).[29] Furthermore, Foucquet categorically asserts that "the satellites of Jupiter were best made for cartography by making use of the method, which relied on the velocity (*liuxing* 流行) of light, which in turn had been determined by Roemer".[30]

It is interesting to read at the beginning of the discussion that Foucquet meant to draw the illustration of the model to show the motion of the first satellite of Jupiter. He does not try to revise Roemer's model of the solar system. Instead, he faithfully explains the model. For this purpose, he says that the Sun 'stands still' (*budong*)[31] without any motions at the centre of the orbit of Jupiter, as well as of that of the Earth.

3. The Cartesian Basis of Foucquet's Physics

Lastly we should like to examine Foucquet's scientific knowledge of physics, as it is reflected in the manuscript which we are discussing. This is crucial for the understanding of how Foucquet recognised the cause of planetary motions. Here we can say that his discussion was clearly based on Cartesian mechanism.

As for the mechanism of planetary motions, he first explains Kepler's magnetic mechanism as the cause of planetary motions. Then he moves on to introduce Boulliau's ellipse, because he followed Riccioli, who had not agreed with the analogy of magnetism used by Kepler.

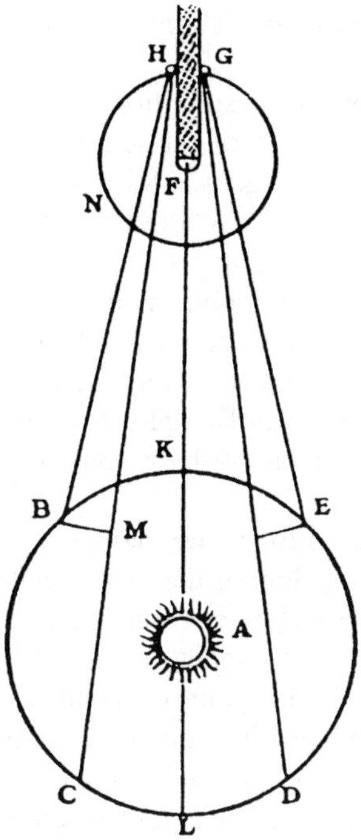

Figure 5 From Huygens, *Treatise on Light*, 1690; English translation by S.P. Thompson, 1912, Dover ed. p. 8.

Foucquet assumes that the Sun is located at the centre of the great circle of planetary orbits, and that it is the essence of the fire (*fuo-zhi-jing*) and can be the source of the motions (*dong-zhi-yuan*).[32] The five planets all receive its movement, because they are in the aether, which transmits the motion from the Sun. It is called the moving power of the aether (*jingqi-zhi-nengli*).

Foucquet then gives the reason that this power causes the non-circular motion of planets.[33] As he tells us, Descartes stated that the motions in the heaven were not perfectly circular.[34] First Foucquet tries to show that, if the centrifugal force works, it should cause circular motion of planets. In order to explain the mechanism, Foucquet cites Descartes' second law of nature from the *Principles of Philosophy*, and explains the planetary motion, based on the Cartesian illustrations, although the figure is missing from the manuscript.

He describes how in the heavens the aether expels the five planets, and says that if the expelling power of aether (*jingqi*) comes straight from the Sun in a uniform manner, then the revolution of the planets should be circular. But, since it works obliquely, so the orbits of the planets should be elliptic.[35] We may attribute his suggestion to Leibniz's theory of planetary motion. Leibniz supposed that "the vortex carrying the planets rotated in spherical layers. And, in order to account for the elliptical orbit, he postulated two motions of the planets: a radial motion from layer to layer and a trans-radial motion in which the planet moved with the same speed as the fluid."[36]

In short, we can reiterate that Foucquet's concern with the cause of the planetary motions centred on Cartesian physics.

Then, as to the propagation of rays in straight line, he continues to discuss the problem in terms of the concept of the medium of subtle aether (*jingqi*), relying on the physical arguments by Descartes as well as Huygens.[37] It is interesting to find that here he mentions the idea of vortex (*xuanquan zhi bo*) in the medium.[38] Interestingly, he transcribes 'physics' by the Chinese term, *gewuqiongli-zhi-xue*.[39]

Lastly, as to the propagation of the light, the manuscript discusses how the luminous body gives impulses to the surrounding aether, and thereby propagates the ray in all directions. We have found that his whole discussion here comprised the complete translation of the *Treatise on Light* by Christiaan Huygens (in 1690), including the discussion of Roemer's determination of the velocity of light, and of longitude, using the first satellite of Jupiter.[40]

We can find this extensive discussion in the chapter of Part 2, Book V, of the *Lifa wenda*, where Foucquet discusses the problem of the determination of longitude by making use of it.[41] We can definitely say that Huygens was a crucial authority for Foucquet in the explanation of this optical problem as well.

The next step we must take is the more detail analysis of the treatise on planetary motions, that is Book V of the manuscript, and the problems of the solar and lunar motions together with the treatise on eclipses which consists of its first three books. Our future work will be concerned with the historical fact of whether or not Foucquet's introduction of Kepler's Laws had any significant effect on the compilation of the *Lixiang kaozheng houbian* in 1742.

Notes and References

1. Oriental and India Office Collections, Or Add. 16634.
2. Borgia Cinese 319(1) & 319(2). Cf. Hashimoto & Jami (1997), in which we have given the table of contents (Table 1).
3. Nissen (1944), p. 32.
4. Yabuuti (1969), p. 171.
5. Wilson, (1989) p. 172.
6. Wilson (1989), p.173.
7. The full title is *La théorie des planetes du Comte de Pagan au tous les orbes celestes sont geometriquement ordonnez, convert le sentiment des astronomes*. Cf. Wilson (1989), p.178.
8. ARSI, Jap. Sin. II 154.
9. *Lifa wenda* V-1-1. In the introduction of the Treatise on Lunar Motion, Foucquet first discusses this topic in detail in Chapter III-1.
10. Book III, Part 1, ff. 72a–b.
11. V-1, ff. 3a–5b. The micrometer is described here.
12. V-1, 8b. Cf. the dark spot on Jupiter produced by the impact of Shoemaker- Levy 9 comet in 1995. As to Cassini's observation, see Tabe et al (1997).
13. V-1, ff. 36a–45b.
14. V-1, ff. 43a–45b.
15. V-1, f. 47b.
16. In the preface to the treatise on planetary motions, we observe that the more common term *tu-yuan*, has been used for the shape of orbits, oval or ellipse, other than circle (V-1, f. i). The term *tuoyuan-xing*, with the hand radical for the character, *tuo*, first appeared in the *Celiang quanyi, quan* 6, in the *Chongzhen lishu*,

where the conic sections are discussed. See the *Xinfa suanshu* edition, *quan* 92, p. 9a, l.2; Taibei reprint version, 1972.
17. V-1, f. 50a.
18. V-1, ff. 51b–52b.
19. Cf. Wilson (1989), pp. 182–183.
20. V-1, ff. 78a–83b.
21. Cf. Wilson (1989), p. 183.
22. Wilson (1970), p. 111.
23. III-1, f. 6b. As to the "semi-Tychonic" world system of Riccioli, see Schofield (1981), and Schofield (1989), p. 40.
24. Debarbat and Wilson (1989), p. 156.
25. III-1, paragraph 14, which has been omitted from the British Library version, the part of which, otherwise, seems to have been copied from the original version. The omission suggests that censorship had taken place.
26. Nissen (1944), p. 32. Later Roemer manufactured in Paris the model based on the Tychonic world system.
27. Brown (1979), pp. 221ff.
28. *Ibid.*, p. 220.
29. V-2, f. 33a. Fig. from Thompson (1912), p. 8.
30. V-2, f. 32b.
31. V-2, f. 33a.
32. V-1, f. 45a.
33. V-1, ff. 43a–45b.
34. Descartes (1644), III, p. 34.
35. V-1, f. 45b.
36. Aiton (1989), p. 10.
37. V-2, f. 43a.
38. V-2, ff. 44a–45a.
39. V-2, f. 32b.
40. English translation by Thompson 1962.
41. V-2, ff. 33a–46a.

Bibliography

Aiton, Eric J. (1989), "The Vortex Theory in Competition with Newtonian Celestial Dynamics, in Taton and Wilson (1989), pp. 3–21.

Brown, Lloyd A. (1979), *The Story of Maps*, reprinted by Dover, New York, first printed in 1949.

Debarbat, S. and Wilson, C. (1989). "The Galilean Satellites of Jupiter from Galileo to Cassini, Roemer and Bradley", in Taton and Wilson (1989), pp. 144–157.

Descartes, René (1644), *Principia Philosophia*.

Hashimoto, K. and Jami, C. (1997), "Kepler's Laws in China: A Missing Link?", *Historia Scientiarum*, Vol. 6/3, pp. 171–185.

Nissen, Andreas (1944), *Ole Roemer*, Copenhagen.

Schofield, C. J. (1981), *Tychonic and Semi-Tychonic World Systems in the Seventeenth Century*, Arno Press, Holland.

Schofield, C. J. (1989), "Tychonic and Semi-Tychonic World Systems", in Taton and Wilson (1989), pp. 32–44.

Tabe, I. et al (1997), "Discovery of a Possible Impact Spot on Jupiter, Recorded in 1690", *Pub. Astron. Soc. Japan*, Vol. 49, pp. L1–L5.

Taton, R. and Wilson, C. Eds. (1989), *Planetary Astronomy from the Renaissance to the Rise of Astrophysics, Part A: Tycho Brahe to Newton. The General History of Astronomy*, Vol. 2A, Cambridge University Press.

Thompson, S. P. (1962), *Treatise on Light by Christiaan Huygens*, English Translation, 1912, reprinted by Dover edition.

Wilson, Curtis (1970), "From Kepler's Laws, so-called, to Universal Gravitation; Empirical Factors", *Archive for History of Exact Sciences*, Vol. 6, pp. 89–170.

Wilson, Curtis (1989), "Predictive Astronomy in the Century after Kepler", in Taton and Wilson (1989), pp. 161–206.

Yabuuti, Kiyoski (1969), *Chugoko no Tenmon Rekiho*, Heibonsha, Tokyo.

Glossary

Cassini	口葛 西尼	Copernicus	歌白尼
Descartes	口葛爾徳	Galileo	加理勒
Huygens	許日尼	Kepler	刻爾伯
La Hire	臘羲爾	Pagan	巴岡
Picard	畢 嘉	Riccioli	利酉理
Römer	洛 默	Tycho	第 谷
Kögler	戴進賢	Foucquet	傅聖澤
Celian quanyi	測量全義	*Chongzhen lishu*	崇禎曆書
Kunyu quantushuo	坤輿全圖説	*Lixiangkaocheng houbian*	曆象考成後編
Lifa wenda	曆法問答	*Shulijinyun*	數理精蘊
Xinfa suanshu	新法算書		
Qiantianjianzheng	欽天監正	Keboer-zhi-fa	刻白爾之法
Liangweige	量 微 格	Danxingxian	蛋形線
Tuoyuan	橢 圓	Xinfa	新 法
Liexian	列 線	Chengxuan	正 弦
Juguang-juhuodian	聚光聚火點	Judian	聚 點
Liuxing	流 行	Budong	不 動
Huozhijing	火 之 精	Dongzhiyuan	動 之 原
Jingqi	精 氣	Jingqi-zhi-nengli	精氣之能力
Gewuqiongli-zhi-xue	格物窮理之學	Xuanquan-zhi-bo	旋圈之波

2.5. Tebbutt vs Russell: Passion, Power and Politics in Nineteenth Century Australian Astronomy

Wayne Orchiston
Australia

1. Introduction

During the second half of the nineteenth century Henry Chamberlain Russell and John Tebbutt were Australia's foremost astronomers. Russell was the Government Astronomer of New South Wales and director of the Sydney Observatory, while Tebbutt, an amateur astronomer, maintained a private observatory at nearby Windsor (for localities mentioned in the text see Figure 1). Initially Russell and Tebbutt collaborated closely, but there was a breakdown in relations during the 1880s and by 1891 they were openly feuding, forcing many leading Australian astronomers to take "sides". When the feud erupted on the pages of *The Observatory*, it became an international issue. The Russell–Tebbutt feud only ended with Russell's death in 1907.

Figure 1 Localities mentioned in the text.

S.M.R. Ansari (ed.), History of Oriental Astronomy, 169–201.

Hetherington (1976), Lankford (1979, 1981), Orchiston (1998a, 1999b) and Rothenberg (1981) have demonstrated that the amateur-professional nexus in astronomy is a particularly fertile area of investigation, addressing as it does the tensions associated with the evolution of a specific science.

This paper, which expands on Orchiston 1998d, discusses the Russell–Tebbutt feud and the ways in which passion, power and politics were brought into play in New South Wales astronomy during the nineteenth century. But to see the feud in full context it is first necessary to summarise salient features of Australian astronomy during the time when Russell and Tebbutt were active.

2. Australian Astronomy During the Second Half of the Nineteenth Century

2.1. *Introduction*

The half-century from 1850 to 1900 was a "golden age" as Australian astronomy rapidly built on its fledgling international reputation established through the pioneering work carried out at the Parramatta Observatory during the 1820s and 30s (see Richardson 1835; Saunders 1990). Four different "fundamental elements" can be identified in this evolutionary process, and each of these is discussed below.

2.2. *Establishment of professional observatories*

Apart from Phillip Parker King (Orchiston 1988f), no-one was particularly active in astronomy in Australia following the demise of the Parramatta Observatory, and it was only in the 1850s that an urgent need for observatories emerged in both Sydney and Melbourne as the lure of the goldfields brought an avalanche of immigrants. The primary demands of an accurate time-service and up-to-date meteorological data were best satisfied by the erection of an observatory.

Thus it was that professional observatories proper were established in the various Australian colonies during the second half of the nineteenth century (see Haynes, Haynes, Malin and McGee 1996; Orchiston 1988a). The first of these was at Williamstown (near Melbourne), in 1853, and the last at Perth, in 1896 (see Table 1). Initially, only Adelaide, Melbourne and Sydney Observatories contained sufficient staff and equipment to warrant ranking on an international scale, but Perth was able to join this trio soon after the turn of the century. Melbourne Observatory only came into existence when the Victorian Government decided to combine the functions of the Flagstaff and Williamstown Observatories.

The only other Australian institution with a substantial nineteenth century involvement in observational astronomy was the New South Wales Lands Department, which even created the official position of "Field Astronomer" in 1886 (Orchiston 1987a).

The work of all of the different "astronomical observatories" in Table 1 went far beyond astronomy and time-keeping to involve – in varying degrees – geomagnetism, meteorology, seismology, tidal studies and trigonometrical survey work. While Adelaide, Melbourne, Perth, Sydney and Williamstown Observatories adopted more wide-ranging policies, the short-lived Flagstaff Observatory was concerned with geomagnetism and meteorology,

Table 1 Australian professional observatories, 1850–1910

Name	Founding year	Main telescope(s)*	Officer(s)-in-charge	Reference(s)
Williamstown	1853	11.4 cm OG	Ellery (1853–1863)	Ellery 1869.
Flagstaff	1858	Small OG	Neumayer (1858–1864)	Perdrix 1990; Weiderkehr 1988.
Melbourne	1863	1.22 m spec 20.3 cm OG 33 cm astr	Ellery (1863–1895) Baracchi (1895–)	Gascoigne 1992; Perdrix 1961, 1970.
Sydney	1858	29.2 cm OG 18.4 cm OG 33 cm astr	Scott (1856–1862) Smalley (1862–1870) Russell (1870–1905) Lenehan (1905–1908) Raymond (1908–)	Orchiston 1988b; Wood 1958.
Adelaide	1874	20.3 cm OG	Todd (1855–1906) Griffiths (1907–1909) Dodwell (1909–)	Edwards 1993, 1994.
Brisbane	1879	Small OG Small tran	MacDonnell (1879–1885) Wragge (1887–1891) McDowall (1891–1902) Spowers (1902–)	Haynes, Haynes and Kitson 1993.
Hobart	1882	Small tran	Shortt (1882–1892) Kingsmill (1893–1909)	Meteorological Department1900.
Perth	1896	30.3 cm spec 33 cm astr	Cooke (1896–1912)	Utting 1989, 1992.

* Key: astr = astrograph; OG = refractor; spec = reflector; tran = transit telescope.

while the observatories in Brisbane and Hobart focussed almost exclusively on meteorology and time-keeping. In addition, all of the observatories in Table 1 had important educational and informational roles to fulfil (see Orchiston 1991a). Because of chronic understaffing, and the emphasis that governments placed on public viewing nights, some of these institutions found their intended research programs severely disrupted.

Research required the right combination of instruments and staff, and in this regard the four major Australian observatories were well served by international standards. The first two directors of Sydney Observatory were Cambridge mathematicians appointed by Airy (Orchiston 1988b, 1998c; Wood 1958), and all of the senior subordinate staff came with excellent newly-acquired academic qualifications or with substantial experience in observational and mathematical astronomy.

A properly-trained professional was also appointed to direct the Adelaide Observatory, in the person of Charles (later Sir Charles) Todd (Edwards 1993; Symes 1976), who had worked at both the Royal Observatory, Greenwich and the Cambridge Observatory. Todd was also the South Australian Superintendent of Telegraphs, and in 1870 added the post of Postmaster General to the list! Todd's assistant and understudy at the Adelaide Observatory was an enthusiastic young local graduate named William Cooke (Hutchison

1980, 1981), who became founding Government Astronomer of Western Australia when the Perth Observatory was opened in 1896, and eventually took charge of the Sydney Observatory.

For a long time, Melbourne was the only major Australian observatory without professionally-trained senior staff. Robert Ellery, founding director at Williamstown Observatory and subsequently at Melbourne Observatory, had trained as a medical practitioner (Gascoigne 1992) and had a background in amateur astronomy, as did his First Assistant, John White. A professionally-trained individual (a surveyor no less) only joined the senior staff in 1882 when Baracchi was appointed.

While the directors at Adelaide, Melbourne, Perth and Sydney also served as the Government Astronomers of their respective colonies, the Brisbane and Hobart Observatories were not so well blessed. From the start, these small observatories were placed in the care of their colony's respective Meteorological Observers, but at some point in time late in the century the Surveyor-General of Queensland took over responsibility for the Brisbane Observatory (see Haynes, Haynes and Kitson 1993).

When it came to astronomical instruments, the major observatories were well equipped, partly due to the high public profile of the 1874 and 1882 transits of Venus (see Ellery 1901) and the pressure that the observatory directors therefore were able to apply to their respective governments for special "equipment funding". The principal refractors at Adelaide, Melbourne and Sydney Observatories were all acquired in this way. However, all were modest by world standards, the largest being the 29.2 cm Schroeder refractor in Sydney. Each observatory also was supplied with a transit telescope, and Melbourne, Perth and Sydney Observatories acquired Grubb astrographs (see Russell 1892a).

Of all the colonial observatory instruments, by far the most impressive was the Great Melbourne Telescope (GMT), which became operational in 1869 (Perdrix 1970). Its 1.22 m speculum mirror by Grubb made this the largest operational equatorially-mounted telescope in the world at the time (Robinson and Grubb 1869), but despite the promise it did not live up to expectations (Gascoigne 1995; Hyde 1987; Perdrix 1992). A large number of nebulae and star clusters were observed, and many of these featured in *Observations of the Southern Nebulae Made With the Great Melbourne Telescope* (Ellery 1885), but drawings based on naked eye observations of these tenuous yet complex objects were rapidly replaced during the last two decades of the nineteenth century by innovations in astronomical photography (Barnard 1898; Lankford 1984; Norman 1938). With severe staff cuts during the depression of the 1890s and increased demands from the International Astrographic Project, the GMT was all but decommissioned, and thereafter was rarely used. From the start, had it been housed in a conventional dome rather than its roll-off roof observatory, and a glass mirror of considerably shorter focal length been installed, then it could have been used for spectroscopic and photographic studies and made a significant contribution to astrophysics. Instead, it proved to be a "white elephant".

The types of research undertaken at Australia's leading government observatories are listed in Table 2. Adelaide, Melbourne, Perth, Sydney and Williamstown Observatories all succeeded in carrying out worthwhile research (see Haynes, Haynes, Malin and McGee 1996) and placing their names before the international astronomical community.

Table 2 Types of astronomical research programs most commonly undertaken by the leading Australian professional observatories, 1850–1910

Transitory Events
Eclipses of the Moon
Eclipses of the Sun
Lunar occultations of planets
Lunar occultations of stars
Phenomena of Jupiter's satellites
Transits of Mercury
Transits of Venus
Short-term Monitoring Projects
Comets (positions and appearance)
Long-term Monitoring Projects
Double stars (separation and position angle)
Planets (appearance)
Sun (surface details)
Search Programs
New Double stars
Sky Survey Work
Star positions

At the Sydney Observatory, Russell and his assistants discovered and measured many new southern double stars (see Innes 1899), and Russell was also one of these who pioneered astronomical photography in Australia (e.g. see Russell 1890a, 1890b, 1891). In Adelaide, Sells and then Cooke assembled a long series of drawings showing Jupiter's changing atmospheric features (Dodwell 1913), while during the 1870s Melbourne Observatory produced what Airy at the time described as the "... best catalogue of stars of the Southern Hemisphere ever published." (*Eleventh Report* ... 1876: 8), and for 20 years, starting in 1874, daily solar photographs were taken (weather permitting) and forwarded to Greenwich (Haynes, Haynes, Malin and McGee 1966). During the late 1880s, Baracchi (1889) and Ellery (1889) carried out stellar spectroscopic observations with the 20.3 cm refractor, while even earlier Le Sueur had used the Great Melbourne Telescope to obtain the spectrum of Eta Argus (now Eta Carinae) and to carry out the first ever spectroscopic observation of an extragalactic object, 30 Doradus (Hearnshaw 1986). In Sydney, Russell (1881b) conducted a spectroscopic examination of the Great Comet of 1881 (C/1881 K1). Staff from Adelaide, Melbourne and Sydney Observatories observed the 1874 transit of Venus (Baracchi 1914; Russell 1892b), and the last two observatories were joined by Perth Observatory in the International Astrographic Project (see Turner 1912).

Research was not the only astronomical contribution made by the staff at the Australian colonial observatories, for both Ellery and Russell were active in grinding, polishing and figuring telescope mirrors. Ellery produced speculum metal mirrors (see Gascoigne 1992), while Russell experimented with glass, in 1880 completing his *piéce de rèsistance*, a mirror with a diameter of 38.1 cm. The associated telescope employed a novel equatorial mounting of Russell's own design, and somewhat reminiscent of that used later for the Great Palomar Telescope (Orchiston 2000b; Orchiston and Bhathal 1982). Russell also developed a range of meteorological instruments (Bhathal 1991).

By the end of the nineteenth century, Australia was viewed – along with South Africa – as a major force in Southern Hemisphere astronomy, and Ellery, Russell and Todd were names well-known to Northern Hemisphere colleagues, particularly those committed to positional astronomy.

2.3. *The growth of popular interest in astronomy*

The period from 1850 to 1910 was blessed with an amazing succession of naked eye comets, three total solar eclipses and two transits of Venus (see Table 3), all potentially visible from Australia. To add to the drama, three of these comets (C/1861 J1, C/1865 B1 and C/1881 K1) were discovered by Australians (see Figure 2), and the two transits not

Table 3 Impressive astronomical objects and events visible from Australia, 1850–1924 (after Orchiston 1997a)

Object/event	Year	Name/type*
Comet	1853	C/1853 G1 (Schweizer)
	1853	C/1853 L1 (Klinkerfues)
	1858–1859	C/1858 L1 (Donati)
	1860	C/1860 M1 (Great Comet)
	1861	C/1861 J1 (Tebbutt = Great Comet)
	1862	109P/Swift-Tuttle
	1864	C/1864 N1 (Tempel)
	1865	C/1865 B1 (Great Southern Comet)
	1874	C/1874 H1 (Coggia)
	1880	C/1880 C1 (Great Southern Comet)
	1881	C/1881 K1 (Tebbutt = Great Comet)
	1881	C/1881 N1 (Shaeberle)
	1882	C/1882 F1 (Wells)
	1882–1883	C/1882 R1 (Great Southern Comet)
	1887	C/1887 B1 (Great Southern Comet)
	1892	C/1892 E1 (Swift)
	1901	C/1901 G1 (Great Comet)
	1910	C/1910 A1 (Great January Comet)
	1910	1P/Halley
Transit of Venus	1874	
	1882	
Total Solar Eclipse	1857	
	1871	
	1910	

* Cometary nomenclature is after Marsden and Williams (1996).

Figure 2 The Great Comet of 1880 attracted considerable public interest (Orchiston Collection).

only attracted Australian observers (amateur and professional) but also teams from Britain and the United States (e.g. see Orchiston and Buchanan 1993).

All of these visually-appealing public spectacles listed in Table 3 demanded description and explanation. Australian professional and amateur astronomers used many means to bring information about these objects and events, and others, before the general public. They established public information services; offered viewing nights at their observatories; prepared press releases, reports, "Letters to the Editor", and even regular "Astronomy" columns for newspapers; published booklets, pamphlets, books and chapters of books on astronomy or aspects of the discipline; manufactured planispheres; presented public lectures on aspects astronomy; and delivered courses on astronomy. I refer to this collective corpus of activities and productions as "public astronomy" (in contradistinction to "research astronomy").

All of the professional observatories were involved in these activities to a greater or lesser extent, but those who made the most significant contributions were Scott and Russell at the Sydney Observatory, Ellery at the Melbourne Observatory, and Cooke at Perth Observatory (see Orchiston 1991a). However, the combined activities of the nation's amateurs brought astronomy to a much wider public audience (see Orchiston 1997a). The

Table 4 Leading Australian amateur astronomers actively involved in the popularisation of astronomy, 1850–1910 (after Orchiston 1997a)

Name	Location	Activities*	Other references
Abbott	Hobart	1,2?,3,5,7?	Orchiston 1992.
Baker	Ballarat	1,2,3?,8	Burk 1986; Davis 1990.
Biggs	Launceston	1,2,3,7	Giordano 1995; Orchiston 1985.
Bone	Castlemaine	1,2,3,7	Orchiston 1986, 1987b.
Dobbie	Adelaide	1,2,3?	Orchiston and Bembrick 1995.
Gale	Sydney	1,3,7	Orchiston, 1988c; Orchiston and Bembrick 1995, 1997.
Innes	Sydney	1,3	Orchiston 1988c; Orchiston, unpublished study.
Macdonnell	Pt Macquarie Sydney	1,2?,3	Baracchi 1914; Orchiston 2001.
Martin	Sydney	1,7	Orchiston and Bhathal 1991.
Roseby	Sydney	1,2,3,7	Orchiston 1988c; Orchiston, unpublished study.
Tebbutt	Windsor	1,2,3,5,7	Bhathal 1993; Orchiston 1988g.
Wooster	Ballarat	1,2,3,4,7	Orchiston, unpublished study.

* Key: 1 = Supplier of astronomical information
2 = Offered public viewing nights
3 = Contributed newspaper reports, "Letters to the Editor"
4 = Contributed newspaper columns
5 = Prepared booklets, books, or chapters of books
6 = Manufactured planispheres
7 = Delivered public lectures
8 = Presented astronomy courses.

leading practitioners are listed in Table 4, but many others also contributed. One non-Australian who helped satisfy the public demand for astronomical information was the noted British populariser, Richard Proctor, and in 1880 he undertook a lecture tour of Australia (see Pitt 1880).

In any review of the achievement of Australia's amateur astronomers, the outstanding contribution that they collectively made to public astronomy cannot be ignored. Through their activities, an understanding or at least heightened public awareness of astronomy reached many more people than the government observatories alone could ever have serviced. In this respect, they supplemented the work of these official institutions, but all the while on a voluntary basis and at considerable personal sacrifice in terms of both time and money. In some towns and cities devoid of professional observatories, amateur astronomers ran their own institutions as small-scale *de facto* city observatories (Orchiston 1989b). Without these individuals, and others who focussed on the popularisation of astronomy through their local news media, the Australian public would have been far less astronomically literate than was indeed the case.

A notable outcome of this substantial amateur involvement in popularisation was the rapid growth in the number of active amateur astronomers during the second half of the nineteenth century. The exponential growth curve for the Sydney region is shown in Figure 3, and it is likely that similar curves typify other major population centres, although the actual numbers in each case were smaller.

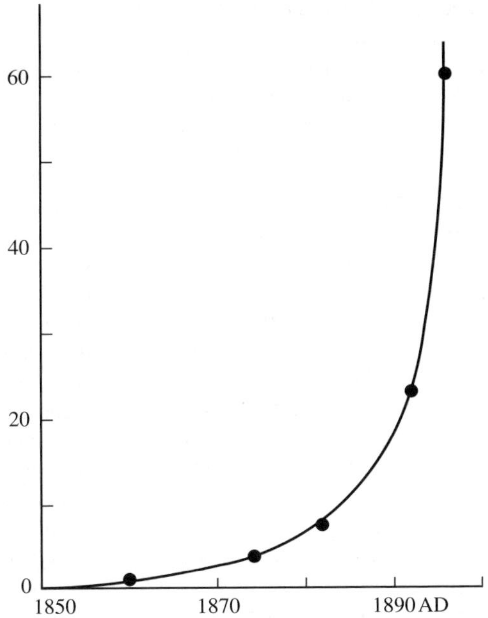

Figure 3 Growth in the number of active amateur astronomers in the Sydney region, 1850–1900.

2.4. *Establishment of significant private observatories*

A particularly important feature of nineteenth century and early twentieth century Australian astronomy was the emergence of a network of private observatories across the nation, established by amateur astronomers involved in research (Haynes, Haynes, Malin and McGee 1996). Research observatories are listed in Table 5, and in the main were distributed along the eastern seaboard of the nation (Figure 1), but with a marked concentration in the Sydney region. Their proliferation between 1880 and 1910 is immediately apparent.

Although the emphasis here is on research astronomy, some of the astronomers listed in this table emulated their professional colleagues and adopted a wider charter of operations. Thus, Abbott, Beattie, Biggs, Bone, Macdonnell and Tebbutt all maintained local time-services, while Abbott, Biggs, Bone and Tebbutt operated meteorological stations. In addition, Biggs was one of Australia's earliest seismologists (*Earth Tremors*, 1984), and Tebbutt studied local floods and tides. About half of the astronomers listed in this table were committed to public astronomy.

All of the observatories represented in Table 5 were furnished with instruments that were similar in aperture to or only marginally smaller than those found in the major Australian government observatories (the Great Melbourne Telescope excepted). Reflectors and refractors were favoured almost equally, and apart from the locally-manufactured reflectors owned by Ross, Innes and Merfield, and Gale's 45.7 cm instrument, all of the reflectors referred to in Table 5 were imported.

Table 5 Significant private research observatories, 1850–1910 (after Orchiston 1989b)

Name of astronomer	Location	Founding decade	Main instruments*	Other references*
Abbott	Hobart	1850s	11.4 cm OG	Orchiston 1992, 1997c.
Tebbutt	Windsor	1860s	11.4, 20.3 cm OG	Orchiston 1982b; White 1979.
Biggs	Launceston	1870s	21.6 cm spec	Orchiston 1985.
Bone	Castlemaine	1870s	20.3 cm OG	Orchiston 1987b, 1997b, 1997c.
Collyer	Sydney	1870s	26.0 cm spec	Baracchi 1914; HHMM.
Macdonnell	Port Macquarie	1870s	15.2 cm OG	HHMM; Orchiston 2001.
(Dr) Wright	Sydney	1870s	21.6 cm spec	Orchiston and Bhathal 1991.
Davidson	Mackay	1880s	15.2 cm OG	Baracchi 1914; Orchiston 1997b.
Morris	Sydney	1880s	21.6 cm spec	Orchiston 1987a.
Gale	Sydney	1890s	15.2 cm OG, 45.7 cm spec	Orchiston 1988c, 1997b; Orchiston and Bembrick 1995, 1997.
Innes	Sydney	1890s	15.6 cm OG, 41.9 cm spec	Orchiston 1988c, 1997b, unpublished study.
Macdonnell	Sydney	1890s	15.2 cm OG	HHMM; Orchiston 2001.
Merfield	Sydney	1890s	17.8 cm spec	Orchiston 1988c, 1997b, unpublished study.
Ross	Melbourne	1890s	30.5 cm spec	Orchiston 1999b; Orchiston and Brewer 1990.
(Mr) Wright	Sydney	1890s	21.6 cm spec	Orchiston 1988c.
Beattie	Sydney	1900s	15.9 cm OG	Orchiston 1997b, unpublished study.
Nangle	Sydney	1900s	16.5 cm OG	Baracchi 1914; Orchiston 1997b.

* Key: OG = refractor; spec = reflector
HHMM = Haynes, Haynes, Malin and McGee 1996.

As Table 6 illustrates, the types of research undertaken also closely mirrored that carried out at the major government observatories, although the amateurs pursued more catholic Short-term Monitoring Projects and Search Programs.

Of all of the amateurs, Tebbutt was Australia's most prolific publisher of scientific papers, followed by Abbott, Merfield, Innes and Gale. All five astronomers, together with Beattie and Nangle, were the equal of their professional colleagues in intellect and mathematical prowess.

Given the calibre of these and other leading Australian amateur astronomers, we should not underestimate their collective impact on research astronomy in Australia. Their combined publications tally far surpassed that of their professional colleagues, and like their government counterparts they published in the same local and overseas journals. Most popular were *Astronomische Nachrichten, Journal of the British Astronomical Association, Monthly Notices of the Royal Astronomical Society* and *The Observatory*, all highly-respected international journals. Through the sheer weight and quality of their publications, amateurs helped to cement Australia's international astronomical reputation.

Table 6 Principal observational programs undertaken by leading Australian amateur astronomers, 1850–1910

Transitory Events
Eclipses of the Moon
Eclipses of the Sun
Lunar occultations of planets
Lunar occultations of stars
Phenomena of Jupiter's satellites
Transits of Mercury
Transits of Venus
Short-term Monitoring Projects
Comets (positions and appearance)
Meteors (fireballs, and shower activity)
Minor planets (positions)
Planets (positions)
Long-term Monitoring Projects
Double stars (separate and position angle)
Planets (appearance)
Sun (surface details)
Variable stars (magnitude variations)
Search Programs
Coloured stars
New double stars
New variable stars (including novae)

Because of the prodigious output of the amateurs and Tebbutt's anomalous position as the nation's foremost astronomer, the distinctions between professionals and serious amateurs were blurred in late nineteenth century Australian astronomy, and it was possible to graduate from amateur to professional ranks as Innes and Merfield demonstrated. This amateur-professional co-operation and collaboration occurred because the forces that emerged to separate the two classes of astronomers in other parts of the world between 1880 and 1920 (e.g. see Hetherington 1976, Lankford 1979, 1981a, 1981b and Rothenberg 1981) simply by-passed Australia altogether at this time. Astrophysics only became fashionable in Australia with the founding of the Commonwealth Solar Observatory in 1924 (see Orchiston 1989a).

The only major international trend to impact on Australian professional astronomy at the time was the International Astrographic Project (Turner 1912), but this did not lead to immediate obvious differences between the nation's amateur and professional astronomers for the government observatories also continued their non-astrographic work, while a number of amateurs experimented successfully with astrophotography. The most notable of these were David Ross of Melbourne (Orchiston and Brewer 1990) and Walter Gale (Reports of the Branches, 1895). Thus, amateur-professional communication was preserved.

Table 7 Australia's earliest formal astronomical groups and societies

Duration	Founder(s)*	Name of group	Reference
1876–1881	Russell (p)	Roy.Soc.N.S.W., Sec.A	Orchiston and Bhathal 1991.
1879–?	Ellery (p)	Roy.Soc.Vic., Astr.Sec.	Orchiston 1998a.
1882–1883	Tebbutt (a)	Austr.Comet Corps	Orchiston 1982a.
1892–	Farr (a)	Roy.Soc.S.A., Astr.Sec.	Waters 1980.
1895–	Gale (a) & Innes (a)	B.A.A., N.S.W. Branch	Orchiston 1988c.
1895–1901	76 (a)	Brisbane Astr. Soc.	Orchiston 1998a.
1897–1907	Ross (a) & Wigmore (a)	B.A.A., Vic. Branch	Orchiston and Perdrix 1990.
1912–1928	Curlewis (p) & Hilton (a)	Astr.Soc.W.A.	WA Astronomical Society 1914

* Key: a = amateur astronomer
p = professional astronomer

2.5. *Formation of astronomical groups and societies*

The formation of organised groups and societies is a critical element in the evolution of any discipline, and this was no more apparent than in nineteenth century Australian astronomy, with its vast geographical distances, physical isolation from the main international centres of astronomical activity, and comparatively small number of astronomers. Australia's earliest astronomical groups and societies are listed in Table 7, and amateurs and professionals played key roles in their formation and operation (see Orchiston 1998a).

In addition to these groups, in 1892 Gale and Innes spent time planning the formation of the nation's first national generalist astronomical society, the Australian Astronomical Society, but in the end they decided not to proceed with this (Orchiston and Bhathal 1984).

Russell and Ellery were the driving forces behind the short-lived Astronomy Sections of the Royal Societies of New South Wales and Victoria, respectively, but amateur astronomers operating alone or in collaboration with professionals were involved in the formation of all of the other groups listed (Ellery 1901; Baracchi 1914). Amateur astronomers also were responsible for founding the Astronomical Section of the Royal Society of South Australia and the Victoria Branch of the British Astronomical Association, but it did not take long for staff from the Adelaide and Melbourne Observatories to exert their influence on these two groups (see Orchiston 1998a).

It was only in the 1890s that Australia's first enduring astronomical groups were formed. These were generalist in focus in that they dealt with all aspects of astronomy (unlike Tebbutt's Comet Corps), and addressed localised rather than national audiences. By this time, Australian astronomy had evolved to the stage where a number of population centres could sustain their own formal groups: there were sufficient active amateur astronomers present; support from both amateurs and professionals was forthcoming; and astronomers of prominence existed who were willing to provide the requisite leadership.

With the possible exception of the Brisbane Astronomical Society, all of the groups listed in Table 7 played an important role in formalising astronomy in Australia, encouraging members to participate in observational programs, and cementing the close relations that for the most part existed between the nation's amateur and professional astronomers. These groups also gave newcomers a structured way of getting started in astronomy, and provided all members with regular meetings, library facilities, and avenues for publication.

3. The Russell–Tebbutt Feud

3.1. Introduction

Russell (Figure 4) and Tebbutt (Figure 5) are regarded by many as Australia's foremost astronomers during the second half of the nineteenth century (see Bhathal 1993). In this section, we provide thumbnail sketches of both men, and then examine the Russell–Tebbutt feud.

Henry Chamberlain Russell (Bhathal 1991; Walsh 1976) was born at West Maitland, New South Wales, in 1836 and came from a distinguished family. His father was the Honourable Bourn Russell, a member of the Legislative Council from 1858 to 1880. After completing his B.A. degree at the University of Sydney in 1859, H.C. Russell was appointed a computer at the Sydney Observatory under Scott (Wood 1958). In 1870 he became Government Astronomer of New South Wales and Director of the Sydney Observatory (Figure 6). With the passage of the years, Russell rose to prominence in both astronomy and meteorology, and contributed significantly to double star astronomy, the International Astrographic Project and the British 1874 transit of Venus program. He also

Figure 4 H.C. Russell, 1836–1907 (Courtesy Sydney Observatory).

Figure 5 Tebbutt, 1834–1916 (Orchiston Collection).

Figure 6 Sydney Observatory (Courtesy Royal Astronomical Society).

observed comets, eclipses, Jovian atmospheric features, and transits of Mercury. Russell invented a number of meteorological instruments, and he also experimented with aspects of telescope design (e.g. see Orchiston 2000a). He played a leading role in the Royal Society of New South Wales for many years, including serving as President (Maiden 1918), and as a founder of the Australasian Association for the Advancement of Science was its inugural President (Hoare 1975; MacLeod 1988). A Fellow of the Royal Society, Russell was a member of the Senate of his *alma mater* for more than 32 years, and was awarded a Companion of the Order of St. Michael and St. George for his long and valuable contribution to New South Wales science. He died in Sydney in 1907.

John Tebbutt (Bhathal 1993; Orchiston 1988g; White 1979) was born at Windsor, near Sydney, in 1834. His father was a successful businessman who turned to farming, and young Tebbutt followed this latter calling. He began systematic observing in 1853, and in 1861 discovered the Great Comet of that year (Orchiston 1998b). The following year, he refused the vacant Sydney Observatory directorship when this was offered to him (Orchiston 1988e, 1998c). Soon after this, he erected his own observatory and a meteorological station (Orchiston 1988d), and as his means improved he expanded this facility (see Figure 7), and furnished it with an outstanding reference library and successively larger refracting and transit telescopes – culminating in the 20.3 cm Grubb telescope. Over the years, Tebbutt used these instruments to observe a wide range of celestial objects and events, with emphasis on comets, double stars, variable stars (see Orchiston 2000b), lunar and solar eclipses, transits of Mercury and Venus, minor planet and planetary positions, Jovian satellite phenomena, and lunar occultations of stars and planets (for a summary see Orchiston 1982b). He also discovered a nova (Ashbrock 1984) and, in 1881, a second great comet (see Orchiston 1981, 1991b, 1999a). In 1905 he received the Royal Astronomical Society's Jackson Gwilt Medal and Gift, for "... his important observations of comets and double stars, and his long continued services to astronomy in Australia, extending over forty years." (cited in Tebbutt 1908a:109). John Tebbutt died in Windsor in 1916, and more than half a century later, in 1973, a lunar crater was named after him, a rare honour for an amateur astronomer.

Russell and Tebbutt were both Fellows of the Royal Astronomical Society and published prolifically, but Tebbutt's lifetime tally of 388 research papers was more than double that

Figure 7 Windsor Observatory in 1880 (Orchiston Collection).

of his Sydney Observatory colleague. Russell was content to address a local audience and approximately half of his papers appeared in the journal of the Royal Society of New South Wales, whereas just 8% of Tebbutt's papers were published in Australian journals (see Table 8). His preferred journals were *Astronomische Nachrichten, Monthly Notices of the Royal Astronomical Society* and *The Observatory*, and these collectively accounted for 83% of all his papers. Whatever the journal, both men tended to publish short, non-analytical data papers, which were the norm at the time for most of those astronomers who espoused positional astronomy.

Both men also scribed books, booklets and pamphlets. In 1887 Tebbutt privately published and distributed a 74-page book about the Windsor Observatory (Tebbutt, 1887), and when advancing years forced him to abandon systematic observing he wrote his 132-page *Astronomical Memoirs* (Tebbutt 1908). These two books were widely distributed within Australia and overseas, as were his Windsor Observatory Annual Reports (which appeared between 1888 and 1903, inclusive), eight volumes of meteorological observations, and a number of pamphlets on astronomy (and, in one case, the astronomy-religion interface). Tebbutt's total output of books, booklets and pamphlets (28) also was more than double that of Russell.

In addition to their observational interests, Russell and Tebbutt also sought to popularise astronomy. Tebbutt was by far the more pro-active when it came to exploiting the print media. His first newspaper report appeared in 1854, and by the end of 1885 these contributions numbered 277 (see Table 9). They continued to appear almost up to the time of his death in 1916 (see "Many Happy Returns" . . . 1915), and although they covered a wide range of subjects comets were his favourite fare. Through these accounts he ". . . was able to share his own joy of astronomy and excitement of discovery with a great many people." (Orchiston 1997a: 48).

Table 8 Journal publications by John Tebbutt, 1862–1915

Journals*	AJo	ANa	ARe	ASP	BAA	NSW	RMN	RMe	Obs	SSR	TOTAL
Totals:	1	149	9	6	20	27	125	1	47	3	388

*Notes: The various journal totals are based in the main on entries in Tebbutt 1908: 118–132 and papers published subsequently.

Journal Key:

AJo = *Astronomical Journal*
ANa = *Astronomische Nachrichten*
ARe = *Astronomical Register*
ASP = *Publications of the Astronomical Society of the Pacific*
BAA = *Journal of the British Astronomical Association*
NSW = *Transactions of the Philosophical Society of New South Wales* plus its successor the *Journal and Proceedings of the Royal Society of New South Wales*
RMN = *Monthly Notices of the Royal Astronomical Society*
RMe = *Memoirs of the Royal Astronomical Society*
Obs = *The Observatory*
SSR = *Southern Science Record* [published in Melbourne]

Table 9 Newspaper articles and reports by John Tebbutt, 1851–1885 (after Tebbutt, 1887)

Discipline	1851–55	1856–60	1861–65	1866–70	1871–75	1876–80	1881–85
Astronomy	1	16	38	23	23	56	85
Meteorology	0	0	8	5	6	4	8
Other	0	0	0	1	1	1	1
Totals:	1	16	46	29	30	61	94

While Tebbutt may have been the more prolific when it came to research and public astronomy, Russell was the superior politician, and this is reflected in his successes with the Royal Society of New South Wales and the Australasian Association for the Advancement of Science. These provided *the* supreme regional and trans-Tasman societal power bases, and Tebbutt's abortive Australian Comet Corps and even the vibrant New South Wales Branch of the British Astronomical Association (of which Tebbutt was founding President) were simply no match.

Furthermore, Russell unashamedly exploited his political acumen for the benefit of the Observatory. As a senior government employee he knew the vagaries of the public service, and had the ear of Parliament. Consequently, over the years he was able to fund the 1874 and 1882 transits of Venus and the 1881 transit of Mercury expeditions; purchase the Schroeder refractor, a new 15.2 cm transit telescope and the Grubb astrograph; add a new wing and dome to the Observatory building; and expand the rural meteorological network (Bhathal 1991). He also gave long service to the University of Sydney, serving for a time as Vice-Chancellor. Consequently, civil honours flowed his way (e.g. the C.M.G.), and he was elected a Fellow of the Royal Society (of London), an honour denied Tebbutt despite his vastly superior research output.

Let us now examine the deteriorating relationship that culminated in the Russell–Tebbutt feud.

3.2. *The deteriorating relationship*

Russell and Tebbutt differed in age by just two years, and at first were close friends. In 1871, the New South Wales Government published a book that included a chapter by Tebbutt which contained the following complimentary remarks about Russell:

> It is matter for congratulation to astronomers that such good use is now being made of the fine equatorial of the Sydney Observatory, and the present director is entitled to the cordial sympathy and co-operation of all lovers of the science. (Tebbutt 1871:12).

It is equally apparent that Tebbutt's sentiments were reciprocated, as the following letter demonstrates:

> Many thanks for the kind way you have spoken to me, manifesting a feeling which I fully reciprocate and which I hope will increase as the years roll over as to our mutual benefit and the advantage of the science to which we have devoted our lives. (Russell 1870b).

This mutual respect and admiration continued throughout the 1870s, both at a personal and a professional level. Each expressed concern about the health of the other and their respective families, and Tebbutt was invited to join the Russell family for meals when visiting Sydney. When Russell's life was threatened during the notorious "letter bomb incident" of 1877 (see Wood 1958), Tebbutt was quick to offer his heartfelt sympathy. Russell's response, reproduced below, says something of his personality and his basic *modus operandi* in inter-personal relations:

> The stories in the paper about my being quite unnerved were all newspaper nonesense [*sic.*] I have felt and indeed still feel that it is a terrible thing to be obliged to recognise the existence of such a diabolical enemy, the more so that I did not before know I had one. It has always been my rule not to offend any one's feelings unless stern duty demanded it, and I thought that I had so far passed quietly through life. I feel now that the kind hand which has so recently preserved me will continue to do so if there be yet work for me to do. (Russell 1877).

When it came to astronomy and meteorology, Russell welcomed and appreciated Tebbutt's observations, and he invited Tebbutt to use the telescopes and the library at the Sydney Observatory. For his part, Tebbut supplied Russell with information on the historic Parramatta Observatory which had closed in 1847, but only after its astronomers had discovered a number of new comets, double stars, star clusters and nebulae, and accumulated observations for the Parramatta Catalogue of stars (Richardson 1835; Saunders 1990).

In 1871 Russell invited Tebbutt to join the expedition that was going to Queensland to observe the total solar eclipse. The following year, he asked Tebbutt to support him in forming a local astronomical group (Russell 1872), and hoped that they would co-operate in their respective 1874 transit of Venus programs (Russell 1874a). About this time, Tebbutt was seeking to install a larger telescope in his Windsor Observatory, and Russell vetted those instruments then available in Sydney. Nor did Russell's assistance with astronomical instrumentation end there, for when about to leave for Europe in 1875 he made the following offer: ". . . if I can do anything for you in the way of Instruments or otherwise I shall be very glad." (Russell 1875). Other example of the amiable relationship could be cited.

However, there were occasional tension points during the 1870s. Tebbutt was proud of his achievements in astronomy and meteorology, and took offence easily if he felt at all slighted. Observations of the 1874 December transit of Venus provided a case in point, and – with hindsight – can be seen to mark the start of a systematic deterioration in the previously amiable relations between the two men. In 1875 January, Russell presented a paper on the transit at a meeting of the Royal Society of New South Wales that Tebbutt did not attend, and an account subsequently appeared in the *Sydney Morning Herald* newspaper which made no mention of Tebbutt's observations. This peeved Tebbutt (1875b), and although it was explained that no offence was intended, he followed up with other letters to the newspaper which professed respect for his "friend" Russell but at the same time promoted his own research and the international status of the Windsor Observatory. He ended up taking the "transit incident" far beyond its original bounds, and raised the interesting spectre of amateur-professional competition in nineteenth century Australian science.

The entire "transit incident" does raise the thorny question of the relative status of Russell and Tebbutt – both in their own eyes and in those of their scientific colleagues, national and international. Tebbutt's superior ranking in New South Wales astronomy was apparent as early as 1862 when he rather than Russell was offered the post of Government Astronomer of New South Wales upon Scott's resignation (see Orchiston 1988d). Then, in 1866, it was Tebbutt who was asked to prepare a report on New South Wales astronomy for the Paris International Exhibition (see Tebbutt 1866), but when this was published (in revised form) by the New South Wales Government in 1871 (see Tebbutt 1871) Russell (1871) contributed a chapter on meteorology for the very same tome. So Tebbutt and Russell had already begun to establish their respective territories in astronomy and meteorology as early as 1870.

By the end of 1880, relations between Tebbutt and Russell were still amiable, even if punctuated by on-going minor irritations, but this all changed in 1881 March when Russell failed to acknowledge Tebbutt's assistance in a research paper. Even the following reassurances from friends like T.H.F. Griffin could not placate him:

> I sympathise with you in the matter of the oversight of your assistance rendered to Mr. Russel [*sic.*] which that gentleman has omitted to acknowledge ... however, as you well point out, the approval which all of your work meets with, & the acknowledgement of its value rendered by the scientific world outside NSW is a far greater recompense than you could receive near home. (Griffin 1881).

This matter of peer recognition and perceived status as an international astronomer was something that was to preoccupy Tebbutt for the remainder of his life. As an obvious corollary, he was also concerned about Russell's astronomical output, and the balance between astronomy and meteorology at the Sydney Observatory. Tebbutt believed that as a publicly-funded institution, the Observatory should contribute equally to both sciences, and he was becoming more and more concerned about the increasing emphasis which Russell was placing on meteorology at the expense of astronomy. This disquiet was echoed by others in the local scientific community.

Next came an important event that would test the level of Russell's regard for Tebbutt without any possible ambiguity, and this was the latter's discovery of Comet C/1881 K1, the Great Comet of 1881. This comet was destined to play a critical role in the history of cometary photography and spectroscopy (see Orchiston 1981, 1999a), but it brought no congratulations from Russell, only concerns:

> For one reason I am rather sorry it has come now, and that is, because it imparts a seeming confirmation to one of those irresponsible and unsigned predictions that have been published in England of late, apparently with the object of frightening the multitude. (Russell 1881a).

This was hardly the response that Tebbutt anticipated, and it would not have been appreciated.

From this critical date, the letters between Russell and Tebbutt became less frequent and the relationship more strained. By the mid-1880s, Russell's letters were formal and to the point: once they had commenced with "My Dear Mr Tebbutt", but now it was a terse "Dear Sir" (e.g. see Russell 1886, 1887a). The deteriorating relationship was also starting to come to the notice of overseas astronomers. In 1887 May 3, Charles Piazzi Smyth wrote Tebbutt a melodramatic letter which ended with an intriguing suggestion:

> But alas the public of N.S.W. are strangely unaware, even after a quarter of a century, that they have in you the rare example of peculiar astronomical genius, which is almost entirely wanting in Anglo-Saxons of the Mother Country, as well as N.S.W., excepting as I said in your instance. If they go on much longer thinking lightly of your orbital calculations – I wonder whether it would be within the range of possibilities that you should come home & do for Great Britain and her Cometic Astronomy what you have been doing so long for that of Australia . . . (Smyth 1887).

Yet there were brief moments of reprieve. In 1886 Tebbutt purchased a 20.3 cm Grubb refractor after Russell referred the agent responsible for the sale to him (Pritchard 1886), and the following year there was a notable warming of relations between the Sydney and Windsor Observatories when Russell was overseas and some of the staff took advantage of his absence to learn more about Tebbutt first hand. It is clear from the warm tone of their letters (e.g. Masters 1887; Pollock 1887) that, with the exception of Russell, staff at the Sydney Observatory had great regard for Tebbutt. This is also evidenced by their responses when Tebbutt distributed copies of his booklet *History and Description of Mr. Tebbutt's Observatory, Windsor, New South Wales* (Tebbutt 1887) towards the end of that year, and even Russell (1887b) was moved to comment that "There are few, if any private observatories in the world which can shew such a record as you."

The year 1888 brought with it an incident which was to dash any prospect of a rapprochement between Russell and Tebbutt and remove once and for all any semblance of civility in their future letters. This was the appearance of Russell's classic historical paper, "Astronomical and meteorological workers in New South Wales, 1788 to 1860" (Russell 1888), which discussed others who were active during the 1850s but omitted any mention whatsoever of John Tebbutt. White (1889) from the Melbourne Observatory was quick to point out that Russell's paper was incomplete without the inclusion of Tebbutt's labours, while Tebbutt (1890) modestly made the same point in a note published in *The Observatory*.

This omission must have been a calculated one, for Russell would have seen Tebbutt's numerous astronomical reports in the Sydney newspapers long before he joined the Sydney Observatory staff in 1859. The 1850s were a critical period in Australian astronomy, falling as they did between the closure of the Parramatta Observatory and the full flowering of the colonial observatories, and Abbott and Tebbutt were the only amateur astronomers of note at this time. Between 1853 and 1860, Tebbutt carried out observations of the Sun, aurorae, meteors, the zodiacal light, lunar occultations, lunar eclipses, Jupiter and its satellites, comets, double stars, and variable stars, and published seventeen different reports in the Sydney newspapers (see Table 9). If anyone did, he certainly deserved a place in Russell's paper. This deliberate snub was later seen by some (e.g. Lenehan 1907c) as the starting point of what was to degenerate into a bitter feud between the two former allies.

As a result of this incident, Tebbutt also reacted negatively towards Russell's friends and colleagues, refusing to join the Australasian Association for the Advancement of Science and contribute to Section A (Astronomy) when this new society was formed in

1888. This group was quickly to become a major force in Australian and New Zealand science (MacLeod 1988), and Tebbutt would suffer by this decision.

As news of the growing tension between Tebbutt and Russell began to circulate throughout the Australian astronomical community some looked to take advantage of the situation by threatening to play one man off against the other for their own political ends (e.g. see Thompson 1889b). Others were eager to express their own concerns about Russell. Amongst these was Chas. Egeson, one of Russell's own employees, who wrote Tebbutt a carefully-worded letter on 1890 March 29:

> It surprises me to think of the amount of work that you have done by private effort. In this Observatory which is unworthy of the name we have telescopes and observers in abundance but there is no result . . .
>
> Our aspirations are like a metal track that guides the intellectual engine and its physical train over the undulating fields of life. To know when to apply the brake and when to steam up is the problem which the driver of the engine has to solve in order to get to the terminus of human destiny. I have known some . . . also known to you, who have wasted life's fuel in simply sounding the steam whistle, that others might hear them, and have in consequence stuck at the foot of the first big hill that presented itself, for want of steam. They are now raking up the last embers in the fire hole in vain endeavour to proceed.

One of the reasons for Tebbutt's growing animosity towards Russell was the Observatory's increasing devotion to meteorology, at the expense of astronomy. Matters came to a head in 1891 when the *Sydney Morning Herald* published statistics on the number of staff employed at the Sydney Observatory in the years 1880 and 1890, total salaries paid in those two years, and the overall cost of running the Observatory for the ten years ended 1890. The total Government allocation over this decade was £41,103, and £24,597 of this (60%) was expended on the Observatory's meteorological functions (*Sydney Morning Herald*, 1891). This brought an immediate critical response from an anonymous writer, but when Tebbutt sent in his own letter this was suppressed. Tebbutt felt that

> . . . as a contributor to the revenue out of which the Observatory is maintained, and a gratuitous promoter of the science of astronomy for upwards of thirty years, I had an undoubted claim to discuss the question raised by the anonymous correspondent referred to, and in my discussion of it I do not think I have in any way transgressed the limits of fair and honest criticism. (Tebbutt 1891b).

Obviously the Editor did not agree, and he even refused to return the letter. But when Tebbutt had a point to make he would not be silenced! To use his own words: ". . . I am determined not to be thus quietly snuffed out." (*ibid.*)

His solution was the obvious one for a gentleman of means in the habit of publishing and distributing his own *Annual Reports* and meteorological monographs: he would publish the "missing letter" and further critical comments himself, as a pamphlet, and distribute this privately.

Late in 1891 September, *The Sydney Observatory and the "Sydney Morning Herald". A Plea for Astronomy in New South Wales* rolled off the printing press in Sydney. The suppressed letter occupied three and a half pages of this little 8-page document, which suggests that quite apart from the political implications, its excessive length and detail may have been a factor (but certainly not the deciding factor) in the original decision to suppress it. There is no logic in reproducing the entire letter here, but the salient points that it makes deserve to be listed:

- The Board of Visitors which was formed in Scott's day was allowed to lapse under Russell's directorship.
- For many years the Observatory has not produced an Annual Report, "... hence I am ignorant of what the Observatory has really done for astronomical science."
- The published astronomical work of the Observatory over the past twenty years is "... quite unworthy of such a highly-equipped institution."
- The Observatory could have made a significant contribution to international astronomy by producing a major catalogue of southern stars, but the Cape Observatory "... has forestalled us of the honour ...".
- Doubtless much observational work has been accomplished at the Observatory since its founding, but what use is this if it is left unpublished.

Tebbutt's concluding paragraph is particularly interesting:

> In conclusion, I may say that by far the greater force of the Sydney Observatory is devoted to the cause of meteorology, to the obvious disadvantage of the sister science. That the efforts of the establishment on behalf of meteorology are praiseworthy and effective no one knows better than myself, and the results will become more and more valuable as the years roll on; but I would suggest the high importance of publishing all that has been done there for astronomy, and also of separating the two departments, so that while the present efforts on behalf of meteorology are continued, a better contribution may be made for the improvement of astronomy.

Supporting the "letter" are additional details of the meteorology–astronomy imbalance, on the basis that although there are those whose meteorological thirst is quenched by the Observatory's weather maps and rain tables, "... there are others who have an astronomical thirst which has equal claims to be satisfied ..." (*ibid.*). Tebbutt then discusses the Observatory's publications since 1862 and points out that

> ... while one linear inch of horizontal space on a library shelf is sufficient for the accommodation of astronomy, about twelve times that space is required for meteorology.

He then compares the published astronomical outputs of the Sydney and Melbourne Observatories, reminding his readers that "It is very well-known that in European astronomical circles the Melbourne Observatory has long been regarded as the chief astronomical establishment in Australia ...". Furthermore,

> Considering its profuse publication of meteorological work, its non-publication of its annual astronomical results, and its sparse communications to the astronomical journals, is it to be wondered at that the [Sydney] Observatory is regarded as chiefly a meteorological establishment, and therefore placed in the second class?

Despite his obvious bias, some of Tebbutt's claims were based on fact. For example, Table 10 lists publications by Russell in astronomy, meteorology and other topics at five-year intervals from 1871 to 1900, and his increased focus on meteorology during the 1880s

Table 10 Henry Russell's publications in astronomy and other disciplines, 1871–1900 (after Orchiston 1988b)

Discipline	Percentage of total publications					
	1871–75	1876–80	1881–85	1886–90	1891–95	1896–1900
Astronomy	63	56	40	23	66	0
Meteorology	26	41	40	68	28	89
Other	11	3	20	9	6	11
Total Publns	19	32	20	22	35	9

is obvious (as is his subsequent return to astronomy during the 1890s). However, it should be borne in mind that a number of the astronomical publications listed actually reported the work of subordinate staff but appeared in Russell's name, a practice which was also prevalent overseas at the time but caused considerable resentment.

What Tebbutt did not consider was the possibility that Russell's meteorological bias at this time may have been warranted. The Observatory was under pressure from politicians, farmers and even the general public to provide meteorological data, and in some cases rural livelihoods depended upon accurate weather forecasts. Sir Charles Todd (1893) from the Adelaide Observatory elaborates:

> That meteorology should have been taken up so energetically [in Australia] and been so liberally supported by the several Colonial Governments, on whose purse, in building up a new nation, there are many claims, is not, however, without a sufficient cause. To successfully occupy and establish industries in new countries, a knowledge of climate and meteorological conditions under which we are to labour is essential to success . . .

In this context, Russell's bias towards meteorology was thoroughly understandable, but ironic given his stated intentions when assuming the post of Government Astronomer in 1870. At that time, he lamented the inordinate amount of time and effort that his predecessor Smalley had put into meteorology and the trigonometric survey of New South Wales, pledging that ". . . I should not like to have the credit of doing as little for Astronomy as he did." (Russell 1870a).

Tebbutt also was well aware that in the nineteenth century most so-called "astronomical" observatories were also involved in trigonometrical surveys and in a range of other scientific disciplines, and in focussing his attack on just astronomy and meteorology he conveniently forgot to mention that staff at the Sydney Observatory were also engaged in geomagnetic and tidal studies. At least Russell had been wise enough to jettison the trigonometrical survey of the colony when he became Government Astronomer of New South Wales (see Orchiston 1987a).

Returning to the chronological narrative: it is apparent that Tebbutt's original letter on its own was far less critical than the pamphlet in dealing with the astronomy–meteorology dichotomy, and had it been allowed to appear in the *Sydney Morning Herald* less damage would have been done and the matter may quickly have died – at least as a public issue.

Instead, it festered for years as the pamphlet found its way onto the desks of Australian and overseas astronomers, and into the libraries of politicians, academics and others of influence. In this way, the Observatory's perceived "ills" reached a wide audience.

As a consequence, many Australian and overseas astronomers wrote to Tebbutt in support of his stand, including Melbourne Observatory's Ellery (1891) who felt that the pamphlet

> . . . will stir up our friend to do a little more standard astrl work and print it. At the same time I am afraid these strictures worry him a bit.
>
> I have a very strong opinion about meteorological work as done at present, and the amount of time and money spent on printing volume upon volume of base observations in various parts of the world is appalling, and the amount of space already required to store these books is equally alarming.

While Tebbutt's brash, frontal attack certainly did not enthuse Russell, it did at least sting him into action on both the observational and publication fronts, thereby achieving the desired effect.

Thus, in 1892 the Government somewhat belated published an attractively-bound and illustrated book about observations of the 1874 transit of Venus made in New South Wales (Russell 1892b). This volume contained more than 70 pages of text and 40 pages of plates, but not one word of the Windsor Observatory, even though Tebbutt not only recorded the event but also published accounts in *Astronomische Nachrichten* and the *Memoirs of the Royal Astronomical Society* (Tebbutt 1875, 1883). All this was well known to Russell, who had even congratulated Tebbutt on his achievement at the time (see Russell 1874b)!

Tebbutt (1893) responded by submitting a long letter of complaint to the *Australian Star*, and this matter soon became international news when it entered the pages of *The Observatory* (Observatories . . . 1893). Russell (1894b) was quick to reply, but the Editors concluded that ". . . after reading his [Russell's] letter, and looking again carefully at the volume in question, we sympathize with Mr. Tebbutt." (*ibid.*). White, from the Melbourne Observatory, wrote to Tebbutt about the "shabby treatment" that he had received from Russell, reassuring him:

> . . . now the subject has been pointedly referred to, you need not worry yourself as to the result. From the experience I have gained during my travels in Europe & America I should infer that Tebbutt was a better known name in the astronomical world than Russell, and I frequently hear surprise that such a quantity of first class work should be done by a private individual in such an out of the way corner of the globe.
>
> It is a pity that Russell is so narrow (in mind I mean, not body) he shows it in many petty ways. (White 1893).

Others wrote in a similar vein.

In yet another snub, Russell purposely omitted any mention of Tebbutt in his "Presidential Report" before the Royal Society of New South Wales in 1892. Although four and a half pages of the published version were devoted to astronomical activities in New South Wales (Russell 1892c), to all intents and purposes Windsor Observatory did not exist!

In order to re-vitalise the Observatory's flagging astronomical interests, Russell became immersed in the Carte du Ciel Project (see Debarbat *et al.* 1988; Gingerich 1992: Chapter 25), and in 1890 September the objectives were received from Howard Grubb (Wood 1958) and the 33 cm Sydney Astrograph (with 26 cm guidescope) was subsequently constructed. Russell (1892a) then published a popular book about the "Sydney Star Camera" (as he called it), but some local astronomers were appalled that public taxes were used to produce this – and the transit of Venus volume – when there was such a backlog of astronomical observations awaiting publication. The South African astronomer, A.W. Roberts (1893), facetiously referred to "... men who issue "picture books" and "Reminiscences" instead of scientific reports...", while *The Observatory* also referred to the transit of Venus book as "... a delightful picture book ... which seems to be a little late in coming." (Notes 1893).

Further friction between Russell and the local astronomical community surfaced when Gale discovered comet C/1894 G1 and Russell's media release led *The Age* newspaper in Melbourne to assign him the credit. The comet also created problems for other staff at the Sydney Observatory, with Sellors prevented from computing the orbital elements (based on his own observations) in official time or publishing them. Subsequently, Russell instructed his staff that no information whatsoever was to be supplied to public enquirers (but particularly any of the local amateur astronomers) without his permission (Sellors 1894). Tebbutt objected to this, but Russell (1894a) was unrepentant.

This deteriorating relationship meant that when the New South Wales Branch of the British Astronomical Association was formed in Sydney at the end of 1894 Russell was passed over as inaugural President and the post was offered to Tebbutt (Orchiston 1988c).

In 1894 and 1895 a severe drought in New South Wales was preoccupying the lives and minds of people, and the efficacy of meteorological forecasts became a topic of public debate, as did prayers for rain. Understandably, it did not take long for Russell and the Sydney Observatory to be drawn into the fray, as evidenced by an article which appeared in the *Windsor and Richmond Gazette* of 1894 September 1 which reads, in part:

> Weather-wise Wragge is really a splendid weather prophet; and though our own Russell don't speak very plainly, it is clear that he is mighty jealous of the Seer of Banana-land. Russell is a jealous person, anyhow; he feared – and therefore hated – poor Egeson – small pumpkins as he was compared to Wragge – and now he sneers at the latter's deductions. The fact is our man is all behind the times – even in his own particular department – that is meteorology. As an astronomer he is infinitely inferior to ... Mr John Tebbutt, of Windsor, who can run rings round the Fort Street man ... Anyhow, Russell has had a pretty long innings; and as he is certainly not up to date, either in astronomy or meteorology, it is about time we paid our money to a first-class man. (Mr. Astronomer Tebbutt 1894).

Clement Wragge was the Government Meteorologist of Queensland, and was popular with the public but was not with the nation's Government Astronomers (see Home and Livingston 1994). Meanwhile, a similar article to the above appeared some months later in the *Redfern Chronicle* newspaper in Sydney and was subsequently reprinted elsewhere. This included the following perceptive comments:

> ... Russell had made it a study – in fact, he was more a meteorologist than an astronomer. And now we have that wretched Wragge saying he isn't. So that our

> Government Astronomer seems just now to be neither one thing nor t'other. He is in a state of suspension . . . Whatever the man is he holds fast to his billet and regularly draws his six or seven hundred a year for doing something or other upon Observatory Hill – among other things being his garden parties to other distinguished Civil Servants and pensioners . . . . No one here attempts to refute Mr. Tebbutt, while even the amateurs we have with us worry Russell's calculations. Of course he wiped out Egeson; but it will require a stronger man to wipe out either Wragge or Tebbutt. (Tebbutt, Wragge, and Russell 1895).

Both letters indicate the degree of public feeling against Russell in some camps, but they also illustrate the power of the press at a time when libel laws were obviously somewhat more lax than they are today. Russell could hardly have relished "trial by media".

Russell took ill with Bright's disease in 1903 October, and went on leave, finally retiring on 1905 February 28. His successor, as Acting Government Astronomer, was H.A. Lenehan, who inherited the controversy over the Observatory's output, but took a somewhat different stance to that adopted by his predecessor. In 1904 he wrote Tebbutt:

> . . . I sincerely trust that if I can be of any service to you or others in observing or doing any thing in my power to serve the ends of scientific men I will have the chance of doing it . . . (Lenehan 1904a).

As a further indication of his attitude, in another letter written this same year, Lenehan (1904b) marks Tebbutt's retirement by congratulating him on a lifetime of achievement in astronomy and meteorology, and concludes:

> I am personally pleased that our intercourse has been so pleasant and in my humble way have ever born testimony to your work as a scientist and friend – a friendship I hope will be continued for many years.

The letter was signed, ". . . your sincere friend and admirer."

Nor did Russell's departure from the Observatory protect him from continued criticism. Thus, on 1906 January 30 Merfield wrote:

> Really I think that one of our recent Govt. astronomers should have been paying into the Govt. funds instead of receiving thereof for the honour of the title he used, as I dont consider that he has advanced the science of astronomy in this state one bit; he has left nothing of value behind him so far as I can see. (Merfield 1906).

By at this stage Merfield was in an excellent position to "see" first hand, for he was employed at the Sydney Observatory, and later he was to describe the equipment there as

> . . . a scrap heap of old iron and brass, for the most part. One good instrument remains, the meridian circle . . .
>
> The late Govt. Astronomer Mr. Russell had some hazy idea that he could improve on the construction of astronomical instruments from such firms as Cooke, Grubb etc etc . . . (Merfield 1909).

As a result, the 29.2 cm Schroeder refractor was a ". . . useless piece of mechanism . . ." (*ibid.*), and Lenehan (1907a) had earlier mentioned that another of the telescopes was non-functional because Russell had dismembered it. To make matters worse, it was soon discovered that there were problems with the Carte du Ciel project in that the settings of more than half of the plates were in error (Merfield 1908).

Russell died on 1907 February 22, and these adverse opinions of him soon reached the ears of his relatives. On 1907 September 30, Lenehan wrote Tebbutt that he had recently met the son, Mr. T. Russell, who "... was boiling at some remarks that he had heard of the unscientific work of his late esteemed father ..." (Lenehan 1907b). Lenehan went on point out that "... the family had a very exalted opinion of his attainments which his work left behind does not justify." This is a useful assessment, coming as it did from one who knew and worked with Russell for many years. Less than two weeks later Lenehan wrote that Russell's widow

> ... is imbued with the false notion that all the animus that existed between yourself & HCR emanated from you. I immediately told her that the trouble was caused in the early stages by HCR ignoring you in his publication on the astronomers & workers in Australia. Unfortunately they look on the goings & writings of the departed through spectacles tinted with great prejudice. (Lenehan 1907c).

We have seen that in reality the breakdown of amiable relations between the two men was a long drawn out affair, which commenced well before the appearance of Russell's 1888 paper (although this certainly served to accelerate the process).

It is equally clear that this relationship deeply upset and irritated Tebbutt. Within a year of Russell's death, he wrote in his *Astronomical Memoirs*:

> Since 1861 I have scarcely ever been directly attacked on astronomical matters; the main opposition locally to my efforts has been by the method of ignoring, a mode of opposition which is difficult to meet. (Tebbutt 1908: 28).

In 1908, one of Tebbutt's aims – the separation of astronomy and meteorology at the Sydney Observatory – was eventually achieved when the Federal Government took over responsibility for the meteorological functions of the state observatories. However, this turned out to be a two-edge sword, for with the loss of funding that had been tied to meteorology the financial survival of these observatories would sooner or later become an issue. Thus, when the Public Service Board carried out an investigation "Into the Working of the Sydney Observatory" in 1909, the research output of that institution was still of concern, and comparisons were still being made with the Windsor Observatory (see Public Service Board 1909) even though Tebbutt had officially retired as an active astronomer several years earlier!

4. Concluding Remarks

For the last two decades of the nineteenth century, the Russell–Tebbutt feud permiated New South Wales astronomy. Underlying it was an on-going quest for power and status, and local astronomical supremacy. Because of the environment in which this occurred, Russell was always going to be the victor, and he worked tirelessly to first establish and then consolidate his position and once entrenched brooked no competition. As Joseph Brooks, the Lands Department's Field Astronomer, so facetiously stated: "... Mr Russell does not like anyone to be 'near the throne'." (Brooks 1897).

In this regard, we are justified in asking whether Russell abused his position to deny Tebbutt natural justice, as the Windsor astronomer has inferred. It is notable that Tebbutt's remarkable solo achievements were never formally recognised by the University, the Royal Society of New South Wales or the Government of New South Wales, and Russell's role in this must be questioned. If anyone had wished to propose Tebbutt for a University of Sydney honorary degree Russell would have been the obvious referee – with the power of veto. Likewise he wielded comparable power through the Royal Society, and as one of the Government's most senior scientists. Although Tebbutt's work was widely recognized and applauded internationally during his lifetime, it was only with his appearance on the new Australian $100 bank note in 1984 that his role in Australian and international science was finally given formal local recognition.

5. Acknowledgements

I am grateful to the following for their assistance in various ways: Dr Richard Dodd and Mike Reid (Carter Observatory); Peter Hingley (RAS Library); Professor Rod Home (University of Melbourne); Dr Nick Lomb (Sydney Observatory); and staff of the Mitchell Library (Sydney). Finally, I should like to thank the IAU for providing a travel grant which made it possible for me to attend the Kyoto General Assembly, where an earlier version of this paper was presented.

6. References

The following abbreviation is used:
TL = *Letters to John Tebbutt*. Mitchell Library, Sydney.

Ashbrook, J., 1984. *The Astronomical Scrapbook. Skywatchers, Pioneers, and Seekers in Astronomy*. Cambridge (Mass.), Sky Publishing Corporation.

Baracchi, P., 1889. Spectra of southern stars observed at the Melbourne Observatory with the McClean direct-vision spectroscope attached to the south equatorial. *Monthly Notices of the Royal Astronomical Society* **50**: 66–71.

——, 1914. Astronomy and geodesy in Australia. **In** Knibbs,G.H. (ed.). *Federal Handbook Prepared in Connection with the Eighty-fourth Meeting of the British Association for the Advancement of Science, Held in Australia, August, 1914*. Melbourne, Government Printer. Pp.326–390.

Barnard, E.E., 1898. The development of photography in astronomy. *Popular Astronomy* **6**: 425–455.

Bhathal, R., 1991. Henry Chamberlain Russell – astronomer, meteorologist and scientific entrepreneur. *Journal and Proceedings of the Royal Society of New South Wales* **124**: 1–21.

——, 1993. *Australian Astronomer John Tebbutt. The Life and World of the Man on the $100 Note*. Sydney, Kangaroo Press.

Brooks, J., 1897. Letter to John Tebbutt, dated 28 April. In TL.

Burk, K., 1986. A history of James Oddie, Captain Henry Evans Baker and the Ballarat Municipal Observatory. *The Oddie-Baker Bull*. **8**: 3–15.

Davis, M.E., 1990. *Henry Evans Baker. A Brave, Able, Original Man*. Brisbane, printed for the author.

Debarbat, S., Eddy, J.A., Eichhorn, H.K., and Upgren, A.R. (eds.), 1988. *Mapping the Sky. Past Heritage and Future Directions*. Dordrecht, Kluwer.

Dodwell, G.F., 1913. *Physical Observations of Jupiter Made at the Adelaide Observatory ... During the Years 1884–1893*. Adelaide, Government Printer.

Earth Tremors. Proceedings of a Seminar Commemorating the Centenary of Instrumental Seismology in Australia. Launceston, Royal Society of Tasmania, Northern Branch (1984).

Edwards, P.G., 1993. Charles Todd and the Adelaide Observatory. *Proceedings of the Astronomical Society of Australia* **10**: 349–354.

——, 1994. The Adelaide Observatory after Todd. *Proceedings of the Astronomical Society of Australia* **11**: 206–210.

Egeson, C., 1890. Letter to John Tebbutt, dated 29 March. In TL.

Eleventh Report of the Board of Visitors to the Observatory, With the Annual Report of the Government Astronomer. Melbourne, Government Printer (1876).

Ellery, R.L.J., 1869. *Astronomical Observations Made at the Williamstown Observatory, in the Years 1861, 1862, and 1863*. Melbourne, Government Printer.

——, 1885. *Observations of the Southern Nebulae Made With the Great Melbourne Telescope from 1869 to 1885. Part 1*. Melbourne, Government Printer.

——, 1889. Preliminary spectroscopic survey of southern stars, made at Melbourne Observatory with a McClean direct-vision spectroscope on the 8-inch equatorial. *Monthly Notices of the Royal Astronomical Society* **49**: 439–445.

——, 1891. Letter to John Tebbutt, dated 29 October. In TL.

——, 1901. A brief history of the beginnings of astronomy in Australia. **In** *Report of the Eighth Meeting of the Australasian Association for the Advancement of Science*. Melbourne, AAAS. Pp.1–17.

Gascoigne, S.C.B., 1992. Robert L.J. Ellery, his life and times. *Proceedings of the Astronomical Society of Australia* **10**: 170–176.

——, 1995. The Great Melbourne Telescope and other 19th century reflectors. *Historical Records of Australian Science* **10**: 223–245.

Gingerich, O. (ed.), 1992. The Great Comet and the "Carte". **In** *The Great Copernicus Chase and Other Adventures in Astronomical History*. Cambridge (Mass.), Sky Publishing Corporation. Pp. 189–194.

Giordano, M., 1995. *Watcher of the Skies. The Story of Alfred Barrett Biggs. Pioneer in Astronomy, Seismology and Telecommunications*. Launceston, Regal Publications.

Griffin, T.H.F., 1881. Letter to John Tebbutt, dated 5 April. In TL.

Haynes, R.D., Haynes, R.A., and Kitson, W.S., 1993. The history of astronomy in Queensland. *Vistas in Astronomy* **36**: 231–252.

Haynes, R., Haynes, R., Malin, D., and McGee, R., 1996. *Explorers of the Southern Sky. A History of Australian Astronomy*. Cambridge, Cambridge University Press.

Hearnshaw, J.B., 1986. *The Analysis of Starlight. One Hundred and Fifty Years of Astronomical Spectroscopy*. Cambridge, Cambridge University Press.

Hetherington, N.S., 1976. Amateur versus professional. The British Astronomical Association and the controversy over canals on Mars. *Journal of the British Astronomical Association* **86**: 303–308.

Hoare, M.E., 1975. The intercolonial science movement in Australasia 1870–1890. *Records of the Australian Academy of Science* **3**: 7–28.

Home, R.W., and Livingston, K.T., 1994. Science and technology in the story of Australian federation: the case of meteorology, 1876–1908. *Historical Records of Australian Science* **10**: 109–127.

Hutchison, D., 1980. William Ernest Cooke: first Government Astronomer. *Early Days* **8**: 93–114.

——, 1981. William Ernest Cooke, astronomer 1863–1947. *Historical Records of Australian Science* **5**: 58–77.

Hyde, W.L., 1987. The calamity of the Great Melbourne Telescope. *Optic News* **13**: 6–11.

Innes, R.T.A., 1899. Reference catalogue of southern double stars. *Annals of the Cape Observatory* **2**(2).

Lankford, J., 1979. Amateur versus professional: the transatlantic debate over the measurement of Jovian longitude. *Journal of the British Astronomical Association* **89**: 574–582.

——, 1981a. Amateurs and astrophysics: a neglected aspect in the development of a scientific speciality. *Social Studies in Science* **11**: 275–303.

——, 1981b. Amateurs versus professionals: the controversy over telescope size in late Victorian science. *Isis* **72**: 11–28.

——, 1984. The impact of photography on astronomy. **In** Gingerich: 16–39.

Lenehan, H.A., 1904a. Letter to John Tebbutt, dated 19 January. In TL.

——, 1904b. Letter to John Tebbutt, dated 13 September. In TL.

——, 1907a. Letter to John Tebbutt, dated 15 September. In TL

——, 1907b. Letter to John Tebbutt, dated 30 September. In TL.

——, 1907c. Letter to John Tebbutt, dated 11 October. In TL.

MacLeod, R., 1988. *The Commonwealth of Science: ANZAAS and the Scientific Enterprise in Australasia, 1888–1988*. Melbourne, Oxford University Press.
Maiden, J.H., 1918. A contribution to the history of the Royal Society of New South Wales, with information in regard to other New South Wales societies. *Journal and Proceedings of the Royal Society of New South Wales* **52**: 215–361.
"Many Happy Returns." Mr. John Tebbutt, F.R.A.S., who celebrates the eighty-first anniversary of his birthday to-day. *Daily Telegraph*, 25 May 1915.
Marsden, B.G., and Williams, G.V., 1996. *Catalogue of Cometary Orbits 1996*. Cambridge (Mass.), Minor Planet Center.
Masters, W.,1887. Letter to John Tebbutt, dated 9 May. In TL.
Merfield, C.J., 1906. Letter to John Tebbutt, dated 30 January. In TL.
——, 1908. Letter to John Tebbutt, dated 5 October. In TL.
Meteorological Department. The Observatory. **In** *Cyclopedia of Tasmania* ... Hobart, Maitland and Krone (1900).
Mr. Astronomer Tebbutt. *Windsor and Richmond Gazette*, 1 September 1894.
Norman, D., 1938. The development of astronomical photography. *Osiris* **5**: 560–594.
Notes. *The Observatory* **16**: 367 (1893).
Observatories. Mr. Tebbutt's, N.S. Wales. *The Observatory* **16**: 361–362 (1893).
Orchiston, W., 1981. Illuminating incidents in Antipodean astronomy: John Tebbutt and the Great Comet of 1881. *Journal of the Astronomical Society of Victoria* **34**: 78–92.
——, 1982a. Illuminating incidents in Antipodean astronomy: John Tebbutt and the abortive Australian association of comet observers. *Journal of the Astronomical Society of Victoria* **35**: 70–83.
——, 1982b. John Tebbutt and the Whakatane Eight Inch Refractor: a review of "The Australian Connection". *Southern Stars* **29**: 215–244.
——, 1985. The contribution of Alfred Barrett Biggs to Australian and Tasmanian astronomy. *Records of the Queen Victoria Museum* **89**: 1–28.
——, 1986. *Observatories of Victoria. 1: Dr William Bone and the Castlemaine Observatory*. Melbourne, Astronomical Society of Victoria (Historical Notes No. 3). Pp. 17.
——, 1987a. Contribution of the Lands Department to the development of astronomy in New South Wales during the nineteenth century. *Australian Journal of Astronomy* **2**: 65–74.
——, 1987b. Dr William Bone, and the role of the amateur observatory in Australian astronomy. *Southern Stars* **32**: 111–128.
——, 1988a. The emergence of professional astronomy in nineteenth century Australia. *Search* **19**: 76–81.
——, 1988b. From research to recreation: the rise and fall of Sydney Observatory. *Vistas in Astronomy* **32**: 49–63.
——, 1988c. Illuminating incidents in Antipodean astronomy: formation of the New South Wales Branch of the British Astronomical Association. *Journal of the British Astronomical Association* **98**: 75–84.
——, 1988d. Illuminating incidents in Antipodean astronomy: John Tebbutt and the founding of the Windsor Observatory. *Journal of the British Astronomical Association* **98**: 287–293.
——, 1988e. Illuminating incidents in Antipodean astronomy: John Tebbutt and the Sydney Observatory directorship of 1862. *Australian Journal of Astronomy* **2**: 149–158.
——, 1988f. Illuminating incidents in Antipodean astronomy: Phillip Parker King and the founding of Sydney Observatory. *Vistas in Astronomy* **32**: 285–301.
——, 1988g. John Tebbutt: an astronomer of note. *Astronomy Now* **2**(5): 11–16.
——, 1989a. Astrophysics and the evolution of twentieth century Australian astronomy. *Search* **20**: 198–202.
——, 1989b. The role of the amateur in early Australian astronomy. *Search* **20**: 16–23.
——, 1991a. Amateur-professional co-operation in astronomy education: a nineteenth century Australian model. *Proceedings of the Astronomical Society of Australia* **9**: 186–188.
——, 1991b. Comet Tebbutt 1881III. *The Observatory* **111**: 313–314.
——, 1992. The contribution of Francis Abbott to Tasmanian and Australian astronomy. *Vistas in Astronomy* **35**: 315–344.

——, 1997a. The role of the amateur in popularising astronomy: an Australian case study. *Australian Journal of Astronomy* **7**: 33–66.

——, 1997b. The role of the large refracting telescope in Australian amateur astronomy: an historical perspective. *Australian Journal of Astronomy* **7**: 89–114.

——, 1997c. The "tyranny of distance" and Antipodean cometary astronomy. *Australian Journal of Astronomy* **7**: 115–126.

——, 1998a. Amateur-professional collaboration in Australian science: the earliest astronomical groups and societies. *Historical Records of Australian Science* **12**: 163–182.

——, 1998b. Illuminating incidents in Antipodean astronomy: John Tebbutt and the Great Comet of 1861. *Irish Astronomical Journal* **25**: 167–178.

——, 1998c. Mission impossible: William Scott and the first Sydney Observatory directorship. *Journal of Astronomical History and Heritage* **1**: 21–43.

——, 1998d. Power and politics in nineteenth century Australian astronomy, in Andersen, J. (Ed.). *Highlights of Astronomy as Presented at the XXIIIrd General Assembly of the IAU*. Dordrecht, Kluwer, pp. 741–743.

——, 1999a. C/1881 K1: A forgotten "Great Comet" of the nineteenth century. *Irish Astronomical Journal* **26**: 33–44.

——, 1999b. Comets and communication: amateur-professional tension in Australian astronomy. *Publications of the Astronomical Society of Australia* **16**: 212–221.

——, 2000a. Illuminating incidents in Antipodean astronomy: H.C. Russell and the origin of the Palomar-type mounting. *Journal of the Antique Telescope Society*, **19**: 13–15.

——, 2000b. John Tebbutt of Windsor, New South Wales: a pioneer southern hemisphere variable star observer. *Irish Astronomical Journal* **27**: 47–54.

——, 2001. William John Macdonnell and the development of astronomy in New South Wales. *Journal of the British Astronomical Association*, **111**: 13–25.

——, and Bembrick, C., 1995. The role of the large reflecting telescope in amateur astronomy: an Australian case study. *Australian Journal of Astronomy* **6**: 53–74.

——, and ——, 1997. The potency of patronage: George Hoskins and the New South Wales Branch of the British Astronomical Association. *Australian Journal of Astronomy* **7**: 1–15.

——, and Bhathal, R., 1982. Illuminating incidents in Antipodean astronomy: the saga of "The Great Sydney Telescope". *British Astronomical Association, New South Wales Branch Bulletin* **612**(Supplement): S1–S9.

——, and ——, 1984. Illuminating incidents in Antipodean astronomy: attempted formation of the Australian Astronomical Society in 1892. *Proceedings of the Astronomical Society of Australia* **5**: 422–425.

——, and ——, 1991. Illuminating incidents in Antipodean astronomy: Section A of the Royal Society of New South Wales. *Australian Journal of Astronomy* **4**: 49–67.

——, and Brewer, A., 1990. David Ross and the development of amateur astronomy in Victoria. *Journal of the British Astronomical Association* **100**: 173–181.

——, and Buchanan, A., 1993. Illuminating incidents in Antipodean astronomy: Campbell Town, and the 1874 transit of Venus. *Australian Journal of Astronomy* **5**: 11–31.

——, and Perdrix, J., 1990. The British Astronomical Association in Australia: a centennial perspective. *Australian Journal of Astronomy* **3**: 163–179.

Perdrix, J., 1961. The Melbourne Observatory. *Journal of the Astronomical Society of Victoria* **14**: 19–25, 45–48, 60–64.

——, 1970. The Great Melbourne Telescope. *Journal of the Astronomical Society of Victoria* **23**: 54–67.

——, 1990. Flagstaff Hill and its observatory. *Australian Journal of Astronomy* **3**: 113–124.

——, 1992. The last great speculum: the 48-inch Great Melbourne Telescope. *Australian Journal of Astronomy* **4**: 149–163.

Pitt, G.W.M., 1880. Letter to John Tebbutt, dated 18 August. In TL.

Pollock, J.A., 1887. Letter to John Tebbutt, dated 2 May. In TL.

Pritchard, C., 1886. Letter to John Tebbutt, dated 21 January. In TL.

Public Service Board,1909. Investigation Under Section 9 of the Public Service Act Into the Working of the Sydney Observatory. MS, Archives Office of New South Wales, Sydney (8/400.2).

Report of the New South Wales Branch (Sydney). *Journal of the British Astronomical Association* **5**: 344–345 (1895).

Richardson, W., 1835. *A Catalogue of 7385 Stars, Chiefly in the Southern Hemisphere, Prepared from Observations Made . . . at the Observatory at Paramatta [sic.] . . .* London, His Majesty's Stationery Office.
Roberts, A.W., 1893. Letter to John Tebbutt, dated 28 November. In TL.
Robinson, T.R., and Grubb, T., 1869. Description of the Great Melbourne Telescope. *Philosophical Transactions of the Royal Society* (London) **159**: 127–161.
Rothenberg, M., 1981. Organisation and control: professionals and amateurs in American astronomy, 1899–1918. *Social Studies in Science* **11**: 305–325.
Russell, H.C., 1870a. Letter to John Tebbutt, dated 27 October. In TL.
——, 1870b. Letter to John Tebbutt, dated 3 December. In TL.
——, 1871. Meteorology in New South Wales. **In** *The Industrial Progress of New South Wales: Being A Report of the Intercolonial Exhibition of 1870, at Sydney; Together With a Variety of Papers Illustrative of the Industrial Resources of the Colony*. Sydney, Government Printer. Pp. 579–615.
——, 1872. Letter to John Tebbutt, dated 24 April. In TL.
——, 1874a. Letter to John Tebbutt, dated 7 March. In TL.
——, 1874b. Letter to John Tebbutt, dated 15 December. In TL.
——, 1875. Letter to John Tebbutt, dated 10 February. In TL.
——, 1877. Letter to John Tebbutt, dated 26 September. In TL.
——, 1881a. Letter to John Tebbutt, dated 25 May. In TL.
——, 1881b. Spectrum and appearance of the recent comet. *Journal and Proceedings of the Royal Society of New South Wales* **15**: 81–86.
——, 1886. Letter to John Tebbutt, dated 20 April. In TL.
——, 1887a. Letter to John Tebbutt, dated 15 January. In TL.
——, 1887b. Letter to John Tebbutt, dated 29 December. In TL.
——, 1888. Astronomical and meteorological workers in New South Wales, 1778 to 1860. **In** Liversidge, A., and Etheridge, R. (eds.). *Report of the First Meeting of the Australasian Association for the Advancement of Science . . .* Sydney, AAAS. Pp.45–94.
——, 1890a. On some celestial photographs recently taken at Sydney Observatory. *Monthly Notices of the Royal Astronomical Society* **51**: 39–43.
——, 1890b. *Photographs of the Milky-Way and Nebeculae at Sydney Observatory in 1890*. Sydney, Government Printer.
——, 1891. Notes on some star photographs recently taken at Sydney Observatory. No. 2. *Monthly Notices of the Royal Astronomical Society* **51**: 494–497.
——, 1892a. *Description of the Star Camera, at the Sydney Observatory*. Sydney, Government Printer.
——, 1892b. *Observations of the Transit of Venus, 9 December, 1874; Made at Stations in New South Wales*. Sydney, Government Printer.
——, 1892c. *Presidential Address*. Sydney, Royal Society of New South Wales.
——, 1894a. Letter to John Tebbutt, dated 12 November. In TL.
——, 1894b. Mr. Tebbutt and the transit of Venus, 1874. *The Observatory* **17**: 81–82.
Saunders, S.D., 1990. Astronomy in Colonial New South Wales: 1788 to 1858. Unpublished Ph.D. Thesis, Department of General Philosophy, University of Sydney.
Sellors, R.P., 1894. Letter to John Tebbutt, dated 5 November. In TL.
Smyth, P. 1887. Letter to John Tebbutt, dated 3 May. In TL.
Sydney Morning Herald, 10 September 1891.
Symes, G.W., 1976. Todd, Sir Charles (1826–1910). **In** Serle, G., and Ward, R. (eds.). *Australian Dictionary of Biography. Volume 6: 1851–1890. R–Z.* Melbourne, Melbourne University Press. Pp. 280–282.
Tebbutt, J., 1866. On the progress and present state of astronomical science in New South Wales. **In** *Catalogue of the Natural and Industrial Products of New South Wales. Forwarded to the Paris Universal Exhibition of 1867, by the New South Wales Exhibition Commissioners*. Sydney, Government Printers. Appendix. Pp. 55–64.
——, 1871. On the progress and present state of astronomical science in New South Wales. **In** *The Industrial Progress of New South Wales: Being A Report of the Intercolonial Exhibition of 1870, at Sydney; Together With a Variety of Papers Illustrative of the Industrial Resources of the Colony*. Sydney, Government Printer. Pp. 617–632.

——, 1875a. Beobachtung des Venusdurchganges in Windsor (N.-S.-Wales). *Astronomische Nachrichten* **85**: 173–176.
——, 1875b. The transit of Venus. *Sydney Morning Herald*, 16 January.
——, 1883. Observations of the transit of Venus, 1874, December 8–9, made at Windsor, N.S. Wales. *Memoirs of the Royal Astronomical Society* **47**: 89–92.
——, 1887. *History and Description of Mr Tebbutt's Observatory*. Sydney, printed for the author.
——, 1888. On the difference of longitude between Mr. Tebbutt's Observatory, Windsor, New South Wales, and the Government Observatories at Sydney and Melbourne. *Monthly Notices of the Royal Astronomical Society* **48**: 313–316.
——, 1890. Astronomical and meteorological workers in N.S. Wales, 1778 to 1860. *The Observatory* **13**: 211–212.
——, 1891a. Our Observatory. *Australian*, 26 September.
——, 1891b. *The Sydney Observatory and the "Sydney Morning Herald". A Plea for Astronomy in New South Wales*. Sydney, printed for the author.
——, 1893. The Government Astronomer and the transit of Venus. *Australian Star*, 20 June.
——, 1908. *Astronomical Memoirs . . .* Sydney, printed for the author.
Tebbutt, Wragge, and Russell. *Windsor and Richmond Gazette*, 21 September 1895.
Thompson, J.P., 1889. Letter to John Tebbutt, dated 2 March. In TL.
Todd, Sir Charles, 1893. Meteorological work in Australia: a review. **In** Tate, R., Rennie, E.H., and Bragg, W.H. (eds.). *Report of the Fifth Meeting of the Australasian Association for the Advancement of Science, Held at Adelaide, South Australia, September, 1893*. Sydney, AAAS. Pp.246–270.
Turner, H.H., 1912. *The Great Star Map. Being a General Account of the International Project Known as the Astrographic Chart*. London, Murray.
Utting, M., 1989. *Cooke's Perth Observatory*. Perth, Perth Observatory.
——, 1992. *Astronomy in Western Australia. Volume 1: 1896 to 1912*. Perth, Perth Observatory.
Walsh, G.P., 1976. Russell, Henry Chamberlain (1836–1907). **In** Serle, G., and Ward, R. (eds.). *Australian Dictionary of Biography. Volume 6: 1851–1890. R–Z*. Melbourne, Melbourne University Press. Pp. 74–75.
Waters, B., 1980. *A Reference History of the Astronomical Society of South Australia Inc. Volume 1. The First Decade 1891–1901*. Adelaide, printed for the author.
Weiderkehr, V.H., 1988. Georg Neumayer und die Grundung eines geophysikalisch-nautischen Observatoriums in Melbourne 1857. **In** Schroder, W. (ed.). *Past, Present and Future Trends in Geophysical Research*. Bremen-Roennebeck, International Commission on History of AIGA. Pp. 111–125.
Western Australian Astronomical Society, 1914. Organisation and membership. *Proceedings of the Western Australian Astronomical Society* **1**: 1–2.
White, E.J., 1889. Letter to John Tebbutt, dated 20 November. In TL.
——, 1893. Letter to John Tebbutt, dated 21 November. In TL.
White, G., 1979. John Tebbutt and the astronomy at the Windsor Observatory. *Proceedings of the Astronomical Society of Australia* **3**: 408–411.
Wood, H., 1958. *Sydney Observatory 1858 to 1958*. Sydney, Government Printer (Sydney Observatory Papers No. 31).

2.6. A New Museum of Astronomy in Korea

Nha Il-Seong
The Nha Il-Seong Museum of Astronomy Yonsei University

My family's long-held dream has come true. This dream was the construction of a museum of astronomy, and now it stands on a small hill top at Yechon Prefecture (λ = 128 deg 28 min E, φ = 36 deg 39 min N) about 200 km south of Seoul, Korea. The site of the museum, 13,200-square meters, is a part of a 35,640-square meter complex named '*Biol-ui-dong-san*', which means '*Hill of Stars*'. The rest of the complex will in the near future have lodgings for visitors. The museum is open to the public from 09:30 to 17:30 every day of the week except Tuesday.

The museum has an observatory with a 40-cm Ritchey Chretien telescope equipped with a photoelectric photometer. Long-term monitorings of the selected long period close interacting binary stars are carried out since 1982, and the observatory opens once a month for the visitors.

At present the museum has a collection of about 150 star maps and about 70 sundials. Star maps are displayed on the first floor of the exhibition hall *Humgyong-gak* and the sundials on the second floor, except for three heavy stone sundails which are on the first floor. These are classified as below.

Star Maps Old and New

a. Photographs of stars painted on walls of ancient graves: Many old graves in the East Asian countries in the period from the 3rd to the 8th century AD have paintings on walls and ceilings. Among those paintings are old star maps with four animals; blue dragon for the east, black tortoise for the north, white tiger for the west, and red bird for the south. Photographs of these paintings arranged in chronological order are useful for the understanding of the early formation of the oriental constellations.

b. Replicas of star maps from the ancient period to the 13th century AD: Old star maps in different civilisations, which are displayed in show-cases, represent the culture of each region. The collected replicas of these maps may help visitors in finding traces of the development of astronomical knowledge.

c. Reconstruction of the stone slab of the 1395-planisphere: In the fourth year (1395 AD) of the first king of the Choson dynasty in Korea, the central astral chart and its history with other inscriptions were engraved on both sides of a large black stone slab (211 cm $\times$ 123 cm $\times$ 12 cm). This stone slab is now exhibited in the Doksu Palace Museum in Seoul. However, due to damage from transportation and water erosion, the inscriptions are partly illegible. The present reconstruction of this historical stone chart offers a clear picture of making a star map in the 14th century (see Figure 2).

d. Rubbings of stone star maps available in China and Korea: Five stone star maps are known to exist; three in China and two in Korea. In order to make a copy of the inscription from the stone slab, a technique of rubbing a paper placing on the stone has been used.

S.M.R. Ansari (ed.), History of Oriental Astronomy, 203–207.

Figure 1 An overview of the museum. The front building is the *Humgyong-gak*, an exhibition hall for star maps and sundials, and the building behind is the observatory, which houses a 40-cm RC telescope.

Figure 2 Public lecture by the author at the star party in the summer of 1999. A large stone slab on a white stone support is the reconstructed stone slab of the 1395 planisphere.

The quality of old rubbings of these stone star maps depends on the dates and skill of the makers.

e. Collection of star maps printed or drawn on papers: Star map making in the old days was not an easy task. Some were printed in large quantities and others by individuals. Chronological arrangement of the collection in show-cases may be seen by visitors.

f. Collection of books in which star maps were printed: Old astronomical books of historical value in which star maps were included are a rare sight for most visitors.

g. Collection of ornamentations: For the most ordinary affairs of life, stars and heavenly symbols decorate ornaments everywhere in the world from old days to the present day. Collection of these, not many at present, will promote visitors' interest in astronomy.

h. Collections of new star maps of various regions: Presentation of star maps helps to guide visitors to where we are now. New star maps such as BD charts, SAO charts and other well-known star atlases, are relevant material for them.

Sundials

a. Photographs of old western and Arabic sundials: Western and Arabic sundials are questionable items for inclusion in the present collection. Therefore, except for a few, only photographs of these are displayed instead.

Figure 3 Several reconstructed stone sundials are investigated by professional visitors. These are displayed on the roof top of the exhibition hall.

b. A reconstruction of a 15th century bronze gnomon and photos of other gnomons: During the reign of the fourth king of the Choson dynasty in Korea, a large bronze gnomon was made and installed on the west side of Ganui-dae, a royal observatory, in Kyongbok Palace in Seoul in 1437. This gnomon was destroyed during the war against the Japanese invasion. A reconstruction, one tenth the size of the original (7 m 45 cm high) is now mounted vertically on a horizontal stone (2 m 61 cm long) on which scales are engraved. A few photographs of gnomons in other regions are also on display.

c. Replicas of sundials made from metals, woods, stones, etc.: A number of sundials made from different materials have been reproduced. Some of these are on display.

d. Rubbings of stone sundials: Large stone sundials are abundant in Korea. Some of these are represented with rubbings.

e. Collection of sundial books: A few old books on sundials written in Latin and German have been collected, along with old Korean and Chinese texts. Dedicated space for these is, however, unfortunately not available at the moment. These books and many other items such as calenders, posters, almanacs, etc. await the construction of an annex building.

Other Activities

Two programs have been created at the Museum; one is the star party and the other is regular public lectures.

a. The star party is designed for enthusiastic star gazers. So far two nights, August 6th and 13th, were attended by 220 and 450 participants, respectively. Each party was scheduled with an afternoon program, 14:00–17:00, and night-time program, 20:00–24:00. In the afternoon, participants visited the exhibition hall and watched solar spectrum, prominence and sunspots. These solar observations were carried out outside, where a Celestron telescope equipped with a thedeostat and a prominence adopter was set up temporarily.

b. There are two different regular public lectures. One is a primary course of four weeks, two hours on every Tuesday evening. Enrolment is limited to about 30 participants, who register in advance. So far about 110 registrations have been made though, about 30 percent of them dropped out.

The contents of this program are as follows:

1st week: A historical review of ancient formations of stellar constellations in two regions, Greece and the Far East.
2nd week: Guidance on how to read star atlases by using the equatorial coordinate system, circumpolar constellations, diurnal motion and corresponding time system.
3rd week: Use of binoculars. Finding the power and field size of binoculars on the star atlas. Outdoor practices both by the naked eye and with binoculars.
4th week: Observations of interesting stellar objects with the 40-cm RC telescope, and taking a group photo.

The second lecture series is the advanced course for those who finish the primary course and those who have equivalent experience. This is a ten-week course, 2–3 hours every Thursday. The first class of 12 is now enrolled.

The contents of these lectures are as following:

1st–4th weeks: Theory of optics and mechanics of cameras. Limiting magnitudes changing diaphragms of a camera. Angular size of a camera. Selection of films. Meteor shower observations. Practices and homework.
5th–10th weeks: Theory of telescopes, refractors and reflectors. Practice of setting a small telescope. Limiting magnitudes on a given night. Techniques of manual gliding for long-exposure photography. Observation techniques and selection of objects. Practices and homework.

2.7. Oriental Astronomy: History of East Asian Mathematical Astronomy Homage to K. Yabuuti and A. Sayili

Shigeru Nakayama

From my point of view, the first professional historians of Chinese mathematics, such as Yoshio Mikami, Li Yen and Ch'ien Pao-ts'ung, were not really writing the history of science. There was no independent disciplinary tradition of Chinese mathematics comparable to the Western Platonic and Euclidian tradition. Chinese mathematics has always been a handmaiden of science or technology, notably providing tools for calendrical astronomy. Mathematics for its own sake, as a kind of intellectual entertainment, existed only in the post-Seki tradition of Japanese mathematics. These historians did not attempt to reconstruct the Chinese tradition – its aims, ideals, presuppositions, values, and biasses – but rather selected out of its remains only the parts interesting from the modern mathematical point of view.

On the other hand, those who began the modern study of ancient Chinese astronomy, such as Henri Maspero, Shinjô Shinzô, Nôda Chûryô, and Chu K'o-chen, were indulging the Orientalist's antiquarian curiosity toward anything Chinese that was old and original, rather than the historian's drive to explain change. Neither of these two groups saw that Chinese exact science evolved as a continuous tradition of computing the ephemerides on which official calendars or almanacs were based. Yabuuti Kiyosi's work made up for this oversight. Instead of simply describing fascinating items from the past, he traced through the calendrical chapters of the dynastic histories the development of astronomical themes, mainly using the analytical tools of modern science.

When I visited Joseph Needham for the first time in 1957, I mentioned Yabuuti's work. Because Needham could not read the Japanese language, he was totally unaware of the discoveries that Yabuuti and his colleagues and students had made about the character of the tradition. The astronomical volume of *Science and Civilisation in China* was already at the stage of galley proofs. It was too late to revise the volume fundamentally to make use of Yabuuti's work. Lacking these insights, Needham simply dismissed calendrical astronomy, the foremost tradition of the exact science in East Asia, as dedicated to the solution of trivial problems.

He had been criticized still earlier, when the first two volumes appeared, because the Japanese contribution to the history of Chinese science was left out of his account. For the remainder of his career, he was aware of this shortcoming, and greatly concerned with ways to have Japanese works translated into English so that he could include them in the later volumes of his survey. He was never able to overcome the obstacles to scholarly translation, so that his writing was always limited by poor access to Japanese publications. Only in some of the volumes written by others could the best scholarship of the whole world be reflected.

In order to pursue the history of calendrical astronomy in my own way, I made a plan, in the 1960s, to work systematically on the solar motion, then on the lunar motion, and

S.M.R. Ansari (ed.), History of Oriental Astronomy, 209–212.

finally on eclipse predictions. Planetary motions were not part of this plan, because they were a peripheral issue in Chinese mathematical astronomy, I concentrated on the crowning achievement of the Chinese tradition, the Season-Granting or *Shou-shih* calendrical system that began in the Yuan dynasty (1279). I collected and examined thoroughly data on the accuracy of observational records of the times of solstices and equinoxes, Eastern and Western, corresponding to the Yuan and earlier times, and investigated possible sources of error.

A by-product was the discovery that Kuo Shou-ching's determination of the times of solstices and equinoxes, which are fundamental in setting up a computational system, was much more exact than earlier determinations in the West. Ptolemy, to fit his adopted value of tropical year length, seems to have cooked the Greek observations, introducing errors of the order of a day. The ninth-century Islamic values recorded by Ibn Yunis fall several hours to half a day late, while the Chinese figures fall within a quarter of an hour for the winter solstice and an hour for the summer solstice. This success is partly due to the giant 40-foot gnomon used, and to Kuo's extra-solstitial observations and reduction methods. It was partly due to the fortuitous circumstance that the solar perigee happened to coincide with the winter solstice ca. 1279, so that the solar shadow-length changed symmetrically before and after the winter solstice.[1]

I was also able to explain the Chinese idea of the secular variation of tropical year length. The Chinese took records of the time of the solstice dated 884, 656, and 523 B.C. as the starting points, and contemporary observations as the endpoints, to calculate the length of the tropical year, Calculation by Newcomb's method shows that these ancient records are two to four days early. Kuo, unaware of this, constantly diminished his constant for year length so that in the course of time it approached the true value. It was natural, because of this, to believe that in ancient times the tropical year had actually been longer and that it would continue to shorten. Not agreeing with the Platonic conviction that celestial phenomena are immutable, the Chinese were inclined toward the more flexible idea that everything, including celestial events, changes constantly.

This idea of changeable heavens was extended by later Japanese astronomers, who believed that every astronomical parameters is subject to periodic change. This conviction colored their understanding of all the data, Eastern and Western, ancient and modern.

After obtaining some highly satisfactory results from my work on the solar motion, I was supposed to move to the next step, the motion of the moon. As those familiar with lunar theory know, it is, however, a far more complicated subject. Since computers were not yet available, I had computed solar motions using an abacus for addition and subtraction and a hand calculator for multiplication and division. When dealing with lunar motion, many more parameters must be considered. In addition to determining sygyzies, and the longer and shorter months and intercalations of a luni-solar civil calendar, the lunar parameters must be correlated with the solar motion, since the synodic month was the major parameter of a luni-solar calendar. Furthermore, additional complications make high accuracy practically unobtainable for many lunar phenomena.

At this point I gave up this endlessly involved analysis and moved into other areas of the history of science. A quarter of a century later, historians of astronomy find sophisticated computer programs readily at hand. Some historians of science of my generation, such as R. Mercier and N. Sivin, have learnt to use computer techniques. I too would like to return to these problems with the cooperation of a young computer adept. A couple of years ago, I participated in a meeting on the history of technical astronomy in Kyoto headed by M. Yano, after my long absence from the field. I was deeply impressed by finding that after a quarter of century since my exit from the field, a new generation of young scholars is emerging who can handle computers easily and are thus qualified to solve hitherto impossible puzzles.

Outside of Japan, however, not many young astronomers are choosing to work on Chinese technical history. None has begun to publish in Europe or the United States in the last decade, and most of those educated in this specialty in China do not have access to careers that allow them to undertake innovative historical research.

Many problems call for study. For instance, in ancient eclipse prediction, there are such parameters as "east–west difference" and "south–north difference" in the case of the *Shou-shih* system, and similar quantities in other systems. These are semi-empirical concepts without clear-cut physical meanings. Studies using a computer should clarify what physical meaning they have, and their role as sources of error for eclipse prediction. In spite of the recent emergence of a group of researchers on calendrical astronomy who work with M. Yano in Kyoto, a systematic research program using computers has barely begun.

Due to the scarcity of new research, my analytical work on the solar motion has not been much cited except by Yabuuti, Sivin and a few others, although it was translated into the Chinese language. On the other hand, another contribution of mine, made almost overnight, has been heavily quoted by such Chinese specialists as Chen Chiu-chin and Chen Mei-tung. It is the analysis of the *Fu-t'ien* calendar.[2]

The *Fu-t'ien* calendar is unique in its algebraic representation of the solar equation of center, which resembles neither the traditional Chinese empirical, pragmatic method of interpolation between observed values at set intervals nor the Hellenistic, Islamic, and Indian schematic, geometric, trigonometric approach. Its representation of equations of second degree was taken over by successive Chinese calendars as well as Uighur calendars in Islamic sources. It provided a basis for the third-degree equations in the *Shou-shih* system.

The compiler of the *Fu-t'ien* calendrical system, Ts'ao Shih-wei, originated in the far west of China. He or his family may have come from Samarkand. If so, this is an interesting instance of astronomical creativity in Central Asia.

Trigonometry as it developed in Islamic countries and Europe is a rigorous method. On the other hand, the Chinese approach links empirical observation and calculation as closely as possible. The Central Asian algebraic representation is intermediate, sacrificing the close integration of the Chinese approach but avoiding the bold jump that explicit trigonometry would require.

When the late Aydin Sayili asked me in 1989 to report on this matter, I sent him a short article entitled "The Emergence of the Third Paradigm for Expressing Astronomical Parameters: Algebraic Function," which was published in *Erdem*.[3] I suggested to him that, since Turkic peoples were active in West China in the eighth century, similar remnants of algebraic astronomical parameters might be found in Turkish astronomical sources.

Four more years passed. Sayili gave me a copy of a new study entitled "Al-Khwarazmi, 'Abdu'l-Hamid ibn Turk and the Place of Central Asia in the History of Science."[4] It includes 100 pages in English and 108 pages in Turkish. I admired his ability to write such an extensive monograph at the age of eighty. Its thesis was largely a direct response to my inquiry about Turks in Central Asia.

In the beginning, he discussed the origin of Arabic algebra as thought by Al-Khwarazmi and others. He discussed the origin of these ideas and their connection with the Turkish origin of the founders of algebra. The middle part of the essay summarized my paper, adduced many sorts of evidence to document the intercourse between Chinese and Turks in the transmission of paper technology, Buddhism, and the decimal place-value system. Where direct proof was absent, he accumulated a great deal of circumstantial evidence. I am grateful for the trouble he took to look deeply into this question, and to find that my short article and inquiry had kindled his scholarly interest so near the end of his life. He passed away in 1994 at the age of eighty, just two weeks after I met him last in Ankara.

I do not think it will be easy to find direct evidence that astronomical parameters were algebraically represented in the area that Turkic culture spanned from the eighth-century Uighur lands to present-day Turkey. They have a common language, which enables international symposia on the history of Turkish culture. There is still hope that they will discover the missing links in the travel of astronomical ideas from Central Asia to China.

Notes

1. "Accuracy of premodern determinations of tropical-year length," *Japanese Studies in the History of Science*, 2: 101–118 (1963), included in *A History of Japanese Astronomy: Chinese Background and Western Impact* (Harvard U.P., 1969), pp. 131–133.
2. Shigeru Nakayama, "The position of the Futian Calendar on the History of East-West Intercourse of Astronomy", in G. Swarup, A. K. Bag and K. S. Shukla (Eds.), *History of Oriental Astronomy, Proceedings of the International Astronomical Union Colloquium No. 91* (Cambridge University Press, 1987), pp. 135–138.
3. *Erdem*, the journal of the Ataturk Culture Center, Vol. 6, no. 18, 1990.
4. *Erdem*, vol. 7, no. 19, Jan. 1991, actually published June 1993.

Part 3

Additional Contributions

3.1. Revisiting An Eighth-Century Chinese Table of Tangents*

Qu Anjing

Northwest University, Xian

1. Introduction

In ancient and medieval China,[1] astronomers always made use of a gnomon to measure the solar motion. Calendar-makers usually gave lengths of the solar shadow at noon on the days of 24 solar seasons (*qi*)[a] in their calendar-making system.[2] Those data were measured at the Imperial Observatory.

Generally speaking, a gnomon is eight Chinese feet (*chi*)[b] in height. By the use of these data, Chinese calendar-makers designed a relative algorithm for calculating the solar shadow at noon of every day.

Among these algorithms, a difference table of the solar shadow appeared in the *Dayan li*,[c3] which was compiled by the Buddhist monk Yi-xing[d] (683–727 AD) in 724 AD.[4]

Scholars have pointed out that the shadow table created by Yi-xing is equivalent to a table of tangents. Because the first western table of tangents appeared 70 years later after Yi-xing's table, they claim that it is the first table of tangents in the history of science.[5] On the other hand, some people argue that there is a slight difference between shadow lengths in the table of the *Dayan li* and the notion of tangent in trigonometry. However, since Yi-xing's innovation is quite different from the tradition of Chinese mathematical astronomy, great attention has been paid to studying and reconstructing this table.

Because the text of the algorithm for constructing the shadow table in the *Dayan li* is too sketchy to reconstruct it, different scholars have made different tables. They believe that there must have some misprints or omissions in Yi-xing's original text.[6]

In this short paper I will not allow myself to become involved in the argument about whether Yi-xing's table should be named a difference table of solar shadow or a table of tangents. My purpose will be to concentrate on the following topic: re-examine the text of Yi-xing, and based on historical material in the *History of Korea*, to reconstruct this table. An application of Yi-xing's table in the 9th century will be also discussed in detail.

2. Reconstruction

Let z represent the zenith distance of the sun. It is given in Chinese degrees (*du*)[e]:

$$1\ du = (360/365.2565)^\circ$$

* Research supported by a special fund of the Education Section of Shaanxi Province, and the China Scholarship Council (97861043).

S.M.R. Ansari (ed.), History of Oriental Astronomy, 215–225.

where 365.2565 days is the length of a sidereal year in the *Dayan li*. Then the length of the solar shadow at noon l is as follows:

$$l = h \text{tg} z \tag{1}$$

where $h = 8$ Chinese feet (*chi*) is the height of the gnomon.

In his *Dayan li*, Yi-xing created an algorithm for constructing a difference table of lengths of the solar shadow $l(z)$, $z = 0, 1, 2, 3, \ldots, 81\ du$.

A new translation of Yi-xing's algorithm in his *Dayan li* is given as follows. The numbers in the left column are the zenith distance given in *du*. Sentences in the right column are explanations of the original text of the *Dayan li*.[7]

	Text in the *Dayan li*	Explanations
0	In the South, at the subsolar point, there is no shadow at noon.	The solar shadow l equals 0 while the zenith distance z equals 0 *du*.
1	For the first *du* of the zenith distance, it takes the initial value 1379. From now on, the differences are increasing by one for each *du*. End at 25 *du*, add up to 26 *fen*.[f]	$z = 1\ du, l = 0.1379\ chi$. $1\ chi = 10000\ fen$. The difference between shadows of $z = 0$ and 1 *du* is the initial value 1379 *fen*. It is the first term of the first difference Δ. From 0 to 25 *du* of zenith distance, the third differences of solar shadow take the same value $\Delta^3 = 1\ fen$. Therefore, let the first term of the second difference $\Delta^2 = 1\ fen$, when $z = 25\ du$, $\Delta^2 = 26$.
25	Increasing further by 2 for each *du*, end at 40 *du*.	From $z = 25$ to 39 *du*, $\Delta^3 = 2$.
40	Increasing further by 6 for each *du*, end at 44 *du*.	From $z = 40$ to 43 *du*, $\Delta^3 = 6$.
44	Increasing by 68 (for the next *du*).	For $z = 44\ du$, let $\Delta^3 = 68$.
45	Increasing further by 2 for each *du*, end at 50 *du*.	From $z = 45$ to 49 *du*, $\Delta^3 = 2$
50	Increasing further by 7 for each *du*, end at 55 *du*.	From $z = 50$ to 54 *du*, $\Delta^3 = 7$.
55	Increasing further by 19 for each *du*, end at 60 *du*.	From $z = 55$ to 59 *du*, $\Delta^3 = 19$.
60	Increasing by 60 (for the next *du*).	For $z = 60\ du$, let $\Delta^3 = 160$.
61	Increasing further by 33 for each *du*, end at 65 *du*.	From $z = 61$ to 64 *du*, $\Delta^3 = 33$.
65	Increasing further by 36 for each *du*, end at 70 *du*.	From $z = 65$ to 69 *du*, $\Delta^3 = 36$.
70	Increasing further by 39 for each *du*, end at 72 *du*.	From $z = 70$ to 71 *du*, $\Delta^3 = 39$.
72	Increasing by 260 (for the next *du*).	For $z = 72\ du$, let $\Delta^3 = 260$.
73	Increasing further 440 for the next *du*.	For $z = 73\ du$, let $\Delta^3 = 440$.
74	Increasing further 1060 for the next *du*.	For $z = 74\ du$, let $\Delta^3 = 1060$.
75	Increasing further 1860 for the next *du*.	For $z = 75\ du$, let $\Delta^3 = 1860$.
76	Increasing further 2840 for the next *du*.	For $z = 76\ du$, let $\Delta^3 = 2840$.
77	Increasing further 4000 for the next *du*.	For $z = 77\ du$, let $\Delta^3 = 4000$.
78	Increasing further 5340 for the next *du*.	For $z = 78\ du$, let $\Delta^3 = 5340$.
	These data are the differences for each *du*. Add up these differences successively to the initial value, convert hundreds to *fen*, and ten to *cun*,[g] the differences of solar shadow for each *du* are here. Add up these differences of solar shadow successively, the length of solar shadow for each *du* of zenith distance are here.	

The original text of Yi-xing's constructive method is found in the fifth section of the *Dayan li*. From the sentences translated above, we know that Yi-xing broke the zenith distance into several parts, and constructed a higher-order difference table on each part successively. Because the text is too sketchy, it is hard to make sense of Yi-xing's method. This is the reason why different scholars reconstructed different tables based on the same source.

For modern scholars, two points of the text are divergent: 1. Comparing with the neighbouring values, the values at 44, 60, and 72 to 78 *du* are the second difference Δ^2 instead of Δ^3;[8] 2. Because the increasing values Δ^3 vary irregularly, a change of the text was made in an article so that the variations of the increasing values Δ^3 seem more rational.[9] Whether or not the text of the *Dayan li* is changed, they have the same problem. That is when the zenith distance $z > 60\ du$; errors in these reconstructed tables are all so great that it is hard to believe that they did complete Yi-xing's table.

A difference table in ancient Chinese mathematical astronomy often took the following form: in the first line, there are names of concerned matters. The beginning value of the first day or the initial *du* (0 *du*) is in the first column of the second line. The value of the next day or *du* is in the next column of the same line. The first term of each higher-order difference-value is in the same line with the beginning value of the first day or initial *du*.

Suppose there is no misprint in the original text of the *Dayan li*; no matter whether the zenith distance z is in an interval or at a single value, the increasing values mentioned in Yi-xing's text always stand for the 3rd difference values Δ^3. Based on the text in the *Dayan li*, a difference table of the solar shadow is reconstructed as Table 1.

When the zenith distance z is $0, 1, 2, \ldots, 81\ du$, the length of the solar shadow l at noon is found in this table.

Values in the sixth column of Table 1 are the theoretical values of $8\text{tg}z$. Because the lengths of a sidereal year in the *Dayan li* are 365.2565 days, 365.2565 *du* in Yi-xing's calendar equals 360°. In this place the unit *du* of China has been converted into the modern degrees as follows:

$$1\ du = (360/365.2565)^\circ$$

Values in the seventh column are the absolute errors, which equal the differences between shadow length l and $8\text{tg}z$.

Values in the eighth column are the relative errors, which equal the absolute errors/$8\text{tg}z$.

The absolute average value of relative errors of Table 1 is 8.7‰. When the zenith distance $z < 60\ du$, the absolute average value of relative errors of Table 1 is 6.6‰; the biggest value of relative error is no more than 11‰. When the zenith distance $z > 60\ du$, the absolute average value of relative errors of Table 1 is 15.6‰; the biggest value of relative error is no more than 36‰.

These analyses show that the stability of Table 1 is much better than any previous reconstructed shadow table based on Yi-xing's text.

People have assumed that there must have some misprints in the text of the *Dayan li*. From the discussion above we see that there is no error!

Table 1 Reconstructed table of tangents in Yi-xing's *Dayan li* (724 AD)

z (*du*)	l (*chi*)		Δ^2	Δ^3	8tg z	Absolute error l – 8tg z	Relative error (l – 8tg z)/8tg z
0	0	1379	1	1	0	0	0
1	0.1379	1380	2	1	0.13763052	0.00026948	0.00195801
2	0.2759	1382	3	1	0.27534291	0.00055709	0.00202327
3	0.4141	1385	4	1	0.41321772	0.00088228	0.00213515
4	0.5526	1389	5	1	0.55133816	0.00126184	0.00228868
5	0.6915	1394	6	1	0.68978652	0.00171348	0.00248407
6	0.8309	1400	7	1	0.82864623	0.00225377	0.00271983
7	0.9709	1407	8	1	0.96800171	0.00289829	0.00299410
8	1.1116	1415	9	1	1.10793858	0.00366142	0.00330471
9	1.2531	1424	10	1	1.24854388	0.00455612	0.00364915
10	1.3955	1434	11	1	1.38990624	0.00559376	0.00402456
11	1.5389	1445	12	1	1.53211620	0.00678380	0.00442773
12	1.6834	1457	13	1	1.67526635	0.00813365	0.00485514
13	1.8291	1470	14	1	1.81945166	0.00964834	0.00530580
14	1.9761	1484	15	1	1.96476971	0.01133029	0.00576673
15	2.1245	1499	16	1	2.11132097	0.01317904	0.00624208
16	2.2744	1515	17	1	2.25920908	0.01519092	0.00672400
17	2.4259	1532	18	1	2.40854119	0.01735881	0.00720719
18	2.5791	1550	19	1	2.55942830	0.01967170	0.00768597
19	2.7341	1569	20	1	2.71198556	0.02211444	0.00815433
20	2.8910	1589	21	1	2.86633271	0.02466729	0.00860587
21	3.0499	1610	22	1	3.02259445	0.02730555	0.00903381
22	3.2109	1632	23	1	3.18090088	0.02999912	0.00943101
23	3.3741	1655	24	1	3.34138801	0.03271199	0.00978994
24	3.5396	1679	25	1	3.50419821	0.03540179	0.01010268
25	3.7075	1704	26	2	3.66948083	0.03801917	0.01036091
26	3.8779	1730	28	2	3.83739272	0.04050728	0.01055594
27	4.0509	1758	30	2	4.00809897	0.04280103	0.01067864
28	4.2267	1788	32	2	4.18177355	0.04492645	0.01074340
29	4.4055	1820	34	2	4.35860012	0.04689988	0.01073575
30	4.5875	1854	36	2	4.53877287	0.04872714	0.01067286
31	4.7729	1890	38	2	4.72249744	0.05040256	0.01057192
32	4.9619	1928	40	2	4.90999198	0.05190802	0.01063064
33	5.1547	1968	42	2	5.10148821	0.05321179	0.01043064
34	5.3515	2010	44	2	5.29723272	0.05426728	0.01024446
35	5.5525	2054	46	2	5.49748830	0.05501170	0.01000670
36	5.7579	2100	48	2	5.70253545	0.05536455	0.00970876
37	5.9679	2148	50	2	5.91267410	0.05522590	0.00934026
38	6.1827	2198	52	2	6.12822542	0.05447458	0.00888913
39	6.4025	2250	54	2	6.34953395	0.05296605	0.00834172
40	6.6275	2304	56	6	6.57696987	0.05053013	0.00768289
41	6.8579	2360	62	6	6.81093163	0.04696837	0.00689603
42	7.0939	2422	68	6	7.05184882	0.04205118	0.00596314
43	7.3361	2490	74	6	7.30018546	0.03591454	0.00491968
44	7.5851	2564	80	68	7.55644368	0.02865632	0.00792300

Table 1 (*Continued*).

z (*du*)	l (*chi*)		Δ^2	Δ^3	8tg z	Absolute error $l - 8\text{tg } z$	Relative error $(l - 8\text{tg } z)/8\text{tg } z$
45	7.8415	2644	148	2	7.82116784	0.02033216	0.00259963
46	8.1059	2792	150	2	8.09494925	0.01095075	0.00135279
47	8.3851	2942	152	2	8.37843151	0.00666849	0.00079591
48	8.6793	3094	154	2	8.67231656	0.00698344	0.00080526
49	8.9887	3248	156	2	8.97737161	0.01132839	0.00126188
50	9.3135	3404	158	7	9.29443709	0.01906291	0.00205100
51	9.6539	3562	165	7	9.62443574	0.02946426	0.00306140
52	10.0101	3727	172	7	9.96838314	0.04171686	0.00418492
53	10.3828	3899	179	7	10.32739983	0.05540018	0.00536439
54	10.7727	4078	186	7	10.70272540	0.06997461	0.00653802
55	11.1805	4264	193	19	11.09573494	0.08476506	0.00763943
56	11.6069	4457	212	19	11.50795824	0.09894176	0.00859768
57	12.0526	4669	231	19	11.94110235	0.11149765	0.00933730
58	12.5195	4900	250	19	12.39707824	0.12242177	0.00987505
59	13.0095	5150	269	19	12.87803232	0.13146768	0.01020868
60	13.5245	5419	288	160	13.38638416	0.13811584	0.01031764
61	14.0664	5707	448	33	13.92487147	0.14152853	0.01016372
62	14.6371	6155	481	33	14.49660446	0.14049554	0.00969162
63	15.2526	6636	514	33	15.10513159	0.14746841	0.00976280
64	15.9162	7150	457	33	15.75451989	0.16168011	0.01026246
65	16.6312	7697	580	36	16.44945345	0.18174655	0.01104879
66	17.4009	8277	616	36	17.19535537	0.20554463	0.01195350
67	18.2286	8893	652	36	17.99853973	0.23006027	0.01278216
68	19.1179	9545	688	36	18.86640272	0.25149728	0.01333043
69	20.0724	10233	724	36	19.80766511	0.26473489	0.01336527
70	21.0957	10957	760	39	20.83268303	0.26301697	0.01262521
71	22.1914	11717	799	39	21.95385049	0.23754951	0.01082040
72	23.3631	12516	838	260	23.18612698	0.17697302	0.00763271
73	24.6147	13354	1098	440	24.54773787	0.06696213	0.00272783
74	25.9501	14452	1538	1060	26.06111764	−0.11101764	−0.00425990
75	27.3953	15990	2598	1860	27.75419980	−0.35889980	−0.01293137
76	28.9943	18588	4458	2840	29.66221246	−0.66791246	−0.02251728
77	30.8531	23046	7298	4000	31.83022630	−0.97712630	−0.03069806
78	33.1577	30344	11298	5340	34.31685090	−1.15915090	−0.03377789
79	36.1921	41642	16638		37.19973210	−1.00763210	−0.02708708
80	40.3563	58280			40.58396462	−0.22766462	−0.00560972
81	46.1843				44.61539850	1.56890150	0.03516502

The left 5 columns in this table are the reconstructed table of tangents of the *Dayan li*. Values in the first column z are the zenith distances of the sun, given in *du*. One *du* equals $(360/365.2565)^\circ$. Values in the second column l are the lengths of solar shadow of a gnomon 8 *chi* in height. Values in columns 3–5 are given in 10^{-4} *chi*. Values in the sixth column 8tg z are the theoretical data, where values z have been converted into modern degrees.

3. Verification

A few years ago, the author found that there is a similar constructional method of such a solar shadow table in the *Xuanming li*.[10] The *Xuanming li*[h] is a calendar-making system compiled by Xu Ang[i] in 822 AD. Most of the algorithms in the *Xuanming li* are adopted from the *Dayan li*. Therefore, the *Xin Tang shu*[j] (*New Tang History*) omitted a lot of the contents of the *Xuanming li*.[11] The solar shadow algorithm of the *Xuanming li* is not found in the version of the *Xin Tang shu*.

Fortunately, an original version of the *Xuanming li* was taken to Korea and Japan in the later Tang dynasty (the ninth century), and was issued for use in Korea and Japan for many years. Nowadays, one can find a much more complete version of the *Xuanming li* in Korea or Japan than in China. In the *History of Korea*, there is a popular edition of this calendar.

In the fifth section of the *Xuanming li*, there is a text for constructing a difference table of lengths of the solar shadow.[12] A translation of the text is as follows:

	Text of the *Xuanming li*	Explanations
0	In the South, at the subsolar point, there is no shadow at noon.	The increasing numbers for each *du* are the third differences of shadow Δ^3. The numbers given in the sentence *add up* are the second differences Δ^2.
1	For the first *du* of zenith distance, it takes the initial value 1379. From now on, the differences are increasing by one for each *du*. End at 25 *du*, add up to 26 *fen*.	
26	From 26 *du* on, increasing by 2 for each *du*. End at 40 *du*, add up to 56 (54) *fen*.	$z = 40\ du$, $\Delta^2 = 56$.
41	From 41 *du* on, increasing by 6 for each *du*. End at 44 *du*, add up to 80 *fen*.	$z = 44\ du$, $\Delta^2 = 80$.
45	At 45 *du* add up to 148.	$z = 45\ du$, $\Delta^2 = 148$.
46	Increasing further by 2 for each *du*. End at 50 *du*, add up to 158 *fen*.	$z = 50\ du$, $\Delta^2 = 158$.
51	From 51 *du* on, increasing by 7 for each *du*. End at 55 *du*, add up to 193 *fen*.	$z = 55\ du$, $\Delta^2 = 193$.
56	From 56 *du* on, increasing by 19(90) for each *du*. End at 60 *du*, add up to 288 *fen*.	$z = 60\ du$, $\Delta^2 = 288$.
61	At 61 *du*, add up to 448 (440).	$z = 61\ du$, $\Delta^2 = 448$.
62	Increasing further by 33 for each *du*.	
66	End at 65 *du*, add up to 580 *fen*.	$z = 65\ du$, $\Delta^2 = 580$.
	From 66 *du* on, increasing by 36 for each *du*. End at 70 *du*, add up to 760 *fen*.	$z = 70\ du$, $\Delta^2 = 760$.
71	From 71 *du* on, incresging by 39 for each *du*. End at 72 *du*, add up to 838 *fen*.	$z = 72\ du$, $\Delta^2 = 838$.
73	At 73 *du*, add up to 1098 (198) *fen*.	$z = 73\ du$, $\Delta^2 = 1098$.
74	At 74 *du*, add up to 1538 *fen*.	$z = 74\ du$, $\Delta^2 = 1538$.
75	At 75 *du*, add up to 2598 *fen*.	$z = 75\ du$, $\Delta^2 = 2598$.

76	At 76 *du*, add up to 4458 *fen*.	$z = 76\ du,\ \Delta^2 = 4458.$
77	At 77 *du*, add up to 7298 *fen*.	$z = 77\ du,\ \Delta^2 = 7298.$
78	At 78 *du*, add up to 11298(54398) *fen*.	$z = 78\ du,\ \Delta^2 = 11298.$
79	At 79 *du*, add up to 16638 *fen*.	$z = 79\ du,\ \Delta^2 = 16638.$
	End at 80 *du*.	
	These data are the differences for each *du*. Add up these differences successively to the initial value, convert hundreds to *fen*, and ten to *cun*, the differences of solar shadow for each *du* are here. Add up these differences of solar shadow successively, the lengths of solar shadow for each *du* of zenith distance are here.	Because the second difference Δ^2 ends at $z = 79\ du$, the length of solar shadow must be up to $z = 81\ du$ in this table.

Comparing the two pieces of text from the *Dayan li* and the *Xuanming li*, it is easy to see that they construct the same difference table.

Although the text in the *Xuanming li* is more clearly formulated than in the *Dayan li*, there is a problem in making sense of it independently. When we study the two pieces of text together, we find that they complement each other to some extent: all of the increasing values Δ^3 are given in the *Dayan li*; and all of the addition values Δ^2 are found in the *Xuanming li*.

According to the text in the *Xuanming li*, it is easy to verify the fact as follows: the increasing values $\Delta^3 = 68, 160, 260\ fen$ in the *Dayan li* are given as 45, 61, and 72 *du*, respectively. Therefore the relevant increasing values $\Delta^3 = 2, 33$, and $440\ fen$ begin from 44, 60, and 73 *du*, respectively.

4. Application

As far as Yi-xing's table is concerned, it seems to us that none of his successors took up this technique for any application elsewhere. In this section, we will give an example of the application of Yi-xing's table.

It is well known that there is a table of lengths of the solar shadow at noon of 24 *qi* in each ancient Chinese calendar-making system. Generally, these shadow lengths are taken from the observations of the calendar-makers. Such a table in the *Xuanming li* is in the fifth section of this calendar. Calculation shows that the shadow lengths of 24 *qi* in this table are the results obtained by the use of Table 1.

In the Tang dynasty, the Imperial Observatory is at Yang-cheng,[k] a place in modern Henan[l] province. The geographical latitude of Yang-cheng is given before the table of 24 *qi*: $\varphi = 34.475\ du$. Because the length of a fourth quadrant in the *Xuanming li* is $91.30\ du$, the zenith distance of Yang-cheng equals $(91.30 - \varphi) = 56.825\ du$.

In the table of 24 *qi*, north polar distances of the sun on the days of 24 *qi* are given. For the *Xuanming li*, there is

The sun's zenith distance $z =$ the sun's north polar distance $p - 56.825\ du$,

Therefore the zenith distance of the sun on the day of 24 *qi* is easy to calculate from its north polar distance p.

Table 2 A difference table of the third order in the *Xuanming li* (822 AD)

North polar distance of the sun (du)	Δ (du)	Δ^2	Δ^3	North polar distance of the sun (du)	Δ (du)	Δ^2	Δ^3
115.20	0.65	−1.6	−0.2	67.40	−0.65	1.6	0.2
114.55	2.25	−1.4	−0.2	68.05	−2.25	1.4	0.2
112.30	3.65	−1.2	−0.2	70.30	−3.65	1.2	0.2
108.65	4.85	−1.0	−0.2	73.95	−4.85	1.0	0.2
103.80	5.85	−0.8		78.80	−5.85	0.8	
97.95	6.65			84.65	−6.65		
91.30				91.30			

Table 3 Solar shadow lengths of the 24 qi in the *Xuanming li*

Names of the 24 qi	Polar distance of the sun (du)	Zenith distance of the sun (du)	Shadow length in *Xuanming li* (*chi*)	Shadow length of calculation (*chi*)
1. Winter Solstice	115.20	58.375	12.7033	12.70325
2. Little Cold 24. Heavy Snow	114.55	57.725	12.3911	12.3911025
3. Severe Cold 23. Little Snow	112.30	55.475	11.3830	11.38304
4. Spring Begins 22. Winter Begins	108.65	51.825	9.9478	9.947765
5. Rain Water 21. Frost Descends	103.80	46.975	8.3781	8.37812
6. Excited Insects 20. Cold Dew	97.95	41.125	6.8874	6.8874
7. Spring Equinox 19. Autumn Equinox	91.30	34.475	5.4470	5.446975
8. Clear & Bright 18. White Dew	84.65	27.825	4.1959	4.195935
9. Grain Rains 17. Limit of Heat	78.80	21.975	3.2069	3.206875
10. Summer Begins 16. Autumn Begins	73.95	17.125	2.4451	2.22505
11. Grain Fills 15. Great Heat	70.30	13.475	1.8989	1.898925
12. Grain in Ear 14. Slight Heat	68.05	11.225	1.5714	1.5714125
13. Summer Solstice	67.40	10.575	1.4780	1.477955

The table of shadow lengths of 24 qi in the *Xuanming li* consists of 6 columns:

The first column is a list of names of 24 qi.

In the second column, there are values of the first difference Δ of the sun's north polar or zenith distance of 24 qi. These values in the original table are given in 10^{-2} du. These data are listed in the second and third columns of Table 3.

Values of the north polar distance of the sun of 24 *qi* are in the third column of the original table. These data are listed in the second column of Table 3. In the original table, values less than 1 *du* are given in 1/84 *du*. Now we have converted them into decimal notation. The shadow lengths of 24 *qi* of the *Xuanming li* form a difference table of the third order.[13]

In the last column of original table, there are lengths of solar shadow at noon of 8-*chi* gnomon of 24 *qi* in Yang-cheng. These lengths are listed in the fourth column of Table 3.

The data in the fifth column of Table 3 are shadow lengths of calculations. As an example, let us see how the length of the solar shadow of the winter solstice was calculated.

In the third column of Table 3, the zenith distance of the sun at the winter solstice is $z = 58.375$ *du*. Refer to Yi-xing's shadow lengths table from Table 1, relevant to the zenith distance $z = 58$ *du*, a shadow length $l(58) = 12.5195$ *chi*.

In the line $z = 58$ of Table 1, the first difference of shadow length is $\Delta = 0.4900$ *chi*. By the use of linear interpolation, an additional shadow length of decimal of zenith distance at the winter solstice 0.375 *du* is $0.375 \times 0.4900 = 0.18375$ *chi*. Then there is the shadow length of the sun at the winter solstice:

$$l(58.375) = 12.5195 + 0.18375 = 12.70325\ chi \approx 12.7033\ chi$$

Comparing the fourth and fifth columns of Table 3, we find that all of the shadow lengths of 24 *qi* in the *Xuanming li* are the results of calculations gained from above procedure.

If we repeated the calculation by the use of any reconstructed shadow length table by previous scholars, we could not obtain the shadow lengths of the *Xuanming li* when the sun's zenith distance $z > 45$ *du*. That shows again that Table 1 is a complete reconstructed table of Yi-xing.

There is also a table of the solar lengths of 24 *qi* in the *Dayan li*. Calculation shows that Yi-xing did not make use of Table 1 to construct his table.

In 724 AD, Yi-xing led a program of wide-ranging field research. All of the ten observation stations were distributed near the meridian 114°E, and ranged in latitude from 170°N to 52°N. Observers measured the sun's shadow length at noon on days of the winter and summer solstices, and the spring and autumn equinoxes.

In their detailed study of data from Yi-xing's program, Joseph Needham and his collaborators considered that only the shadow lengths at the summer solstice seemed to be the observed data. Values at the winter solstice and equinoxes were calculated from the summer solstice with techniques such as a table of tangents.[14] Their inferences are correct to some extent. Such a technique which Yi-xing used for calculating some values of his meridian survey is the difference table of solar shadow listed in Table 1.

Let w and s express the lengths of the solar shadow at the winter and the summer solstice, respectively. By the use of Table 1, the zenith distances z_w and z_s of the sun can be calculated. Suppose the latitude of the observation station is φ, there are:

$$\varphi = (z_w + z_s)/2$$

$$\varphi = z_w - \varepsilon$$

$$\varphi = z_s + \varepsilon$$

where ε is the obliquity of the ecliptic. In the *Dayan li*, it is 23.9 *du*. From one of the above expressions, we can calculate the sun's shadow length at the equinoxes with Table 1.

Calculations show that in some places the shadow lengths at the winter and the summer solstices were all observed values. In the other places, only one of the values at solstices was the observed datum. By the use of one of the above expressions, all values of shadow lengths at the actual equinox (*ding chunqiufen*)[m] recorded by Yi-xing were calculated from the data of the summer or the winter solstice with Table 1. There is no doubt that Yi-xing had made use of Table 1 for calculating many values of his meridian survey.

5. Conclusion

By analyzing and comparing the texts in the *Dayan li* and the *Xuanming li*, the difference table of the solar shadow in the *Dayan li* is reconstructed as in Table 1. The absolute average value of relative error of Yi-xing's shadow table is 8.7‰.

It is clear that the reconstructed table is a piecewise difference table. When the zenith distance $z < 74\,du$, it consists of nine difference tables of the third order. When $z > 72\,du$, the shadow lengths of Table 1 forms a difference table of the fifth order. This is the highest difference table found in the ancient Chinese mathematical astronomy so far.

Because the shadow lengths of the sun of 24 *qi* are the fundamental data for each traditional Chinese calendar, it was believed that these lengths were certainly the results of measurement. This paper hopes to prove that this is not the truth.

Using Yi-xing's shadow table as in Table 1, according to the sun's north polar distance or zenith distance, the shadow lengths of the sun of 24 *qi* in the *Xuanming li* are calculated. When the 24 *qi* are broken into four parts by points of two equinoxes and two solstices, it is noticeable that each part of the sun's north polar distance or zenith distance consists of a difference table of the third order.[15]

This fact shows that the data of zenith distance of the sun at 24 *qi* in the *Xuanming li* are calculated results derived using an algorithm.

Therefore we may conclude that those shadow lengths of the sun at 24 *qi* which are adopted in the *Xuanming li* have no relevance to measurement to any extent! The significance is that the calendar-maker of the ninth century had made conscious efforts to make use of a theoretical method instead of direct observations only. This is great progress in traditional Chinese mathematical astronomy. There is no doubt that Yi-xing's difference table of solar shadow had a big influence on the calendar-making system in the Tang dynasty.

Notes

1. Medieval China refers to the period between the 2nd and 15th century.
2. In a traditional Chinese calendar-making system, similar to the zodiacal signs, the ecliptic is broken into 24 parts of equal size. People named the time when the sun passes through the 24 nodes together *qi*. For instance, the winter and the summer solstices, and the spring and the autumn equinoxes are 4 of the 24 *qi*.
3. Chinese people call a calendar-making system *li*.
4. Yi-xing is rendered as I-hsing (see Cullen) in many western materials.
5. See Cullen, Liu and Zhao.

6. Based on Yi-xing's algorithm, two different tables have been reconstructed by Cullen, Liu and Zhao.
7. See Ouyang, pp. 659–660.
8. See Cullen.
9. See Liu And Zhao.
10. See Yu.
11. See Ouyang.
12. See Yu, pp. 88–91.
13. See Table 2.
14. See Beer.
15. See Table 3.

References

A. Beer, Ho Ping-yu, Lu Gwei-djen, J. Needham, E.g. Pulleyblank, and G. Thompson. "An 8th-Century Meridian Line: YI-XING's Chain of Gnomons and the Pre-history of the Metrid System." *Vistas in Astronomy*, Vol. 4 (1961), pp. 3–28

C. Cullen. "An Eighth Century Chinese Table of Tangents." *Chinese Science*, Vol. 5 (1982), pp. 1–33

Liu Jinyi and Zhao Chengqiu. "The World's Earlist Table of Tangents, A Contribution Made by Yi-xing of the Tang Dynasty." *Ziran kexueshi yanjiu*[n] (*Studies in the History of Natural Sciences*). Vol. 5, No. 4 (1986), pp. 298–309

Ouyang Xiu. *Xin Tang Shu* (*New History of Tang Dynasty*). Beijing: Zhonghua shuju[o] (Chinese Book House), 1975, pp. 745–770, 659–660

Yu Jinglao[p] (ed.). *Hanguo kexue jishushi ziliao daxi*[q] (*Historical Materials of Science and Technology of Korea – Astronomy*), Vol. 2. Seoul: Lijiang[r] Press, 1986, pp. 88–91

Glossary

a	qi	气	*b*	chi	尺
c	Dayan li	大衍历	*d*	Yi-xing	一行
e	du	度	*f*	fen	分
g	cun	寸	*h*	Xuanming li	宣明历
i	Xu Ang	徐昂	*j*	Xin Tang shu	新唐书
k	Yang-cheng	阳城	*l*	Henan	河南
m	ding chunqiufen	定春秋分	*n*	Ziran kexueshi yanjiu	自然科学史研究
o	Zhonghua shuju	中华书局	*p*	Yu Jinglao	俞景老
q	Hanguo kexue jishushi ziliao daxi	韩国科学技术史资料大系	*r*	Lijiang	骊江

3.2. Recent Advances in the Studies of History of Astronomy in China

Li Qibin* **and Chen Meidong****

**Beijing Astronomical Observatory, CAS*

***Institute for History of Natural Science, CAS*

Abstract

Advances in studies of the various aspects of the history of Chinese astronomy in about the last ten years are reviewed, namely, the ancient records and their modern application, the astronomical chronology, the calendar system, the ancient astronomers, astronomical atlases and instruments, and the origin and development of astronomy. A large number of papers on the history of astronomy published recently in Chinese language are introduced briefly.

1. Introduction

Since 1957, when the Institute of History of Natural Science was established led by Professors Xi Zezong and Bo Shuren, a group of astronomers has been actively studying the history of Chinese astronomy and has made quite a considerable contribution in this area. In 1974, the Chinese Academy of Sciences decided to co-ordinate the studies of the history of Chinese astronomy. Since then a number of astronomers working in observatories have been involved in this endeavour. In 1978, the Chinese Astronomical Society as well as the Society for the History of Natural Science separately established commissions on the history of astronomy, both of which have been promoting the domestic and international exchanges and cooperation between astronomers. Since 1993 biannual conferences devoted to the study of the history of oriental astronomy (HOA) have been held by astronomers of China, Japan and Korea. More than 500 papers have been published on HOA in the last ten years, most of which are written in the Chinese language.

2. Collection of Ancient Records and Documents and their Modern Application

A general compilation of ancient astronomical records (Zhuang and Wang, 1988) was published by the Beijing Astronomical Observatory. In this book, more than ten thousand records have been collected from more than 150,000 ancient documents. They are concerned with records of sunspots, aurora, meteorites, meteors and meteor showers, comets, novae and supernovae, solar and lunar eclipses, and occultation.

Some scholars have worked on the modern application based on the records collected. Xu Zhentao and Jiang Yaotiao (1990) carefully collated and analyzed the records of sunspots. They also studied the terminology in early sunspot records and analyzed the pattern of solar activities in the historical period. Li Qibin (1987, 1992) gave the fuzzy possibility of historical supernovae and their sketchy positions.

S.M.R. Ansari (ed.), History of Oriental Astronomy, 227–235.

Zhang and Han (1996) calculated that the average increment in the length of the day was about 1.5 ms/century, based on the timing records of 32 solar eclipses in ancient China. Zhang Peiyu (1993) discussed the precise identification of the lunar eclipse records. Han and Zhang (1994) obtained the value of ΔT as 1.51 ms/century from the records of 48 lunar eclipses ranging from the third to the thirteenth century. Han (1994, 1997) also investigated the Japanese records of six solar eclipses and a record of an annular solar eclipse, and gave the value of ΔT as 1.6 ms/century. Liu Ciyuan (1991) suggested a time window method to analyze the data of lunar occultations recorded in ancient documents and for studying the secular variation of the earth's rotation. He obtained the same results with data derived from Babylonian lunar eclipses and Chinese solar eclipses from the seventh century BC to the second century AD, from Arab solar and lunar eclipses, and from European total solar eclipses from the eighth to the twelfth century AD. He obtained some different result with data derived from western data in the fifth century AD. Liu Ciyuan (1992) gave a statistical analysis of ancient Chinese data of lunar occultations and of close approaches.

Chen Y.-N. and Huang Yi-Long (1994) reconsidered the records of the guest star of AD185. They found that the description of that event did not tally with the features of the supernova, so they suggested that the guest star of AD185 might have been a comet. Huang Yi-Long (1991) investigated records of "Mars staying at Xin Mansion" in ancient Chinese documents and found that most of those phenomena had not in fact happened. Based on his investigation of the records of those events, he analyzed the characteristics of astrology in China.

The Institute for Archeology in Beijing (1989) collected and collated the astronomical relics and documents in which Xi Zezong, Xia Nai and other scholars studied the ancient star maps, and investigated the problems of recording the unearthed relics. Bo Shuren (1997) completed a collection including 99 ancient classic books on astronomy covering 10,000 pages. Deng Wenkuan (1996) collated the documents on astronomy and the calendar saved from the Dunhuang grottos. The astronomical archives of the Qing dynasty (1645–1910) were collected by Beijing First Archive House and Beijing Planetarium (Cui Zhenhua and Zhang Shucai, 1997).

3. Astronomical Chronology

Due to the absence of sufficient chronological evidence, historical events in China cannot be dated before 840 BC. Recently Chinese astronomers have paid more attention to finding evidence of timing from the astronomical records. Some astronomical phenomena recorded in ancient documents could be considered as evidence for dating historical events.

The significance of the ancient records on "double dawn at Zheng during the first year of King Yi", described in ancient documents, has been emphasized, since Liu Chaoyang (1944) pointed out that the appearance of double dawn might be attributed to solar eclipse just before sunrise. Kevin Pang identified the event as due to the solar eclipse on April 21, 889 BC. Besides discussing further the various facts relating to how the dawn was formed twice, Liu Ciyuan and Zhou Xiaolu (1999) organized an expedition team to observe the solar eclipse on 11 sites in northwest China on March 9, 1997. Liu concluded that the double dawn was caused by the solar eclipse in the early morning, and deduced the first year of King Yi to be 899 BC.

Y.C. Chang (1978) had tried to identify the comet that appeared in that year as Halley's Comet and dated the event to 1057 BC. Since then different results for this event have been given by other authors. Recently Jiang Xiaoyuan (1999) has pointed out that the records of comets should be disregarded in the studies of chronology because it is difficult to identify comets as Halley's Comet. Instead of cometary events, he listed seven records of astronomical phenomena such as the position of the Sun, Moon, Jupiter and Mercury, and also the phases of the Moon recorded on different dates during the war between Emperors Wu and Zhou. He emphasized the importance of consistent coincidence between the recorded series of astronomical phenomena and computer-generated ones. He concluded that the year in which Emperor Wu started the conquest was 1044 BC and that Emperor Zhou was defeated in 1045 BC.

Zhang Peiyu (1999) examined the solar and lunar eclipses in the inscriptions on bones and tortoise shells of Shang dynasty. He rejected the record of "three flames occulting the sun" as being due to any solar eclipse. Identifying the dates of five recorded lunar eclipses, he then suggested that the period of Emperor Wuding was about 1250–1192 BC and confined the period during which Emperor Wu defeated Emperor Zhou to 1050–1020 BC.

4. Astronomical Atlases and Catalogues

Pan Nai (1989) presented in his monograph a systematic exposition of stellar observation in ancient China. As to the origin of the lunar mansions, he concluded that the position of determinant stars should be based on the observations carried out before the sixth century BC. A set of stellar positions in the *Shi's Star Catalogue* was determined in about 450 BC, and another set was observed around 170 AD. He also discussed Yang Weide's *Star Catalogue* and Guo Shoujing's *Star Catalogue*, Chen Zuo's *Constellations* and the *Constellations in Verse* written in Shui-Tang dynasties, *Dunhuang Star Atlas*, Su Song's *Astronomical Atlas* carved in stone and preserved in Suzhou, and the *General Star Atlas of both Northern and Southern Celestial Hemisphere*. According to Sun Xiaochun (1994), the epoch of *Shi's Star Catalogue* should be considered as being around 78 BC. Hu Weijia (1998) examined the uncertainty in the method used in deducing the time of observations and concluded that *Shi's Star Catalogue* would have been completed not long before the Tang dynasty. Wang Dechang *et al.* (1989) studied the star map carved on a stone tablet preserved in Changshu. Sun Xiaochun and Kistmaker (1997) introduced the constellations and stellar observation. Chen Meidong (1989) studied the constellations of Chen Zuo's and Su Song's maps. Hu Lingui and Zhong Wanli (1991) studied the stellar map excavated from the tomb of the Han dynasty in Xi'an. In the painted vault of the tomb, more than 80 star-markings of the 28 lunar mansions were drawn and the stars were linked to form constellations. Feng Shi (1990) studied the Beidou-Dragon-Tiger figures unearthed in Puyang, Henan province, which were painted more than 6000 years ago. Sun Xiaochun (1995) studied the star atlas and catalogue of Chongzhen Calendar.

Yi Shitong (1989) and Pan Nai (1989) studied the star atlas preserved in Longfu Temple in Beijing. The temple was built in 1453. A large stellar atlas, the diameter of which was 1.97 m, was drawn on the ceiling of the temple. The positions of 1420 stars tally with those given in the *Constellations in Verse*, written in the Shui-Tang dynasty. Wang Dechang *et al.* (1989) studied the star atlas carved on the stone preserved in Changshu, Jiangsu province. The star atlas drawn in 1506 seems to follow the *Astronomical Atlas* preserved in Suzhou. However, 284 constellations and 1466 stars were drawn, 4 constellations and 33 stars more than in the Suzhou atlas.

Chen Meidong (1996) completed a collection of studies on ancient star charts, covering 18 papers and 111 plates, and with Bo Shuren, summarized the results of studies in this field.

5. Ancient Astronomical Instruments, Astronomers and Calendar

Hua Tongxu (1991) completed his monograph on clepsydra in China, while Li Zhichao and Chen Yu (1993) wrote about the water-driven globe of Zhang Heng. Li Zhichao (1997) has written a book on water-driven globes. He discussed the working principle and structure of the globe made by Yixing and Liang Linzan. Hu Weijia (1994) discussed the mechanism of the water-driven globe and star-mapping method in the *Xin Yi Xiang Fa Yao*. Li Qibin (1997) investigated the structure of the gnomon and simulated the device for observing the image of the rod and the sun used by Guo Shoujing in Dengfeng ancient observatory in the thirteenth century. He explained geometrically the pinhole imaging principle of the device. It was supposed that the pinhole imaging principle was employed by Guo Shoujing for the first time in the world.

A collected biography of Chinese scientists was published (Du Shiran, 1992, 1993) in which the life and achievements of more than fifty Chinese astronomers and fourteen foreign astronomers who had worked in China are sketched.

Chen Chen-Yih, Xi Zezong and Jao Tsung-i (1994) compared the music and astronomy of China and Babylon.

Deng Wenkuan (1996) collated the astronomical texts and calendar saved from the Dunhuang grottos. Xi Zezong and Deng Wenkuan (1989) discussed the method for dating the fragmented Dunhuang calendar. Qu Anjing, Ji Zhigang and Wang Rongbin (1994) investigated the application of interpolation and polynomial function used in ancient calendar calculations. They found a set of constants in calendar calculations and simulated the method of deducing those constants.

6. Origin and Development of Astronomy

Jiang Xiaoyuan (1991) investigated the special status of astronomy in China, stressing the relationship of astronomy to politics, culture, ethics and architecture in his monograph on the origin of astronomy. He also carried out a comparative study of Chinese, Indian and Babylonian astronomy, and investigated the exchanges between China, India and Babylon in the field of astronomy.

Xi Zezong (1989) studied the place of astronomy in social culture. Chen Jiujing (1996) completed a monograph on the history of Chinese Muslim astronomy

Bo Shuren (1989) claimed that the theory of spherical heaven is more advanced than the theory of canopy-heaven. According to Chen Zhengyi and Xi Zezong (1991), the combination of theory and observations in Chenzi's mathematical model for explaining the celestial phenomena is a significant contribution to the history of sciences. Jiang Xiaoyuan (1991) re-examined the theory of canopy-heaven. Xu Fengxian (1994) investigated ancient astronomers' conception of anomalistic astronomical phenomena. Shi Yunli (1992) reviewed the controversy between the theories on the earth's motion in ancient China. Chen Meidong (1997) reviewed the arguments on the spin-left or spin-right of the sun, moon and five bright planets. Hu Tiezhu (1990) compared the characteristics of Chinese and Indian astronomy. Lu Dalong (1997) found the relationship between the *Li Xiang Kao Chen Hou Pian* with Newton's theory of lunar motion.

Liu Nanwei, Wu, Li Qibin and co-authors (1989) investigated the history of nautical astronomy in ancient China. In the fifteenth century, Zheng He led a huge fleet which sailed to the Middle East and east Africa through the Indian Ocean. The navigator in Zheng He's fleet determined the position of fleet in real time by observing the angle of elevation of the bright stars. The authors identified the guide stars in *Zheng He's Guiding Map* and the method of observations.

7. Comprehensive Research

Chen Zhungui completed the last volume of *History of Chinese Astronomy* in 1989. In this volume, astronomical instruments, the astronomical concepts of ancient astronomers, and the history of modern astronomy are dealt with.

The History of Chinese Astronomy Series, a comprehensive set of monographs on HOA, were completed in 1998, edited by Bo Shuren and containing contributions from 20 authors (1999). The 10,100-page series consists of 12 volumes covering various subjects: astronomers (Chen Jiujijng), calendars (Zhang Peiyu), astronomical concepts (Chen Meidong), astrology (Lu Yang), astrometry (Wu Shouxian), astronomical instruments (Quan Hejun), astronomical institutes and education (Chen Xiaozhong), astronomy of minorities (Chen Jiujing), propagation of modern astronomy into China (Cui Zhenhua and Du Shenyun), modern application of ancient records (Zhuang Weifeng), terminology of ancient Chinese astronomy (Xu Zhengtao), history of modern astronomy (Miao Yongkuan and Xiao Naiyuan). The results of these studies in the history of Chinese astronomy in this century are summarized in this series. Some of the topics dealt with have never been systematically covered in any monograph.

Jiang Xiaoyuan (1995) completed his book: *Astrology in History*. Chen Jiujing (1996) completed his monograph: *Chinese Muslim Astronomy*. Cui Zhenhua and Chen Dan (1993) published their book: *World History of Astronomy*. Xuan Huancan (1992) published the book: *History of Astronomy*. Chen Meidong (1994) completed the book: *Chronicle of Events in Natural Sciences – Astronomy*.

8. Concluding Remarks

In the last decade, the ancient astronomical records in documents were systematically collected and collated. A large number of astronomical records were excavated and studied. Based on the material collected, the researches on application to modern astronomy progressed. Some works examined the chronology of the era earlier than 841 BC by using ancient astronomical records in both the excavated relics and documents, which led to fruitful results. Besides this, a series of monographs on history of Chinese astronomy published recently has summarized the results of studies, which will be helpful for further research work in this field.

In future works, besides the textual analysis of both documents and excavated relics, the researches on the application of ancient records in modern astronomy will be further developed. The translation of Chinese astronomical records and research works into foreign languages would be given more attention to provide foreign astronomers with original materials of ancient astronomical records and to promote exchanges between foreign and Chinese astronomers.

References

Papers with * are in English, papers without are in Chinese language

Bo Shuren (1989), Concerning the Gai Tian theory in Zhou Bi Arithmetical Classics of the Gnomon and Circular Path, *Researches on History of Natural Sciences*, Vol. 8, No. 4, pp. 297–305.

Bo Shuren, Ed. (1997), *Collection of Classics in Sciences and Technologies – Astronomy*, Henan Education Press, 10157 pp.

Bo Shuren, Ed. (1999), *Series of the History of Chinese Astronomy*, Hebei Science and Technology Press, Shijiazhuang, in press.

*Chen-Chen-Yih, Xi Zezong and Jao Tsung-i (1994), A Comparative study of Acoustics and Astronomy in Babylon and in China, prior to and during the Time of Marquis Yi set-bells, in *Two-tone Set-Bells of Marquis Yi*, Ed. Chen Chen-Yih, World Scientific, Singapore, pp. 297–366.

Chen Jiujing (1993), *Collected Papers of Chen Jiujin*, Heilongjiang Education Press, Harbin, 420 pp.

Chen Jiujing (1996), *History of Chinese Muslim Astronomy*, Guangxi Science Technology Press, Nanning, 376 pp.

Chen Meidong (1989), Historical Development of Chen Zuo's Constellation, *Collected Papers on History of Science and Technology*, Vol. 16, pp. 77–91.

Chen Meidong (1994), *Chronicle of Events in Natural Sciences – Astronomy*, Liaoning Education Press, Shengyang, 124 pp.

Chen Meidong, Ed. (1996), *Collection of Studies on Ancient Star Maps*, Liaoning Education, Shengyang, 321pp.

Chen Meidong (1997), Controversy between the Spin-left and Spin-right Theories of the Sun, Moon and Five Bright Planets, *Researches on History of Natural Sciences*, Vol. 16, No. 2, pp. 147–160.

*Chen Y.-N. and Huang Yi-Long (1994), *Nature*, Vol. 371, No. 6496, pp. 398–399.

Chen Zhengyi and Xi Zezong (1991), Chenzi's Model and the Early Measurement of the Sun, in *On the History of Science of Ancient China*, published by Institute for Humanities, Kyoto University, Kyoto, pp. 367–383.

Chen Zhungui (1989), *History of Chinese Astronomy*, Vol. 4, Shanghai People's Publishing House, Shanghai, 631 pp.

Cui Zhenghua and Chen Dan (1993), *World History of Astronomy*, Jiling Education Press, Changchun, 601 pp.

Cui Zhenhua and Zhang Shucai, Eds. (1997), *Collection on Astronomical Archive of Qin Dynasty*, Elephant Press, Henan, 486 pp.

Deng Wenkuan (1996), *Collected Collations of the Astronomical Texts and Calendar Saved in Dunhuang Grottos*, Jiangsu Classics Press, Nanjing, 767 pp.

Du Shiran (1992, 1993), *Collected Biography of Chinese Scientists*, Science Press, Beijing, Vol. 1, pp. 22–665, Vol. 2, pp. 667–1338.

Feng Shi (1990), Studies on the Early Constellation Maps of China, *Researches on Natural Science History*, Vol. 19, No. 2, pp. 108–118.

*Han Yanben (1994), ΔT derived from Japanese Ancient Records of an Annular Solar Eclipse in AD1183, *Astrophysics Reports*, No. 25, pp. 68–72.

*Han Yanben (1997), Variation of the Earth's Rotation Derived from Japanese Records of Ancient Solar Eclipses, *Acta Geophysica Sinica*, Vol. 40, No. 2, pp. 285–288.

*Han Yanben and Zhang Peiyu (1994), ΔT Derived from the Latest Textual Criticism of Timing Records of Lunar Eclipses in Ancient China, *Chinese Science Bulletin*, Vol.39, No. 21, pp. 1847–1848.

Hu Lingui and Zhong Wanli (1991), *Xi'an Jiaotong University West Han Dynasty Mural Tomb*, Xi'an Jiaotong University Press, Xi'an, pp. 28–48.

Hu Tiezhu (1990), The Calculation of Five Planets in Dayan Calendar and Surya-Siddhanta, *Master Dissertation*, The Institute for the History of Natural Sciences, pp. 1–28.

Hu Weijia (1994), On the Escapement and Star Mapping in the *Xin Yi Xiang Fa Yao, Researches on History of Natural Sciences*, Vol. 13, No. 3, pp. 244–253.

Hu Weijia (1998), On the Catalogues of Twenty Eight Lunar Mansions and Shi Shen's Constellations in the Astronomical Book of the Tang Dynasty, *Researches on History of Natural Sciences*, Vol. 17, No. 2, pp. 139–155.

Hua Tongxu (1991), *On the Chinese Clepsydra*, Anhui Science and Technology Press, Hefei, 238 pp.

Huang Yi-Long (1991), Counterfeit Astronomical Records in Ancient China – A Case Study on the Celestial Phenomenon "Mars Staying at Xin Mansion", *Researches on History of Natural Sciences*, Vol. 10, No. 2, pp. 120–131.

(The) Institute for Archaeology, Ed. (1989), *Collected Papers of Chinese Ancient Astronomical Relics*, Culture Relics Press, Beijing, 511pp.

Jiang Xiaoyuan (1991), *The Truth of Ancient Chinese Knowledge of Heavens*, Liaoning Education Press, Shengyang, 417pp.

Jiang Xiaoyuan (1995), *Astrology in History*, Shanghai Science and Technology Press, Shanghai, 397pp.

Jiang Xiaoyuan (1999), Astronomical Phenomena Described in *Guo Yu* and their Occurring Years When Emperor Wu Conquered the Emperor Zhou, *Researches on History of Natural Sciences*, Vol. 18, No. 3 (in press).

*Li Qibin (1987), A Recent Study on the Historical Novae and Supernovae, in *High Energy Astrophysics, Proceedings of the Second Workshop of the Max-Planck-Gesellschaft and the Chinese Academy of Sciences*, G. Boerner, Ed., Springer-Verlag, pp. 2–25.

*Li Qibin (1992), A Sketch of the Position of Historical Supernovae in Chinese Ancient Records, in *Supernovae and their Remnants*, Li *et al.*, Eds., International Academic Publishers, Beijing, pp. 57–64.

*Li Qibin (1997), Application of Pinhole Imaging Principle in Guo Shoujing's Measurements of Tropical Year, in *Oriental Astronomy from Guo Shoujing to King Sejong*, Nha I.-S. and F.R. Stephenson, Eds., Yonsei University Press, Seoul, pp. 229–236.

Li Zhichao (1997), *Study on Water Driven Instruments*, China Science and Technology Press, Hefei, 195pp.

Li Zhichao and Chen Yu (1993), A Textual Research and Reconstruction of Water-driven Globe of Zhang Heng, *Researches on History of Natural Sciences*, Vol. 12, No. 2, pp. 120–127.

Liu Chaoyang (1944), An Elementary Study of Solar and Lunar Eclipses in the Period from the Yin Dynasty to the Beginning of Zhou Dynasty, *Series of Publications for the Study of Chinese Culture*, Vol. 4, pp. 1–10.

Liu Ciyuan (1991), Ancient Chinese Lunar Motion Records and Secular Variation of Earth Rotation, *Ph.D. dissertation*, Shaanxi Observatory, Lintong, Shaanxi, 72 pp.

Liu Ciyuan (1992), A Statistical Analysis on Ancient Chinese data of Lunar Occultation and Close Approaches, *Researches on History of Natural Sciences*, Vol. 11, No. 4, pp. 299–306.

Liu Ciyuan and Zhou Xiaolu (1999), Observation of Skylight in Northern Xinjiang during Solar Eclipse on March 9, 1997, in *Total Solar Eclipse and Environment*, Zhang H-q. *et al.*, Eds., Science Press, Beijing, in press.

Liu Nanwei, Wu Zhongzheng, Li Qibin, Li Jing and Jiang Xiquan (1989), *Ancient Nautical Astronomy in China*, Guangzhou Branch of Popular Science Press, Guangzhou, 140 pp.

Lu Dalong (1997), The Gui Mao Yuan Li and Issac Newton's Theory of the Moon, *Researches on the History of Natural Sciences*, Vol. 16, No. 4, pp. 329–336.

Pan Nai (1989), *A History of Stellar Observation in China*, Xuelin Press, Shanghai, 485 pp.

Qu Anjing, Ji Zhigang and Wang Rongbin (1994), *An Investigation of the Ancient Mathematics and Astronomy*, Xibei University Press, Xi'an, 322 pp.

Shi Yunli (1992), The Concept of Ancient about the Earth Moving towards Four Directions, *Study on Dialectics of Nature*, No. 4, pp. 43–49.

Sun Xiaochun (1994), On the Constellation of Shi Shen's School during Han Dynasty, *Researches on History of Natural Sciences*, Vol. 13, No. 2, pp. 123–138.

Sun Xiaochun (1995), The Star Atlas and Catalogue in Chongzhen Calendar, *Researches on History of Natural Sciences*, Vol. 14, No. 4, pp. 323–330.

*Sun Xiaochun and Kistmaker, Jacob (1997), *The Chinese Sky during the Han – Constellating Stars and Society*, Brill, Leiden, 240 pp.

Wang Dechang, Che Yixiong and Huang Buqing (1989), The Star Atlas Carved in Stone Preserved in Changshu, in *Collection of Studies on Astronomical Relics*, The Institute for Archaeology, Ed., Wenwu Press, Beijing, pp. 381–390.

Xi Zezong (1989), The Place of Astronomy in Chinese Traditional Culture, *Sciences* (China), Vol. 41, No. 2, pp. 90–96.

Xi Zezong and Deng Wenkuan (1989), A Dating of the Fragmentary Calendar from Dunhuang, *Publications of the Historical Museum of China*, No. 12, pp. 12–22.

Xu Fengxian (1994), View on Abnormal Celestial Phenomena in Ancient China, *Researches on History of Natural Sciences*, Vol. 13, No. 3, pp. 201–208.

Xu Zhentao and Jiang Yaotiao (1990), *The Study of the Ancient Records in China and their Modern Applications*, Nanjing Univ. Press, 386 pp.

Xuan Huancan (1992), *History of Astronomy*, High Education Press, Beijing, 406pp.

Y.C. Chang (1978), The Tendency in Orbital Evolution of Halley Comet and its Ancient History, *Acta Astronomica Sinica*, Vol. 19, No. 1, pp. 109–118.

Yi Shitong (1989), Star Atlas Preserved in Longfu Temple, in *Collected Papers on Chinese Unearthed Cultural Relics*, The Institute for Archaeology, Ed., Liaoning Education Press, Shenyang, pp. 395–400.

Zhang Peiyu (1993), A Study on the Identification and Accuracy of the Ancient Records of Lunar Eclipse in China, *Acta Astronomica Sinica*, Vol. 34, No. 1, pp. 63–79.

Zhang Peiyu (1999), The Solar and Lunar Eclipse Recorded on Turtle Bone and the Era of King Wuding of Shang Dynasty, *Culture Relics*, No. 3, pp. 66–76.

Zhang Peiyu and Han Yanben (1996), Variation of the Earth Rotation derived from Timing Records of Solar Eclipse in Ancient China, *Chinese Science Bulletin*, Vol. 41, No. 6, pp. 525–526.

Zhuang Weifeng and Wang Lixing, Eds. (1988), *Compilations of the Ancient Astronomical Records in China*, Jiangsu Science and Technology Press, Nanjing, 1100 pp.

Glossary

Bo Shuren	薄树人	Han Yanben	韩延本
Che Yixiong	车一雄	Hu Lingui	呼林贵
Chen Cheng-yil	程贞一	Hu Tiezhu	胡铁珠
Chen Dan	陈丹	Hu Weijia	胡伟佳
Chen Jiujin	陈久金	Hua Tongxu	华同旭
Chen Meidong	陈美东	Huang Buqing	黄步青
Chen Xiaozhong	陈晓中	Huang Yi-Long	黄一农
Chen Yu	陈宇	Ji Zhigang	纪志刚
Chen Zhungui	陈遵妫	Jiang Yaotiao	蒋窈窕
Cheng Zhengyi	程贞一	Jiang Xiaoyuan	江晓原
Cui Zhenhua	崔振华	Jiang Xiquan	姜锡全
Deng Wenkuan	邓文宽	Li Jing	李竞
Du Shengyun	杜升云	Li Qibin	李启斌
Du Shiran	杜石然	Li Zhichao	李志超
Feng Shi	冯时	Liu Chaoyang	刘朝阳

Liu Ciyuan	刘次源	Wu Shouxian	吴守贤
Liu Nanwei	刘南威	Wu Zhongzheng	吴钟铮
Lu Dalong	鲁大龙	Xi Zezong	席泽宗
Lu Yang	卢央	Xia Nai	夏鼐
Miao Yongkuan	苗永宽	Xu Fengxian	徐凤先
Pan Nai	潘鼐	Xu Zhentao	徐振韬
Qu Anjing	曲安京	Xuan Huancan	宣焕灿
Quan Hejun	全和钧	Y.C.Chang	张钰哲
Rao Zongyi	饶宗颐	Yi Shitong	伊世同
Shi Yunli	石云里	Zhang Peiyu	张培瑜
Sun Xiaochun	孙晓淳	Zhang Shucai	张书才
Wang Dechang	王德昌	Zhong Wanli	钟万励
Wang Lixing	王立兴	Zhou Xiaolu	周小鲁
Wang Rongbin	王荣彬	Zhuang Weifeng	庄威凤

3.3. Eclipse Records in Early Korean History: The Koryo-sa

F. Richard Stephenson
University of Durham

1. Introduction

In recent years the author has been much concerned with the accuracy and reliability of the astronomical records in Chinese and Korean history. These records (along with their Japanese counterparts) have formed the basis of investigations of past occurrences of such diverse events as aurorae, sunspots, comets, meteor showers and supernovae. Key questions relate to (i) the completeness of the East Asian records of celestial phenomena; (ii) the accuracy of the reported dates; and (iii) the reliability of the individual records.

Lately my attention has been drawn to the *Koryo-sa*, and in particular the eclipse records in this work. The *Koryo-sa* (covering the period from AD 918 to 1392) is the only true dynastic history of Korea and follows the pattern of a typical Chinese official history. Astronomical records are conveniently assembled in an astronomical treatise. This is, unfortunately, not the case at other periods in Korean history: namely the Three Kingdoms Period and the Yi Dynasty. Among celestial phenomena, eclipses have the special attraction that using recent studies of Earth's past rotation it is possible to readily and accurately compute both the dates and local circumstances of all such events (whether of the Sun or Moon) over many centuries. An investigation of the completeness and reliability of the eclipse records in the *Koryo-sa* enables an assessment to be made of the quality of the other astronomical data in the same history.

2. Eclipse Records in the Koryo-sa

The *Koryo-sa* (History of Koryo) was compiled by a team of scholars under the direction of Chong In-ji. It was completed in AD 1451, nearly 60 years after the fall of the Koryo dynasty. Although the history covers the entire Koryo dynasty (from AD 918 to 1392), it contains very few reports of celestial phenomena from before AD 1000. However, from the start of the reign of King Hyonjong (AD 1010) the astronomical records are fairly consistent. Following the pattern of a typical Chinese dynastic history, these records are grouped in the astronomical treatise (chaps. 47–49) of the *Koryo-sa*. Many of these same observations are also found in the annals (chaps. 1–46) of the history, but the astronomical treatise is a more convenient source, as well as providing more detailed information. As is typical of the astronomical chapters of a Chinese dynastic history, the treatise is presumably based on the records of the court astronomers at the capital – mainly Songdo (Kaesong), but for a short period around AD 1250 at the island refuge of Kangwha. Several Korean eclipse reports directly mention the official astronomers.

S.M.R. Ansari (ed.), History of Oriental Astronomy, 237–243.

2.1. Solar eclipses

Reports of solar eclipses are contained in the first section of the astronomical treatise (part of chap. 47), along with references to other daylight phenomena such as sunspots and solar haloes. Lunar eclipses are found in the later section of the history (the remainder of chap. 47, together with the whole of chaps. 48 and 49) along with other lunar and planetary observations, and also reports of comets, meteors and aurorae – all in strictly chronological order. The date of each event is given in a standard form: year (within the appropriate king's reign), lunar month and cyclical day number. In the case of solar eclipses, the date is almost invariably specified as the first day of the lunar month; there are only two exceptions (in AD 1049 and 1059), when the event occurred on the last day of a month. However, for lunar eclipses, the day of the month is never given. I have employed a computer program which I developed to convert dates expressed on the lunar calendar (including the cyclical day) direct to the Julian calendar.

In all, 135 solar eclipses are recorded in the astronomical treatise of the *Koryo-sa*. These range in date from AD 1012 to 1391. Most reports note only the mere occurrence of an eclipse. However, on five occasions (AD 1245, 1321, 1361, 1366 and 1390) it is asserted that the eclipse was total (but without any descriptive details). Total obscurations of the Sun are extremely rare at any one place, and computation shows that none of these five events could have actually been total at the capital, although the eclipse of AD 1245 was probably annular there (Stephenson, 1997). Possibly an original statement that an eclipse was "almost total" has been abbreviated to imply totality. In AD 1016, an eclipse ceremony is described in which the king avoided entering the palace, wore plain robes and provided food relief because of the eclipse.

There are also many examples of prediction among the accounts of solar eclipses. These include nine direct allusions to unsuccessful prediction, for instance: "the Sun should have been eclipsed but was not eclipsed" or "the Grand Astrologer reported that the Sun should have been eclipsed but it was not eclipsed". In 1320 the Yuan ambassador warned the Korean court of an eclipse and recommended the cessation of celebrations and the wearing of plain clothes by the officials. However, the text adds that the expected eclipse did not appear. On the contrary, in 1289 the official astronomers failed to predict an eclipse and were punished.

There are twelve further references to predicted eclipses when an anticipated event was not seen on account of cloud or rain. Typical entries are of the following form: "the Sun should have been eclipsed but on account of rain it was not seen". In the absence of any allusion to a prediction, I have initially presumed that an actual observation is implied, but – as computation reveals – there are many exceptions to this rule (see below).

2.2. Lunar eclipses

As many as 2115 lunar eclipses are recorded in the astronomical treatise of the *Koryo-sa*, substantially more than for their solar counterparts. This to some extent reflects the greater frequency of lunar eclipses at any one place. The records cover the period from AD 1009 to 1392. Once again, most reports note only the occurrence of an eclipse. However, on 20 occasions it is stated that an eclipse was total (but without any descriptive details). Even before analysing these records, it can be inferred that the record of total eclipses is very incomplete; about 30 per cent of all lunar eclipses in any given period are total. Eclipse ceremonies (similar to the solar rite in AD 1246) are described briefly in AD 1125 and 1157. In each case it is stated that the king, dressed in plain robes, performed a ritual because of the eclipse. Presumably similar ceremonies occurred at other eclipses, but these are not recorded.

As in the case of solar eclipses, there are many examples of prediction among the accounts of lunar eclipses. There are only two direct allusions to unsuccessful prediction (AD 1030 and 1071), when it is reported that "the Moon should have been eclipsed but it was not eclipsed". However, on five further occasions (AD 1026, 1078, 1156, 1196 and 1216) it is noted that the court astronomers failed to predict an eclipse which was actually observed. On the second of these occasions (1078), the Sung ambassador was present, causing some embarrassment to the Korean court. In 1196 the head of the astronomical bureau was punished for his negligence. Unfavourable weather was said to prevent the visibility of as many as 20 anticipated eclipses. On three of these dates (1113, 1229 and 1227) it is specifically stated that the astronomers had predicted an eclipse but it was not seen on account of cloud or rain. On the remaining dates it is simply asserted that the Moon was eclipsed but cloud or rain prevented visibility. As for solar eclipses, in the absence of any allusion to a prediction, I have initially supposed that an actual observation is implied.

3. Eclipse Computations

The eclipse computations in this paper are made using programs designed by the author. These programs incorporate the results of recent studies on Earth's past rotation (Stephenson and Morrison, 1995). The programs are designed to compute the dates and local circumstances (magnitudes, local times and altitudes) of all eclipses visible at a given location over any selected period. Computations were made for Songdo (modern name Kaesong), the Korean capital for most of the period; the distance between this site and the 13th century island refuge of Kangwha is not significant.

3.1. Computation of solar eclipses

The local circumstances of all solar eclipses which were visible in Korea between AD 1009 (the date of the earliest eclipse observation reported in the *Koryo-sa* and 1392 (the end of the Koryo dynasty) have been computed by the author and compared with the record. A summary of the results is as follows: commencing with predicted events.

Of the ten direct allusions to predictions, only on one occasion (AD 1320) was an eclipse actually visible in Korea. Computation shows that several of the remainder were visible only in the Arctic or equatorial regions. Even the eclipse of AD 1320 only reached a magnitude of about 0.05 and would probably pass unnoticed. Of the twelve anticipated eclipses which were said to be obscured by cloud or rain, four were invisible in Korea. Hence these four reports must relate to faulty predictions. Most of the remaining eight eclipses in this category could have been seen in Korea if the weather had been fine. In view of the low success rate of eclipse prediction during the Koryo dynasty (see also below), it may reasonably be inferred that most of the eclipses which were said to be clouded out at the capital were reported by provincial observers, although this is never mentioned in the record.

Computation further reveals that even for those records where there is no allusion to prediction, a significant proportion cannot have been observed. As many as 16 eclipses in this category were invisible in Korea; as for the events which were said to be predicted, some of these eclipses were only visible in the arctic or equatorial regions. In addition, the recorded eclipse of AD 1357 did not take place. Clearly, the prediction of solar eclipses in Koryo was still at a low level of attainment, although with a single exception (AD 1357), all of the expected eclipses were visible somewhere on the Earth's surface.

Deducting predictions, whether reported or inferred, of the original total of 135 solar eclipses, no more than 106 can have been *observed* at the Korean capital. This latter total falls significantly short of the number actually visible in Korea during the period from AD 1009 to 1392 – namely 136. Based on these figures, the Koryo solar eclipse records are only about 70 per cent complete. Some of the missing eclipse reports could well be due to unfavourable weather, but it also seems likely that others had been lost by the time that the *Koryo-sa* was compiled. For instance, there are significant lacunae during some reigns, notably around AD 1200 – see Fig 1a. The eclipse records from the reign of King Munjong (AD 1047–1083) are particularly curious. Although nine visible and three invisible eclipses are reported – all without any details – as many as seven more escaped notice. It is significant that during the last 50 years of the Koryo dynasty (the reigns of Kongminwang, Sin-u and Kongyangwang), every observable eclipse was reported – evidence of the marked astrological importance of eclipses to a declining dynasty.

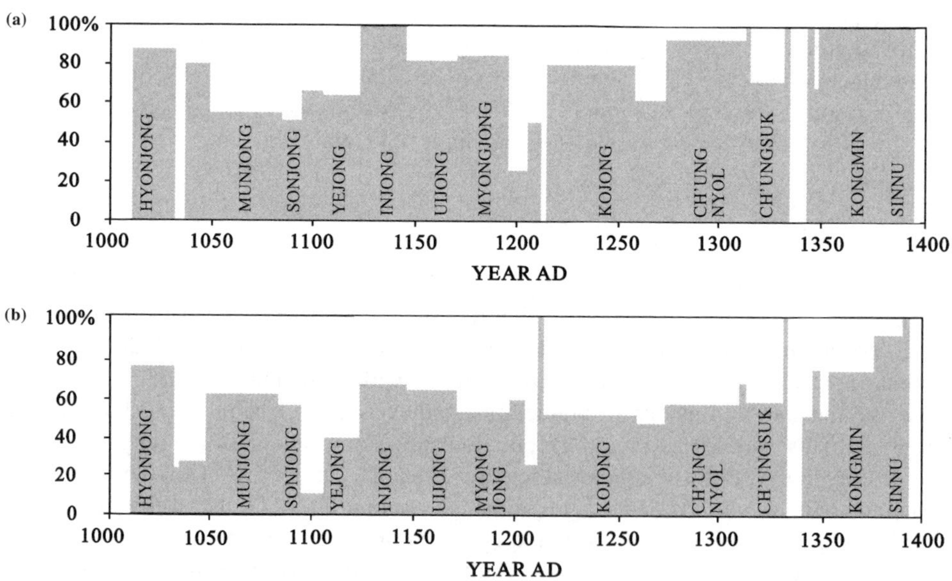

Figure 1 (a) Koryo-sa solar eclipse records: completeness. (b) Koryo-sa lunar eclipse records: completeness.

Finally in this section, it is noteworthy that almost all dates of solar eclipses are accurately recorded. Discrepancies between the reported dates (when concerted to the Julian calendar) and the corresponding computed dates seldom exceed one day.

3.2. *Computation of lunar eclipses*

As in the case of solar eclipses, the local circumstances of all lunar eclipses which were visible in Korea between 1009 (the date of the earliest observation reported in the *Koryo-sa* and 1392 were computed and compared with the record. The following is a summary of the results, commencing with predicted events.

For the two direct allusions to unsuccessful prediction (AD 1030 and 1113), computation indicates that a partial eclipse should have actually been visible. However, when the astronomers were said to have failed to predict an observed eclipse, computation confirms visibility in each case. On most of the 20 occasions when an expected eclipse was said to be invisible on account of cloud, the anticipated event indeed took place, but there are also several unsuccessful predictions. Of these, the eclipses of AD 1009, 1082, 1255, 1319 and 1374 were invisible in Korea, while on three further occasions (1322, 1372 and 1379) the Moon only entered the Earth's penumbral shadow.

The number of recorded eclipses which prove to be penumbral (twelve in all, including the three reported predictions) is surprising. Five of these occurred in the 11th century (AD 1011, 1032, 1040, 1058 and 1087) after which there are no further examples until the 14th century (AD 1322, 1350, 1361, 1372, 1373, 1379, 1390). Present-day experience shows that even the largest penumbral eclipses can easily pass unnoticed by the unaided eye. In addition, the numerous Babylonian records of lunar eclipses are devoid of a single allusion to a penumbral eclipse (Stephenson, 1997). Although three of the eclipses reported in the *Koryo-sa* were marginally umbral (penumbral magnitudes ranging from 1.07 to 1.12), cloud was said to prevent visibility on two of these occasions. The mean magnitude of the remaining nine penumbral eclipses was quite small: only 0.83. It may thus be confidently inferred that all twelve were predicted, rather than observed. In particular, numerous larger penumbral eclipses were overlooked during the same interval of time.

Discounting events which were stated to have been predicted and not seen, there remain 187 records for which observation might reasonably be inferred. However, computation indicates that of these eclipses, only 170 would actually be visible in Korea; the remaining 17 must represent failed predictions. Of these predictions, apart from penumbral eclipses, in three instances (AD 1012, 1106, 1120) no eclipse took place, while a further five eclipses (1037, 1050, 1145, 1199 and 1222) would be invisible in Korea. Either the extant records are deficient or the astronomers reported a significant number of non-events. On the other hand, computation does indicate that all the eclipses recorded as total were indeed so, although many total eclipses were not recorded as such.

Computation further demonstrates that the extant records are far from complete; many eclipses are overlooked in the *Koryo-sa* records. In the period from AD 1010 to 1392, some 340 lunar eclipses would be visible in Korea (weather permitting). However, only 188 of these events (or 55 per cent) are actually recorded – a total which itself includes 18 reported predictions. Ignoring these predictions, this total reduces to only 50 per cent, a fraction which is substantially less than for solar eclipses. Unfavourable weather can only have been a partial factor in accounting for the missing lunar eclipse reports. Possibly the official astronomers were less vigilant during the night.

A summary of the completeness of the lunar eclipse records reign by reign is given in Fig 1b. During some reigns, whole series of eclipses are overlooked, suggesting deficiencies in the preserved records rather than in the level of original observation. Thus in the reign of King Sukchong (AD 1096–1105), although 11 eclipses should have been visible, only a single event (the very last in Sukchong's reign) is actually recorded. Similar remarks apply in the reign of King Huijong (AD 1205–1211), when only two of the eight eclipses were reported and even one of these was a prediction. During the reign of King Myongjong (AD 1171–1197), nine of the first 12 lunar eclipses (up to AD 1183) were recorded, but after that date there is a sequence of omissions from AD 1185 to 1191, during which 8 eclipses which should have been visible are not mentioned. On the other hand, the efficiency in the last half century of the Koryo Dynasty is relatively high: around 80 per cent. Either the degree of preservation of records in this period is much better or the astronomers were more efficient; as noted above for solar eclipses, near the end of a dynasty which was clearly in decline much more interest would be taken in celestial portents.

The accuracy of the recorded dates of solar eclipses has been shown to be very high. However, this would be greatly assisted by the consistent occurrence of these events on the first day of a lunar month. No such check is possible on the dates of other celestial phenomena, including lunar eclipses. Nevertheless, the dating accuracy for eclipses of the Moon is also extremely high, both for observations and predictions – apart from the three instances when no eclipse occurred on or near the stated date. There is only a single example of an error of more than one day: the recorded date of the eclipse of 1082 Nov 8 is actually 3 days too early.

An interesting aspect concerns the time of night when the calendar date was changed. In East Asia, although the civil day began at midnight, it has been suggested that the astronomical day began in the early hours of the following morning (Kiang, 1972). This is confirmed by the lunar eclipse records in the *Koryo-sa*. Although the reported dates of most lunar eclipses in this work give the evening date, a substantial number carry the date of the following morning. For example, the recorded date of the eclipse which occurred on the night of AD 1012 Aug 4/5 corresponds to Aug 4, while for the eclipse of 1013 Jan 29/30 the reported date is equivalent to Jan 30. There are as many as 31 instances of this latter convention, which was especially prevalent from AD 1030 to 1090 and again from 1360 to the end of the dynasty. Computation shows that such morning dates are most frequent when the Moon set eclipsed; i.e. observation commenced near dawn. However, it is clear that no set rule was adopted.

4. Conclusion

Investigation of the eclipse records in the *Koryo-sa* reveals some striking features, notably the presence of many predictions which judging from the reports alone are indistinguishable from observations. Dates of both solar and lunar events are almost invariably correct to within one day, which strongly suggests that the dates of other celestial events reported in the *Koryo-sa* are usually accurate. However, the completeness of the records is very uneven, which in part is probably due to the loss of material by the time the *Koryo-sa* was compiled. These conclusions lead to the suggestion that reports of other celestial phenomena in the *Koryo-sa* – e.g. comets or meteor showers – might be similarly incomplete. For a general discussion of statistics of astronomical records in Korea from AD 1200 to 1450, see Kim Yonggi (1997).

The fact that solar eclipse records are much more complete than their lunar counterparts could well be due to the great astrological significance of such events. This discrepancy is especially marked in view of the fact that lunar eclipses are much easier to detect with the unaided eye than solar obscurations.

No doubt Korean records of celestial phenomena other than eclipses are affected by similar factors. However, the false sightings of eclipses probably all owe their origin to faulty predictions. Prediction of other celestial events other than eclipses was seldom practised throughout East Asia. This suggests a significantly higher degree of reliability for the reports of these phenomena, if not completeness.

References

Kiang, T., 1972. "The past orbit of Halley's Comet". *Memoirs of the Royal Astronomical Society*, **76**, pp. 27–66.

Kim Yonggi, 1997. "Statistics of astronomical records in Korea during AD 1200–1450". In *Oriental astronomy from Guo Shoujing to King Sejong* (ed. Nha Il-Seong and F.R. Stephenson). Yonsei University Press, Seoul, pp. 143–158.

Stephenson, F.R., 1997. *Historical Eclipses and Earth's Rotation*. Cambridge University Press.

Stephenson, F.R. and Morrison, L.V., 1995. "Long-term fluctuations in the Earth's rotation: 700 BC to AD 1990". *Philosophical Transactions of the Royal Society of London, Series A*, **351**, pp. 165–202.

FULL PROGRAMME OF THE JOINT DISCUSSION

Monday, August 25 1997

Oriental Astronomy during the Ancient and Medieval Period

Session 1 (9:00–10:30)	**Chair: S.M.R. Ansari (India)**
On the Earliest Stage of Chinese Astronomy: Three Hypotheses	Y. Maeyama (Germany)
Islamic Astronomy in China: Two New Sources for the Islamic Calendar *Huihui-li*	B. van Dalen (Netherlands) and M. Yano (Japan)
An Arabic Commentary on al-Ṭūsī's *al-Tadhkirah* and its Sanskrit Translation	T. Kusuba (Japan)
Coffee break (10:30–11:00)	

Session 2 (11:00–12:30)	**Chair: M. Yano (Japan)**
Indian Astronomy in Ancient China	J. Xiao (Yuan (China)
Three Korean Star Maps of the 18th Century	Il-Seong Nha (Korea)
A New Museum of Astronomy in Korea	Il-Seong Nha (Korea)
Knowledge of Starry Sky and Agricultural Activities in Indonesia: A Review	B. Hidayat (Indonesia)
The Projection Method in Chinese, Korean and Japanese Star Maps	K. Miyajima (Japan)

Session 3 (14:30–15:30)	**Chair: D. Pingree (USA)**
On the Obliquity of the Ecliptic	K.-Y. Chen (USA)
A Note on the Vedāṅga Astronomy	Y. Ôhashi (Japan)
Spherical Trigonometry in the Astronomy of the Medieval Kerala School	K. Plofker
Astronomical Dating and Statistical Analysis of Ancient Eclipse Data,	K.D. Pang, K.K. Yau and H.-H.Chou (USA)
Coffee break (15:30–16:00)	

Modern Astronomy in the Orient

Session 4 (16:00–17:30)	**Chair: S. Dick (USA)**
Dṛikapakṣasāraṇī: A Sanskrit Version of de La Hire's *Tablulae Astronomicae*	D. Pingree (USA)
Modern Astronomy in Indo-Persian Sources	S. M. R. Ansari
Takamine and Saha: Contacts with Western Astrophysics	D. H. DeVorkin (USA)
Astronomy Education in the East	S. Isobe (Japan)

Tuesday August 26, 1997

Session 1 (9:00–10:30)	**Chair: S. Débarbat (France)**
Kepler's Laws in China as Observed in *Lifa Wenda*	K. Hashimoto (Japan)
Contemporary Astronomy in Iran	Y. Sobouti (Iran)
Power and Politics in Nineteenth Century Australian Astronomy	W. Orchiston (New Zealand)
Status of Astronomy in Uzbekistan	S. Ehgamberdiev (Uzbekistan)

Poster Session

Astronomical Observations in Asia from Delisle's Manuscripts preserved in the Paris Observatory Library	S. Débarbat (France)

About the Contributors

Ansari, S. M. Razaullah (born in 1932) is a former Professor of Physics, Aligarh Muslim University (Aligarh/India). By training he is a theoretical/mathematical physicist. He secured his B.Sc. (Honours) and M.Sc. degrees in Physics from the University of Delhi, and his D.Sc. (Dr. rer. nat.) from Eberhard Karl University at Tübingen (Germany), with a dissertation on a problem of mathematical physics. During his stay in Germany (1959–69) he had been working as a Fellow of Alexander Von Humboldt Foundation (Bonn), and as Research Fellow and Research Associate of the German Council of Research. There, he also specialized in the history of Islamic science in general and the history of Indian and Islamic mathematics in particular.

Ansari has researched in solar physics and interstellar matter. Later he shifted to the history of exact science in Medieval and Modern India. His pioneering paper is: "Introduction of Modern European Astronomy in India during 18th–19th Centuries", published in *Indian Journal of History of Science* (Vol. 20, 1985), which was also published as a monograph. Thereafter, he shifted to history of astronomy in Medieval India. He has completed a critically edited Persian Text of Astronomical Tables of Maharaja Swai Jai Singh – the *Zīj-i Muḥammad Shāhī* – a project sponsored by the Indian National Science Academy (New Delhi).

Ansari has to his credit about 60 research papers, the above-mentioned monograph and this edited volume on *History of Oriental Astronomy.* He is quite known on the international level through his participation in the International Congresses of History of Science (of IUHPS), which he has been attending since 1977. Notably, he has also held a number of international offices, namely, Member of the IUHPS Executive Council as an assessor (1989–93, 1994–97), President of IUHPS Commission for Science and Technology in Islamic Civilization (1993–97), President of IUHPS-IAU Joint Inter-Union Commission for History of Astronomy (1997–2001), President of the IAU Commission for History of Astronomy (1994–97), and he is the current President of the IUHPS Commission for History of Ancient and Medieval Astronomy (2001–2005) and the Indian Society for History of Astronomy.

Moreover, Ansari was elected a Fellow of the Royal Astronomical Society (1972), a member of the International Astronomical Union (1973), and is a full (effective) member of the International Academy of History of Science (Paris), since 1997. He is also a life member of the Astronomical Society of India and of the Indian Society of History of Mathematics. Further, he is a former member of the Indian National Commission for History of Science and has been on the Editorial Board of *Indian Journal of History of Science* (New Delhi), editor of the international journal: *Studies in History of Medicine and Science* (New Delhi) for 1984–2000, and consulting editor of the *Science Net Work Series*, published by Birkhauser Verlag, Basel. Currently he is on the editorial board of *Gaṇit Bhārtī*, the Bulletin of the Indian Society of History of Mathematics.

Besides giving finishing touches to his *Critical Edition* of the Persian text of Jai Singh Astronomical Tables (with English translation and commentary), Ansari is also working on a Monograph: *History of Islamic Science in Medieval India.* He intends also to edit the

Proceedings of the Symposium: "*Astronomical Heritage of the Non-European Cultural Areas*", organised by him at the 21st ICHS (Mexico, 2001). The following are his recent publications:

> "Transmission of Arabic–Islamic Astronomy to Medieval India", *Archives Internationales d'Histoire des Sciences*, Vol. 45 (1995), No.135, pp. 273–297.
>
> "Promotion of Astronomy by Indian Monarchs During the 15th–18th Centuries", in *Astronomy in the Time of King Sejong*, Proceedings of an International Conference to Commemorate the 600th Anniversary of His Birth, (held on September 1st, 1997, Daejeon, Korea), Eds. Kyung J. Sim and Changbom Park, Korean Astronomical Observatory, Daejeon (Korea), 2001, pp. 58–75.
>
> "Islamic Astronomy in India During 16th–18th Centuries and its Interaction with the Traditional Astronomy", *Proceedings of the Conference to Commemorate the 500th Anniversary of Tantrasanhagarha by Nilakantha Somayaji,* held in Chennai, March 11–13, 2000, Ed. S. Sriram, (Indian Institute of Advance Studies), Simla (India), 2002, pp. 145–156.
>
> "Practical Astronomy in Indo–Persian Writings", Paper presented at the Joint Discussion on Applied Historical Astronomy, held during the 24th General Assembly of the International Astronomical Union, and organized by the Commission 41 (History of Astronomy), in Manchester (UK), Aug. 11, 2000, *Indian Journal of History of Science*, Vol. 37, No. 3 (2002), pp. 255–265.

E-mail: Raza.Ansari@gmx.net

Chen, Kwan-Yu was born in Shanghai, China. After graduation from Nanyang Model Middle School, he went to the USA for a higher education. He received a Bachelor of Science degree and a Master of Science degree in mechanical engineering from the Illinois Institute of Technology. He then studied astronomy and received his Ph.D. in astronomy from the University of Pennsylvania. He worked as a design engineer at International Harvester 1956–57), a research assistant at the National Radio Astronomy Observatory (summer 1960), an assistant instructor in mathematics at the University of Pennsylvania (1960–63), and an instructor in astronomy at the Pennsylvania State University (1962–63).

He began his teaching and research career at the University of Florida in 1963, first as an assistant professor, then an associate professor and a professor. His principal research interest lay in the study of close binary stars. During the academic year 1974–75, he was a visiting professor at Adam Mickiewicz University in Poznan, Poland. He retired in 1999, and is now a professor emeritus at the University of Florida. He continues his work in the field of history of astronomy. He is a member of the International Astronomical Union, and has joined Commissions 41 and 42. His most recent publications are:

> "The East–West Interaction: A Background to Oriental Astronomy in the Thirteen–Fifteenth Centuries", in *Oriental Astronomy from Guo Shoujing to King Sejong*, Eds. I.-S. Nha and F. R. Stephenson, Yonsei University Press, Seoul, 1997, pp. 87–104.
>
> "A Study of the Motions of Rahu and Ketu", in *Proceedings of The Third International Conference on Oriental Astronomy*, held Oct. 27–30, 1998, Ed. M. Hirai, Fukuoka University of Education Press, Fukuoka, 2000, pp. 93–96 (with E.-H. Lee).

E-mail: chen@astro.ufl.edu

Chou, Hung-hsiang studied at the University of Hong Kong under the guidance of Professor Tung Tso Pin, a pioneer in Shang dynasty (1600–1100 B.C.) oracle bone research. He published *Imperial Records of the Shang and Yin Dynasty Kings* in 1958. He later received a Ph.D. from the Australian National University, and has been a Professor at the University of California, Los Angeles since.

His important publications are:

Oracle Bone Collections in the United States, University of California Press, Berkeley and Los Angeles, 1976.

"Chinese Oracle Bones", *Scientific American*, Vol. 240 (1979), pp. 135–149.

"Computer Analysis of Some Ancient Chinese Sunrise Eclipse Records to Determine the Earth's Rotation Rate", *Vistas in Astronomy*, Vol. 31 (1988), pp. 833–847, (with Kevin D. Pang).

"Shang Dynasty Oracle Bone Solar and Lunar Eclipse Records: Astronomical Dating and Statistical Analysis", *Proceedings of the International Symposium Commemorating the 100th Anniversary of the Discovery of Shang Dynasty Oracle Bones*, Anyang, China, Institute of History, Chinese Academy of Social Sciences, Beijing, 2002 (with Kevin D. Pang).

Dalen, Benno van received his Ph.D. in 1993 from the Mathematics Faculty of Utrecht University (Netherlands). In his dissertation he developed statistical methods for analysing mathematical properties of ancient and medieval astronomical tables and applied these methods in order to draw historical conclusions about the relationships of tables and astronomical handbooks. In recent years he has been working on two projects: a new version of Kennedy's "Survey of Islamic Astronomical Tables" of 1956 which includes information on more than 100 astronomical handbooks that have since become known, and a study of an Islamic astronomical handbook prepared by Muslims in Yuan China (late thirteenth-century) and translated into Chinese 100 years later. The former project is currently being supported by the Deutsche Forschungsgemeinschaft (DFG), the latter has been supported by the Japanese Society for the Promotion of Science (JSPS) and the Dibner Institute (Cambridge MA). His most important recent publications are as follows:

"A Statistical Method for Recovering Unknown Parameters from Medieval Astronomical Tables", *Centaurus* 32 (1989), pp. 85–45.

"On Ptolemy's Table for the Equation of Time", *Centaurus* 37 (1994), pp. 97–153.

"Al-Khwārizmī"s Astronomical Tables Revisited: Analysis of the Equation of Time", in: *From Baghdad to Barcelona. Studies on the Islamic Exact Sciences in Honour of Prof. Juan Vernet* (Eds. Josep Casulleras and Julio Samsó), Barcelona (Instituto Millás Vallicrosa) 1996, pp. 195–252.

"A Non-Ptolemaic Islamic Star Table in Chinese", in *Sic itur ad astra. Studien zur Geschichte der Mathematik und Naturwissenschaften. Festschrift für Paul Kunitzsch zum 70. Geburtstag* (Eds. Menso Folkerts and Richard P. Lorch), Wiesbaden (Harrassowitz) 2000, pp. 147–176.

"Origin of the Mean Motion Tables of Jai Singh", *Indian Journal of History of Science*, Vol. 35, No.1 (2000), pp. 41–66.

E-mail: dalen@em.uni-frankfurt.de

DeVorkin David is curator, history of astronomy and the space sciences, at the National Air and Space Museum, Smithsonian Institution, Washington DC, USA. He got his M.Phil. degree (in Astronomy) from Yale and his Ph.D. degree in the history of astronomy from the University of Leicester for his 1978 thesis: "An Astronomical Symbiosis: Stellar Evolution and Spectral Classification in Early Astrophysics, 1860–1910".

DeVorkin has written extensively, mainly on topics concerning astrophysics and astronomy. Specifically, he has to his credit about 110 publications, consisting of nine monographs, chapters in books, entries in encyclopaedias, and articles published in well-known peer-reviewed journals. His recent publications are:

Henry Norris Russell: Dean of American Astronomers, Princeton University Press, Princeton, 2000.

Beyond Earth: Mapping the Universe, National Geographic Society, Washington, D.C., 2002. This is an edited collection of essays based on a new exhibition he has curated, entitled "Explore the Universe".

"Hybrid Studies: Solar System Astronomy in America – Essay Review of Ron Doel, *Solar System Astronomy*", *Studies in the History and Philosophy of Modern Physics*, Vol. 1, No. 31 (2000), pp. 99–103.

"Who Speaks for Astronomy? How astronomers responded to government funding after World War II", *Historical Studies in the Physical and Biological Sciences*, Vol. 31, pt 1 (2000), pp. 55–92.

"Quantum Physics and the Stars V: Physicists at Mount Wilson Prior to 1922", *Journal for the History of Astronomy,* Vol. 31 (2000), pp. 301–321.

"Internationalism, Kapteyn, and the Dutch Pipeline", in P. C. van der Kruit and K. van Berkel, Eds., *The Legacy of J. C. Kapteyn*, Kluwer Academic, Dordrecht, 2000, pp. 129–150.

"'Space Science', 'Astronomy and Astrophysics', and 'Optical Telescopes'", in *Reader's Guide to the History of Science*, Ed. A. Hessenbruch, Fitzroy Dearborn, London, 2001.

"Menzel at Princeton", *J. History of Astronomy*, Vol. 33 (2002), pp. 119–131.

"Was Pluto Discovered? Contrasting Standards of Practice in Mathematical Astronomy and the New Astrophysics", for *Curtis Wilson Festschrift Volume*, Dibner Institute for the History of Science, MIT, Cambridge, expected in 2003.

E-mail: david.devorkin@nasm.si.edu

Hashimoto, Keizo (born in Kobe in 1941) was educated in Science and Astrophysics at Kyoto University, and in History and Philosophy of Science at Cambridge University. He had also the privilege of studying Chinese History of Science under the supervision of Prof. Joseph Needham, and originally of Prof. K. Yabuuti at Kyoto University. At Cambridge, he was awarded his Ph.D. degree for his dissertation entitled *Chinese Acceptance of European Astronomy: 1629-1635*, which was later published as a book (see the first publication below).

Hashimoto has worked at the Research Institute for Humanities, Kyoto University, and at Needham Research Institute, Cambridge. He is now Professor of History and Philosophy of Science and Technology, belonging to Faculty of Sociology, Kansai University, Osaka (Japan). His main field of current research is the history of astronomy and science, particularly in China. His interest also covers the philosophy of science as well as the history of technology. He has so far published numerous papers and books concerning Chinese astronomy, science and technology in Japanese, also several in English. His main publications in English are:

> *Hsü Kuang-Ch'i and Astronomical Reform: The Process of the Chinese Acceptance of Western Astronomy 1629–1635*, Kansai University Press, Osaka, 1988.
>
> *East Asian Science: Tradition and Beyond*, Papers from the Seventh International Conference on the History of Science in East Asia, Kyoto, Aug. 2–7, 1993, Eds. K. Hashimoto, C. Jami and L. Skar, Kansai University Press, 1995.

E-mail: keizo@ipcku.kansai-u.ac.jp

Maeyama, Yas is a Professor in the Institute for History of Science, Goethe University, Frankfurt (Germany). Although he graduated as an engineer, he shifted later to history of science and secured his Ph.D. (Dr. rer. nat.) in 1971 under the supervision of the late Prof. Willi Hartner for his dissertation "Hypothesen zur Planetentheorie des 17. Jahrhunderts". Since then his researches focused on solar theories, Babylonian lunar theory, Greek and Chinese stellar observations (star catalogues) etc. Besides editing (with W. G. Saltzer) Willi Hartner's *Festschrift: Prismata* (Wiesbaden, 1977), and Collected Works: *Oriens-Occidens*, Vol. II (Hildesheim, 1984), Maeyama's major publications are the following:

> "Ancient Stellar Observations: Timocharis, Aristyllus, Hipparchus, Ptolemy – the Dates and Accuracies", *Centaurus*, Vol. 27 (1984), pp. 280–310.
>
> "Astronomical Periods in the Solar System", in T. Sasao *et al.* (Eds.), *Progress and Future Observational Possibilities*, Tokyo 1994, pp. 295–305.
>
> "The Stellar Reference-Points in Ancient China and the Evolution of Positional Astronomy", in *East Asian Science: Tradition and Beyond*, Eds. K. Hashimoto, C. Jami and L. Skar, Kansai University Press, Osaka, 1995, pp. 385–394.
>
> "Determination of the Sun's Orbit: Hipparchus, Ptolemy, al-Battānī, Copernicus, Tycho Brahe, *Archive of History of Exact Sciences*, Vol. 53 (1998), pp. 1–49.
>
> "Zur geozentrischen Planetenbewegung: Methoden zum Studium der Astronomiegeschichte", *Mathesis*: *Festschrift für M. Schramm*, Ed. R.Thiele, Diepholz, Berlin, 2000, pp. 282–297.

E-mail: Maeyama@em.uni-frankfurt.de

Miyajima, Kazuhiko (born in 1946) graduated from the Department of Astrophysics, Kyoto University, in 1969 and entered graduate school there. He studied History of Chinese Astronomy under Prof. Kiyosi Yabuuti. He was employed by Doshisha University as a teacher of history of science and that of modern astronomy. He is presently Professor in Science and Engineering Research Institute, Doshisha University, Kyoto, Japan. Besides his numerous writings in Japanese, he is the author of the following publications in English:

> "Wang Xichan's Model of the Solar Systems in His Xiao'an Xinfa", in Kim Yung Sik and Francesca Bray (Eds.): *Current Perspectives in the History of Science in East Asia*, Seoul National University Press, Seoul, 1999, pp. 347–352.

> "Japanese Celestial Cartography before the Meiji Period", in *Cartography in the Traditional East and South-East Asian Societies*, Book 2 of *The History of Cartography*, Eds., J. Harley and David Woodward, The University of Chicago Press, 1994, Chicago, pp. 579–603.

> "On the Constellation Plate of the Changdok Palace", *Journal of the Korean History of Science Society*, Vol. 5, No.1 (1983), pp. 139–147, translated into Korean and with Japanese summary.

> "New Identification of Islamic Astronomical Instruments Described in *Yuan-shi*", in *Science and Skills in Asia*, edited by the "Publication Committee", Kyoto 1982, pp. 407–427.

Nakayama, Shigeru (born in 1928) graduated from the University of Tokyo in 1951 with an astrophysics major. Since then, his interest shifted to the history of science, a more humanistic aspect of science. He visited the USA as a Fulbright Fellow and was conferred his Ph.D. degree in the History of Science and Learning (history of universities) at Harvard in 1959. He had the privilege to have teachers such as K. Yabuuti, Thomas Kuhn and Joseph Needham. His dissertation was later published as a book: *A History of Japanese Astronomy: Chinese Background and Western Impact*, Harvard University Press, Cambridge (Mass.) 1969, 329 pp.

In 1960, he joined the Faculty of Arts and Sciences, University of Tokyo, until his retirement in 1989. In 1989, his post was shifted from the University of Tokyo to the School of International Business Administration, Kanagawa University, where he had been Professor of STS Centre. He is now Professor Emeritus retired, but still teaches graduate students. During his research career, his interest shifted gradually from the history of traditional East Asian sciences to more contemporaneous science policy studies. He took the Kyoto University Scientific Mission to Iran, Afghanistan and Pakistan in 1959, and has been a visiting Professor to Harvard (1972–73); to CSIR, New Delhi (1977–78); UC, Berkeley (1981–82) and to the Japanese Research Centre at Beijing (1990).

Besides being a member of several learned societies, Nakayama is also a member of IAU, and a full (effective) member of the International Academy of History of Science (Paris), of which he has been also a Vice-President.

Besides his above-mentioned *classical* book, another of his important publications in English is: "Japanese Scientific Thought" in *Dictionary of Scientific Biography*, Ed. Charles C. Gillispie, Vol. XV, Charles Scribner's, New York, 1978, pp. 728–758. He has to his credit more than 30 books in Japanese and 7 books in English. His most recent publications of interest are:

> *A Social History of Science and Technology in Contemporary Japan*, Vol. 1, Trans Pacific Press, Melbourne (Australia), 2001, 632 pp.

Science, Technology and Society in Contemporary Japan, Cambridge University Press, 1999, 226 pp. (with M. Low and H. Yoshioka).

"History of Astronomy Before and After the Computer", *Proceedings of the Third International Conference on Oriental Astronomy*, held October 27–30, 1998, edited by Masanori Hirai, Fukuoka University of Education, Fukuoka, Japan, pp. 35–38.

"The Spread of Chinese Science into East Asia", in *Current Perspectives in the History of Science in East Asia*, edited by Y. S. Kim and F. Bray, Seoul National University Press, Seoul, 1998, pp. 13–20.

"The Chinese 'Cyclic' View of History vs. Japanese 'Progress'", in *The Idea of Progress*, edited by A. Burgen, P. McLaughlin and J. Mittelstraß, Walter de Gruyter, Berlin, 1997, pp. 65–75.

For further details of his works, see his web page:
http://homepage2.nifty.com/shigeru-histsci/
E-mail: KFH00250@nifty.com

Nha, Il-Seong (born in 1932) secured his B.Sc. and M.Sc. (Physics) degrees from Yonsei University Seoul (Korea) and Ph.D. (Astronomy) in 1971 from the University of Pennsylvania, Philadelphia, USA, where he later worked as a Research Associate. After his return to Korea, he was an Associate Professor at the Department of Astronomy and Meteorology, Yonsei University, Seoul (1974–1977) and Professor (1978–1998). During 1981–1989, he acted as the Director of Yonsei University Observatory and since 1989 he is Professor Emeritus at Yonsei University, Seoul, and also George L. Paik, Endowment Professor for 2002–2003.

Nha is well known on a national and international level. He has also been Visiting Professor at the University of Florida (January–April 1981), Universitat Nacionale Autonomy de Mexico (Summer 1981) and recently at the Pedagogical University of Krakow. Further, he is a member of the Board of Trustees of King Sejong Memorial Society (since 1991) and of the Korean–German Literature Institute (since 1991), and a member of the Editorial Board of the *Journal of Astronomical History and Heritage* (since 1998). He is also ex-President of the Korean Astronomical Society, and presently Chairman of the Working Group of Historical Instruments of IAU Commission 41.

Nha began his astronomical career at the Flower and Cook Observatory and his professional interest was originally in the photometric studies of eclipsing binary stars. Later he shifted to the history of Korean astronomy, particularly to the restoration of ancient star maps and sundials. Moreover, he is one of the founders of the International Conferences of Oriental Astronomy, first held in 1993, and his crowning achievement is the establishment of "The Nha Il-Seong Museum of Astronomy" in Yecheon (Korea) in 1999, by means of his own resources.

Nha is the author and co-author of about 20 books and two Conference Proceedings. In the following only a few of his recent publications are listed:

"*History of Korean Astronomy*", Seoul National University Press, Seoul, 2000.

"*An Atlas of O-C Diagrams of Eclipsing Binary Stars*", Parts 1–6, Pedagogical University Press, Krakow , 2001, (with J. M. Kreiner and C. H. Kim).

"*Story of Solar and Lunar Eclipses*", King Sejong Memorial Society, Seoul, 2002, (with J. B. Lee).

E-mail: SLISNHA@chollian.net

Ohashi, Yukio (born in 1955) obtained his B.Sc. (Physics) from Saitama University (Japan) in 1979, and his M.A. (Chinese Culture) also from Saitama University in 1981. He studied the history of Indian astronomy under the guidance of Prof. K. S. Shukla at Lucknow University (India) from 1983 to 1987, and obtained his Ph.D. (History of Mathematics) from Lucknow University in 1992. He completed the postgraduate (doctorate) course of Hitotsubashi University (Japan) in Social Studies (Social History of the East) in 1989. Some of his recent papers are:

"Originality and Dependence of Traditional Astronomies in the East", in Chan, Alan K.L. *et al.* (Eds.), *Historical Perspectives on East Asian Science, Technology and Medicine*, Singapore University Press and World Scientific, Singapore, 2002, pp. 394–405.

"Early History of the Water-Driven Clock in the East", *Technology and History*, Korean Society for the History of Technology and Industry, Vol. 1, No.1, 2000, pp. 37–67.

"Remarks on the Origin of Indo-Tibetan Astronomy", in Selin, H. (Ed.): *Astronomy Across Cultures*, Kluwer Academic, Dordrecht, 2000, pp. 341–369.

"Historical Significance of Mathematical Astronomy in Later-Han China", in Kim, Yung Sik and Francesca Bray (Eds.): *Current Perspectives in the History of Science in East Asia*, Seoul National University Press, Seoul, 1999, pp. 259–263.

"Preliminary Remarks on the Origin of 'Mori' and 'Mieri' in Chinese Calendars", in Tatsuhiko Kobayashi *et al.* (Eds.), Proceedings of the Fourth International Symposium on the History of Mathematics and Mathematical Education using Chinese Characters, Maebashi, 1999, pp. 97–102.

"The Cylindrical Sundial in India, Supplement to *Indian Journal of History of Science*, Vol. 33, No. 4 (1998), pp. S147–S205.

"Early History of the Astrolabes in India", *Indian Journal of History of Science*, Vol. 32, No. 3 (1997), pp. 194–295.

E-mail: yukio-ohashi@dk.pdx.ne.jp

Orchiston, Wayne is a Research Associate at the Anglo-Australian Observatory, and Archivist at the Australia Telescope National Facility, both of which are in Sydney. He has B.A. Honours and Ph.D. degrees from the University of Sydney, and formerly worked at the Division of Radiophysics, C.S.I.R.O. (Sydney); at Sydney Observatory (part-time); headed the Astronomy Group at Victoria College (now Deakin University) in Melbourne; and was Executive Director of the National Observatory of New Zealand, in Wellington. In 1985 he was a Visiting Fellow at Mt. Stromlo and Siding Spring Observatories in Canberra.

He is one of the Editors of the Journal of Astronomical History and Heritage; a Fellow of the Royal Astronomical Society; Secretary of IAU Commission 41 (History of Astronomy); and also Secretary of the IAU–IUHPS (DHS Division) Inter-Union Commission for History of Astronomy since 2001.

His research interests lie mainly in the history of the Cook voyages; Australian and New Zealand astronomy; historic transits of Venus; early radio astronomy; astronomical education; and meteoritics. He has more than 160 astronomy publications, including astronomical biographies in the multi-volume *Dictionary of New Zealand Biography*, edited by Orange, C. *et al.*, Auckland University Press and Department of Internal Affairs. Auckland. To note are the following recent books and articles:

Nautical Astronomy in New Zealand: The Voyages of James Cook, Carter Observatory Wellington, 1998, 131 pp.

"Australian Aboriginal, Polynesian and Maori Astronomy", in Walker, C. (Ed.), *Astronomy Before The Telescope*, British Museum, London, 1996, pp. 318–328.

"From the South Seas to the Sun: The Astronomy of Cook's Voyages", in Lincoln, M. (Ed.), *Science and Exploration in the Pacific. European Voyages to the Southern Oceans in the Eighteenth Century*, Boydell Press in association with the National Maritime Museum, Woodbridge, 1998, pp. 55–72.

"A Polynesian Astronomical Perspective: The Maori of New Zealand", in Selin, H. (Ed.), *Astronomy Across Cultures: The History of Non-Western Astronomy*, Kluwer Academic, Dordrecht, 2000, pp. 161–196.

"Southern Hemisphere Observations", in Andersen, J. (Ed.), *Highlights of Astronomy, papers presented at the XXIVth General Assembly of the IAU*, Astronomical Society of the Pacific San Francisco, 2002, in press.

E-mail: wo@aaoepp.aao.gov.au

Pang, Kevin D. received a Ph.D. from UCLA in 1970, and has worked in NASA planetary exploration and Earth applications programs since. His roles include a Co-Investigatorship in Voyager to Jupiter and Saturn, and Viking landings on Mars; and Principal Investigator-ship on several projects. He was also a consultant to the RAND Corp. He has published six papers in the prestigious *Science* and *Nature*.

Pang's historical research was recognized by Pollock and Dudley awards. He served a member of a National Science Foundation commission to China in 1985 to investigate historical floods. His major publications are:

"Extraordinary floods in early Chinese history and their absolute dates", *J. Hydrol.* Vol. 96 (1987), pp. 139–155.

"Postglacial rebound and other influences on the Earth's secular rotation rate from analysis of ancient eclipse records", in *Dynamics of the Ice Age Earth: A Modern Perspective*, P. Wu, Ed., Trans Tech, Zurich, Switzerland, 1999, pp. 459–488.

"Absolute chronology of the Xia, Shang and Zhou dynasties by dating 17 eclipses, in *21st Century Chinese Astronomy Conference*, K. S. Cheng and K. L. Chan, Eds., World Scientific, Singapore, 1997, pp. 523–526.

"The legacies of eruptions", *Sciences*, Vol. 31 (1991), pp. 30–33.

"Astronomical evidence for the *Bamboo Annals*' chronology of early Xia", *Early China* Vol. 15 (1990), pp. 87–95, (with David S. Nivison).

"Ancient observations link sun's brightness and Earth's climate", *Eos (Transactions of the American Geophysical Union)*, 2002, in press.

E-mail: KevinDPang@aol.com

Pingree, David (born in 1923) is University Professor at Brown University and holds the Chair of its Department of the History of Mathematics, succeeding the grand legendary figure of Historian of Science in Antiquity, Otto Neuegebaur. He has written extensively on the transmission of the exact sciences from one culture to another in antiquity and the medieval period. His researches focus on astronomy, mathematics, astrology, divination, and astral magic in ancient, medieval, and Renaissance cultures in West and South Asia and Europe.

His several hundred publications are well known to historians of science, namely, books, chapters in books, and articles. For a full bibliography see the forthcoming *Festschrift*: *Ketuprakāśa: Studies in the History of the Pre-modern Exact Sciences in Honour of David Pingree*, edited by C. Burnett, J. Hogendijk, and K. Plofker, to be published shortly by E. J. Brill, Leiden. A sample of his most famous and classical works are listed in the following:

The Pañcāsiddhāntikā of Varāhamihira, 2 Vols., Danish Royal Academy, København, 1970–1971, (with O. Neugebauer).

Census of the Exact Sciences in Sanskrit, Series A, Vols. 1–5: American Philosophical Society, Philadelphia, 1970–1994.

The Astronomical Works of Gregory Chioniades, Vol. 1, *The Zīj al-'Alā'ī*, Part 1, Text, translation and Commentary, Part 2, Tables; Series: *Corpus des Astronomes Byzantins*, Vol. II, General Editor: Anne Tihon, J. C. Gieben, Amsterdam, 1985–1986.

Jyotiḥśāstra, Astral and Mathematical Literature, Vol. VI, Fasc. 4 of the series: *A History of Indian Literature*, edited by Jan Gonda, Otto Harrassowitz, Wiesbaden, 1981.

Astral Sciences in Mesopotamia, E. J. Brill, Leiden, 1999 (with H. Hunger).

E-mail: C/o Kim Plofker <Kim_Plofker@brown.edu>

Plofker, Kim is currently a Postdoctoral Fellow at the Dibner Institute for History of Science and Technology at MIT, where she works on medieval Indian and Islamic mathematics and astronomy. She is a Resident Faculty Fellow at Brown University, where she has resumed teaching in the Department of the History of Mathematics since the Fall 2000.

Her Ph.D. dissertation was supervised by Prof. David Pingree (Brown University); the title being: "Mathematical Approximation by Transformation of Sine Functions in Medieval Sanskrit Astronomical Texts". Her recent and forthcoming publications are:

"The 'Error' in the Indian 'Taylor Series Approximation' to the Sine", *Historia Mathematica*, Vol. 28 (2001), pp. 283–295.

"The Tithicintāmanji of Gaṇeśa", *SCIAMVS*, Vol. 2 (2001), pp. 251–289.

"The Astrolabe and Spherical Trigonometry in Medieval India", *Journal for the History of Astronomy*, Vol. 31 (2000), pp. 38–54.

"Use and Transmission of Iterative Approximations in India and the Islamic World", to appear in the Proceedings of the Conference: *From China to Paris, 2000 Years of Mathematical Transmission*, held in Bellagio, Italy, May 2000, edited by Joseph Dauben, Yvonne Dold-Samplonius, Menso Folkerts, and Benno van Dalen. Series: Boethius Texte und Abhandlungen zur Geschichte der Mathematik und der Naturwissenschaften, Band 46, Publisher: Franz Steiner Verlag Stuttgart. 2002.

"Derivation and Revelation: The Legitimacy of Mathematical Models in Indian Cosmology", to appear in *Mathematics and the Divine*, edited by Teun Koetsier and Luc Bergmans, Elsevier, (with S. Ikeyama).

E-mail: Kim_Plofker@brown.edu

Qu Anjing (born in 1962) got his B.A. (pure mathematics) in 1984, and Ph.D. (history of science) in 1994 from Northwest University (NWU), China. He is a full Professor in Mathematics at NWU since 1997. Presently he is the Director of the Center for the History of Mathematics and Sciences of NWU, and a supervisor of several Ph.D. candidates in the history of science.

Anjing has been recently elected as a member of the Executive Committee of the International Commission for the History of Mathematics for the period 2002–2005. He is serving as a Vice-President of the Chinese Society for the History of Mathematics for 2002–2006. Notably, he has been a visiting scholar at the Needham Research Institute Cambridge, UK (1994–1995); Harvard University, USA (1999–2000); and Kyoto Sangyo University, Japan (2001–2002).

Anjing's research focuses on the history of mathematical science. About 60 academic papers have been published in Chinese, English, Japanese and Italian. The following are the recent books in Chinese, along with a few papers.

New Studies on the Zhoubi Suanjing, Shanxi People's Press, Xian, 2002.

A Concise History of Science and Technology in Ancient China – Mathematics, Liaoning Education Press, Shenyang 2000 (Edited by Qu Anjing).

Explorations of Mathematical Astronomy in Ancient China, Northwest University Press, Xian, 1994 (Edited by Qu Anjing, Z. Ji and R. Wang).

"Revisit to the Solar Eclipse theory in Ancient China", *Studies in the History of Natural Sciences*, Vol. 21, No. 2 (2001), pp. 97–114.

"The Third Approach to the History of Mathematics in China", *Proceedings of the International Congress of Mathematicians – 2002*, Vol. III, Higher Education Press, Beijing, 2002, pp. 947–958.

"Continued Fractions, Astronomical Cycles, and the Problem of the Derivation of $\pi = 355/113$ in Medieval China", *Archive for the History of Exact Sciences* (to be published).

E-mail: qaj@sein.sxgb.com.cn

Stephenson, F. Richard (born in 1941) holds degrees of B.Sc. Honours in Physics (University of Durham), M.Sc. (Geophysics), Ph.D. (Geophysics) and D.Sc. (Astronomy), all of the University of Newcastle-upon-Tyne.

Following research fellowships at Newcastle University (1973–1979) and Liverpool University (1979–1982), Stephenson came in 1982 to Durham University where he is still based. In 1989/90 and separately 1990/91, he was Senior Resident Research Fellow at Jet Propulsion Laboratory, Pasadena, USA. His current position is Professorial Fellow in Astronomy in the Department of Physics, University of Durham.

He is currently President of IAU Commission 41 and a member of IAU Commission 19. In 1992, he was awarded the Jackson-Gwilt Medal by the Royal Astronomical Society and the Tompion Gold Medal by the Worshipful Company of Clockmakers, London. He is also a Freeman of the City of London.

Stephenson's main research interest is Applied Historical Astronomy: the application of early astronomical observations to problems in modern astronomy and geophysics. Among his numerous publications, his most recent books are:

> *Oriental Astronomy from Guo Shoujing to King Sejong*, Yonsei University Press, Seoul, 1997, Co-editor Nha Il-Seong.
>
> *Historical Eclipses and Earth's Rotation*, Cambridge University Press, Cambridge, England, 1997.
>
> *Historical Supernovae and their Remnants*, Clarendon Press, Oxford, 2002, Co-author David A. Green.

E-mail: F.R.Stephenson@durham.ac.uk

Yano, Michio (born in 1944 in Kyoto) is a full professor of Kyoto Sangyo University (KSU) since 1984, and is presently Professor at its Department of Cultural Studies. He earned his Master's degree from the Department of Indian Philosophy (KSU), completed his Ph.D. course work at the Department of Sanskrit (KSU), and finally was awarded his D.Litt. degree by the same University in 1996.

Yano has studied and taught at the Department of History of Mathematics of Brown University, and is the Academic Consultant to the Indira Gandhi National Centre for the Arts (New Delhi, India) since 1996. His major fields of research are the history of Indian and Islamic exact sciences. Recently he has devised a computer program of traditional Indian calendar pañcānga. He is a full (effective) member of the International Academy of History of Science (Paris) since 1997, is on the editorial boards of *Historia Scientiarum*, *Historia Mathematica*, and *Archieves Internationales d'Histoire des Sciences*, and presently Chief Editor of the *SCIAMVS*, an international journal for exact sciences in antiquity and the middle ages.

Besides writing numerous articles in Japanese, he has to his credit the following books and papers published recently in English:

> *Abū Ma'shar: The Abbreviation of the Introduction to Astrology*, together with the Medieval Latin translation of Adelard of Bath (with Ch. Burnett and K. Yamamoto), E. J. Brill, 1994.

Kushyār Ibn Labbān's Introduction to Astrology, in *Studia Culturae Islamicae*, Vol. 62, Institute for the Study of Languages and Cultures of Asia and Africa, Kyoto, 1997.

"Japanese Contribution to the History of Chinese Science", *Historia Scientiarum*, Vol. 6, No. 2 (1996), pp. 123–158 (with H. Kawahara).

"Distance of Planets in Indian Astronomy", *in Oriental Astronomy from Guo Shoujing to King Sejong*, edited by I.-S. Nha and F. R. Stephenson, Yonsei University Press, 1997, pp. 113–120.

"Al-Kindī on Finding Buried Treasure", *Arabic Sciences and Philosophy*, Cambridge University Press, Vol. 7 (1997), pp. 57–90. (with Ch. Burnett and K. Yamamoto).

"Tables of Planetary Latitute in the *Huihui li*", in *Current Perspectives in the History of Science in East Asia*, edited by Yung Sik Kim and Francesca Bray, Seoul National University, 1999, pp. 307–315.

"Yabuuti Kiyosi as a Historian of Exact Sciences", *East Asian Science, Technology, and Medicine*, No. 18 (2001), pp. 13–19.

E-mail: Yanom@cc.kyoto-su.ac.jp

Yau, Kevin K. received his Ph.D. from the University of Durham, England, in 1988. The title of his thesis is: "An Investigation of Some Contemporary Problems in Astronomy and Astrophysics by Way of Early Astronomical Records". He has been working as a Member of the Technical Staff at the Jet Propulsion Laboratory. He is a Member of the Flight Team in the Cassini Project to Saturn.

His important publications are:

"A Revised Catalogue of Far Eastern Observations of Sunspots (165 BC to AD 1918)", *Quarterly J. Royal Astron. Soc.* Vol. 29 (1988), pp. 175–197.

"Far Eastern Observations of Halley's Comet: 240 BC to AD 1368", *J. British Interplanetary Soc.* Vol. 38 (1985), pp. 195–216 (with F. R. Stephenson).

"The Past and Future Motions of Comet P/Swift-Tuttle", *Mon. Not. Roy. Astron. Soc.* Vol. 266 (1994), pp. 305–316.

"Meteorite Falls in China and Some Related Human Casualty Events", *Meteoritics*, Vol. 29 (1994), pp. 864–871.

"The Earth's Paleo-Rotation, Postglacial Rebound and Lower Mantle Viscosity From Analysis Of Ancient Chinese Eclipse Records", *Pure & Applied Geophysics*, Vol. 145 (1995), pp. 459-485 (with Kevin D. Pang).

"The Need for More Accurate 4000-Year Ephemerides, Based on Lunar and Spacecraft Ranging, Ancient Eclipse and Planetary Data", in *Dynamics, Ephemerides and Astrometry of the Solar System*, S. Ferraz-Mello *et al.*, Eds., Kluwer, Dordrecht, 1996, pp. 113–116 (with Kevin D. Pang).

Index

Previously published in Astrophysics and Space Science Library book series:

- **Volume 273**: **Lunar Gravimetry**
  Author: Rune Floberghagen
  Hardbound, ISBN 1-4020-0544-X, May 2002
- **Volume 272**: **Merging Processes in Galaxy Clusters**
  Editors: L. Feretti, I.M. Gioia, G. Giovannini
  Hardbound, ISBN 1-4020-0531-8, May 2002
- **Volume 271**: **Astronomy-inspired Atomic and Molecular Physics**
  Author: A.R.P. Rau
  Hardbound, ISBN 1-4020-0467-2, March 2002
- **Volume 269**: **Mechanics of Turbulence of Multicomponent Gases**
  Authors: Mikhail Ya. Marov, Aleksander V. Kolesnichenko
  Hardbound, ISBN 1-4020-0103-7, December 2001
- **Volume 268**: **Multielement System Design in Astronomy and Radio Science**
  Authors: Lazarus E. Kopilovich, Leonid G. Sodin
  Hardbound, ISBN 1-4020-0069-3, November 2001
- **Volume 267: The Nature of Unidentified Galactic High-Energy Gamma-Ray Sources**
  Editors: Alberto Carramiñana, Olaf Reimer, David J. Thompson
  Hardbound, ISBN 1-4020-0010-3, October 2001
- **Volume 266: Organizations and Strategies in Astronomy II**
  Editor: André Heck
  Hardbound, ISBN 0-7923-7172-0, October 2001
- **Volume 265: Post-AGB Objects as a Phase of Stellar Evolution**
  Editors: R. Szczerba, S.K. Górny
  Hardbound, ISBN 0-7923-7145-3, July 2001
- **Volume 264: The Influence of Binaries on Stellar Population Studies**
  Editor: Dany Vanbeveren
  Hardbound, ISBN 0-7923-7104-6, July 2001
- **Volume 262: Whistler Phenomena Short Impulse Propagation**
  Authors: Csaba Ferencz, Orsolya E. Ferencz, Dániel Hamar, János Lichtenberger
  Hardbound, ISBN 0-7923-6995-5, June 2001
- **Volume 261: Collisional Processes in the Solar System**
  Editors: Mikhail Ya. Marov, Hans Rickman
  Hardbound, ISBN 0-7923-6946-7, May 2001
- **Volume 260: Solar Cosmic Rays**
  Author: Leonty I. Miroshnichenko
  Hardbound, ISBN 0-7923-6928-9, May 2001
- **Volume 259: The Dynamic Sun**
  Editors: Arnold Hanslmeier, Mauro Messerotti, Astrid Veronig
  Hardbound, ISBN 0-7923-6915-7, May 2001
- **Volume 258: Electrohydrodynamics in Dusty and Dirty Plasmas Gravito-Electrodynamics and EHD**
  Author: Hiroshi Kikuchi
  Hardbound, ISBN 0-7923-6822-3, June 2001
- **Volume 257: Stellar Pulsation - Nonlinear Studies**
  Editors: Mine Takeuti, Dimitar D. Sasselov
  Hardbound, ISBN 0-7923-6818-5, March 2001
- **Volume 256: Organizations and Strategies in Astronomy**
  Editor: André Heck
  Hardbound, ISBN 0-7923-6671-9, November 2000
- **Volume 255: The Evolution of the Milky Way Stars versus Clusters**
  Editors: Francesca Matteucci, Franco Giovannelli
  Hardbound, ISBN 0-7923-6679-4, January 2001

- **Volume 254: Stellar Astrophysics**
  Editors: K.S. Cheng, Hoi Fung Chau, Kwing Lam Chan, Kam Ching Leung
  Hardbound, ISBN 0-7923-6659-X, November 2000
- **Volume 253: The Chemical Evolution of the Galaxy**
  Author: Francesca Matteucci
  Hardbound, ISBN 0-7923-6552-6, May 2001
- **Volume 252: Optical Detectors for Astronomy II**
  State-of-the-art at the Turn of the Millennium
  Editors: Paola Amico, James W. Beletic
  Hardbound, ISBN 0-7923-6536-4, December 2000
- **Volume 251: Cosmic Plasma Physics**
  Author: Boris V. Somov
  Hardbound, ISBN 0-7923-6512-7, September 2000
- **Volume 250: Information Handling in Astronomy**
  Editor: André Heck
  Hardbound, ISBN 0-7923-6494-5, October 2000
- **Volume 249: The Neutral Upper Atmosphere**
  Author: S.N. Ghosh
  Hardbound, ISBN 0-7923-6434-1, (in production)
- **Volume 247: Large Scale Structure Formation**
  Editors: Reza Mansouri, Robert Brandenberger
  Hardbound, ISBN 0-7923-6411-2, August 2000
- **Volume 246: The Legacy of J.C. Kapteyn**
  Studies on Kapteyn and the Development of Modern Astronomy
  Editors: Piet C. van der Kruit, Klaas van Berkel
  Hardbound, ISBN 0-7923-6393-0, August 2000
- **Volume 245: Waves in Dusty Space Plasmas**
  Author: Frank Verheest
  Hardbound, ISBN 0-7923-6232-2, April 2000
- **Volume 244: The Universe Visions and Perspectives**
  Editors: Naresh Dadhich, Ajit Kembhavi
  Hardbound, ISBN 0-7923-6210-1, August 2000
- **Volume 243: Solar Polarization**
  Editors: K.N. Nagendra, Jan Olof Stenflo
  Hardbound, ISBN 0-7923-5814-7, July 1999
- **Volume 242: Cosmic Perspectives in Space Physics**
  Author: Sukumar Biswas
  Hardbound, ISBN 0-7923-5813-9, June 2000
- **Volume 241: Millimeter-Wave Astronomy: Molecular Chemistry & Physics in Space**
  Editors: W.F. Wall, Alberto Carramiñana, Luis Carrasco, P.F. Goldsmith
  Hardbound, ISBN 0-7923-5581-4, May 1999
- **Volume 240: Numerical Astrophysics**
  Editors: Shoken M. Miyama, Kohji Tomisaka, Tomoyuki Hanawa
  Hardbound, ISBN 0-7923-5566-0, March 1999
- **Volume 239: Motions in the Solar Atmosphere**
  Editors: Arnold Hanslmeier, Mauro Messerotti
  Hardbound, ISBN 0-7923-5507-5, February 1999
- **Volume 238: Substorms-4**
  Editors: S. Kokubun, Y. Kamide
  Hardbound, ISBN 0-7923-5465-6, March 1999
- **Volume 237: Post-Hipparcos Cosmic Candles**
  Editors: André Heck, Filippina Caputo
  Hardbound, ISBN 0-7923-5348-X, December 1998

- **Volume 236: Laboratory Astrophysics and Space Research**
  Editors: P. Ehrenfreund, C. Krafft, H. Kochan, V. Pirronello
  Hardbound, ISBN 0-7923-5338-2, December 1998
- **Volume 235: Astrophysical Plasmas and Fluids**
  Author: Vinod Krishan
  Hardbound, ISBN 0-7923-5312-9, January 1999
  Paperback, ISBN 0-7923-5490-7, January 1999
- **Volume 234: Observational Evidence for Black Holes in the Universe**
  Editor: Sandip K. Chakrabarti
  Hardbound, ISBN 0-7923-5298-X, November 1998
- **Volume 233: B[e] Stars**
  Editors: Anne Marie Hubert, Carlos Jaschek
  Hardbound, ISBN 0-7923-5208-4, September 1998
- **Volume 232: The Brightest Binaries**
  Authors: Dany Vanbeveren, W. van Rensbergen, C.W.H. de Loore
  Hardbound, ISBN 0-7923-5155-X, July 1998
- **Volume 231: The Evolving Universe**
  Selected Topics on Large-Scale Structure and on the Properties of Galaxies
  Editor: Donald Hamilton
  Hardbound, ISBN 0-7923-5074-X, July 1998
- **Volume 230: The Impact of Near-Infrared Sky Surveys on Galactic and Extragalactic Astronomy**
  Editor: N. Epchtein
  Hardbound, ISBN 0-7923-5025-1, June 1998
- **Volume 229: Observational Plasma Astrophysics: Five Years of Yohkoh and Beyond**
  Editors: Tetsuya Watanabe, Takeo Kosugi, Alphonse C. Sterling
  Hardbound, ISBN 0-7923-4985-7, March 1998
- **Volume 228: Optical Detectors for Astronomy**
  Editors: James W. Beletic, Paola Amico
  Hardbound, ISBN 0-7923-4925-3, April 1998
- **Volume 227: Solar System Ices**
  Editors: B. Schmitt, C. de Bergh, M. Festou
  Hardbound, ISBN 0-7923-4902-4, January 1998
- **Volume 226: Observational Cosmology with the New Radio Surveys**
  Editors: M.N. Bremer, N. Jackson, I. Pérez-Fournon
  Hardbound, ISBN 0-7923-4885-0, February 1998
- **Volume 225: SCORe'96: Solar Convection and Oscillations and their Relationship**
  Editors: F.P. Pijpers, Jørgen Christensen-Dalsgaard, C.S. Rosenthal
  Hardbound, ISBN 0-7923-4852-4, January 1998
- **Volume 224: Electronic Publishing for Physics and Astronomy**
  Editor: André Heck
  Hardbound, ISBN 0-7923-4820-6, September 1997
- **Volume 223: Visual Double Stars: Formation, Dynamics and Evolutionary Tracks**
  Editors: J.A. Docobo, A. Elipe, H. McAlister
  Hardbound, ISBN 0-7923-4793-5, November 1997
- **Volume 222: Remembering Edith Alice Müller**
  Editors: Immo Appenzeller, Yves Chmielewski, Jean-Claude Pecker, Ramiro de la Reza, Gustav Tammann, Patrick A. Wayman
  Hardbound, ISBN 0-7923-4789-7, February 1998
- **Volume 220: The Three Galileos: The Man, The Spacecraft, The Telescope**
  Editors: Cesare Barbieri, Jürgen H. Rahe†, Torrence V. Johnson, Anita M. Sohus
  Hardbound, ISBN 0-7923-4861-3, December 1997

- **Volume 219:** **The Interstellar Medium in Galaxies**
  Editor: J.M. van der Hulst
  Hardbound, ISBN 0-7923-4676-9, October 1997
- **Volume 218:** **Astronomical Time Series**
  Editors: Dan Maoz, Amiel Sternberg, Elia M. Leibowitz
  Hardbound, ISBN 0-7923-4706-4, August 1997
- **Volume 217:** **Nonequilibrium Processes in the Planetary and Cometary Atmospheres: Theory and Applications**
  Authors: Mikhail Ya. Marov, Valery I. Shematovich, Dmitry V. Bisikalo, Jean-Claude Gérard
  Hardbound, ISBN 0-7923-4686-6, September 1997
- **Volume 216:** **Magnetohydrodynamics in Binary Stars**
  Author: C.G. Campbell
  Hardbound, ISBN 0-7923-4606-8, August 1997
- **Volume 215:** **Infrared Space Interferometry: Astrophysics & the Study of Earth-like Planets**
  Editors: C. Eiroa, A. Alberdi, Harley A. Thronson Jr., T. de Graauw, C.J. Schalinski
  Hardbound, ISBN 0-7923-4598-3, July 1997
- **Volume 214:** **White Dwarfs**
  Editors: J. Isern, M. Hernanz, E. García-Berro
  Hardbound, ISBN 0-7923-4585-1, May 1997
- **Volume 213:** **The Letters and Papers of Jan Hendrik Oort as archived in the University Library, Leiden**
  Author: J.K. Katgert-Merkelijn
  Hardbound, ISBN 0-7923-4542-8, May 1997
- **Volume 212:** **Wide-Field Spectroscopy**
  Editors: E. Kontizas, M. Kontizas, D.H. Morgan, G.P. Vettolani
  Hardbound, ISBN 0-7923-4518-5, April 1997
- **Volume 211:** **Gravitation and Cosmology**
  Editors: Sanjeev Dhurandhar, Thanu Padmanabhan
  Hardbound, ISBN 0-7923-4478-2, April 1997
- **Volume 210:** **The Impact of Large Scale Near-IR Sky Surveys**
  Editors: F. Garzón, N. Epchtein, A. Omont, B. Burton, P. Persi
  Hardbound, ISBN 0-7923-4434-0, February 1997
- **Volume 209:** **New Extragalactic Perspectives in the New South Africa**
  Editors: David L. Block, J. Mayo Greenberg
  Hardbound, ISBN 0-7923-4223-2, October 1996
- **Volume 208:** **Cataclysmic Variables and Related Objects**
  Editors: A. Evans, Janet H. Wood
  Hardbound, ISBN 0-7923-4195-3, September 1996
- **Volume 207:** **The Westerbork Observatory, Continuing Adventure in Radio Astronomy**
  Editors: Ernst Raimond, René Genee
  Hardbound, ISBN 0-7923-4150-3, September 1996
- **Volume 206:** **Cold Gas at High Redshift**
  Editors: M.N. Bremer, P.P. van der Werf, H.J.A. Röttgering, C.L. Carilli
  Hardbound, ISBN 0-7923-4135-X, August 1996
- **Volume 205:** **Cataclysmic Variables**
  Editors: A. Bianchini, M. Della Valle, M. Orio
  Hardbound, ISBN 0-7923-3676-3, November 1995
- **Volume 204:** **Radiation in Astrophysical Plasmas**
  Author: V.V. Zheleznyakov
  Hardbound, ISBN 0-7923-3907-X, February 1996
- **Volume 203:** **Information & On-Line Data in Astronomy**
  Editors: Daniel Egret, Miguel A. Albrecht
  Hardbound, ISBN 0-7923-3659-3, September 1995

- **Volume 202: The Diffuse Interstellar Bands**
  Editors: A.G.G.M. Tielens, T.P. Snow
  Hardbound, ISBN 0-7923-3629-1, October 1995
- **Volume 201: Modulational Interactions in Plasmas**
  Authors: Sergey V. Vladimirov, Vadim N. Tsytovich, Sergey I. Popel, Fotekh Kh. Khakimov
  Hardbound, ISBN 0-7923-3487-6, June 1995
- **Volume 200: Polarization Spectroscopy of Ionized Gases**
  Authors: Sergei A. Kazantsev, Jean-Claude Henoux
  Hardbound, ISBN 0-7923-3474-4, June 1995
- **Volume 199: The Nature of Solar Prominences**
  Author: Einar Tandberg-Hanssen
  Hardbound, ISBN 0-7923-3374-8, February 1995
- **Volume 198: Magnetic Fields of Celestial Bodies**
  Author: Ye Shi-hui
  Hardbound, ISBN 0-7923-3028-5, July 1994
- **Volume 193: Dusty and Self-Gravitational Plasmas in Space**
  Authors: Pavel Bliokh, Victor Sinitsin, Victoria Yaroshenko
  Hardbound, ISBN 0-7923-3022-6, September 1995
- **Volume 191: Fundamentals of Cosmic Electrodynamics**
  Author: Boris V. Somov
  Hardbound, ISBN 0-7923-2919-8, July 1994
- **Volume 190: Infrared Astronomy with Arrays The Next Generation**
  Editor: Ian S. McLean
  Hardbound, ISBN 0-7923-2778-0, April 1994
- **Volume 189: Solar Magnetic Fields Polarized Radiation Diagnostics**
  Author: Jan Olof Stenflo
  Hardbound, ISBN 0-7923-2793-4, March 1994
- **Volume 188: The Environment and Evolution of Galaxies**
  Authors: J. Michael Shull, Harley A. Thronson Jr.
  Hardbound, ISBN 0-7923-2541-9, October 1993
  Paperback, ISBN 0-7923-2542-7, October 1993
- **Volume 187: Frontiers of Space and Ground–Based Astronomy The Astrophysics of the 21st Century**
  Editors: Willem Wamsteker, Malcolm S. Longair, Yoji Kondo
  Hardbound, ISBN 0-7923-2527-3, August 1994
- **Volume 186: Stellar Jets and Bipolar Outflows**
  Editors: L. Errico, Alberto A. Vittone
  Hardbound, ISBN 0-7923-2521-4, October 1993
- **Volume 185: Stability of Collisionless Stellar Systems Mechanisms for the Dynamical Structure of Galaxies**
  Author: P.L. Palmer
  Hardbound, ISBN 0-7923-2455-2, October 1994
- **Volume 184: Plasma Astrophysics Kinetic Processes in Solar and Stellar Coronae**
  Author: Arnold O. Benz
  Hardbound, ISBN 0-7923-2429-3, September 1993
- **Volume 183: Physics of Solar and Stellar Coronae: G.S. Vaiana Memorial Symposium**
  Editors: Jeffrey L. Linsky, Salvatore Serio
  Hardbound, ISBN 0-7923-2346-7, August 1993
- **Volume 182: Intelligent Information Retrieval: The Case of Astronomy and Related Space Science**
  Editors: André Heck, Fionn Murtagh
  Hardbound, ISBN 0-7923-2295-9, June 1993

- **Volume 181: Extraterrestrial Dust Laboratory Studies of Interplanetary Dust**
  Author: Kazuo Yamakoshi
  Hardbound, ISBN 0-7923-2294-0, February 1995
- **Volume 180: The Center, Bulge, and Disk of the Milky Way**
  Editor: Leo Blitz
  Hardbound, ISBN 0-7923-1913-3, August 1992
- **Volume 179: Structure and Evolution of Single and Binary Stars**
  Authors: C.W.H. de Loore, C. Doom
  Hardbound, ISBN 0-7923-1768-8, May 1992
  Paperback, ISBN 0-7923-1844-7, May 1992
- **Volume 178: Morphological and Physical Classification of Galaxies**
  Editors: G. Longo, M. Capaccioli, G. Busarello
  Hardbound, ISBN 0-7923-1712-2, May 1992
- **Volume 177: The Realm of Interacting Binary Stars**
  Editors: J. Sahade, G.E. McCluskey, Yoji Kondo
  Hardbound, ISBN 0-7923-1675-4, December 1992
- **Volume 176: The Andromeda Galaxy**
  Author: Paul Hodge
  Hardbound, ISBN 0-7923-1654-1, June 1992
- **Volume 175: Astronomical Photometry, A Guide**
  Authors: Christiaan Sterken, J. Manfroid
  Hardbound, ISBN 0-7923-1653-3, April 1992
  Paperback, ISBN 0-7923-1776-9, April 1992
- **Volume 174: Digitised Optical Sky Surveys**
  Editors: Harvey T. MacGillivray, Eve B. Thomson
  Hardbound, ISBN 0-7923-1642-8, March 1992
- **Volume 173: Origin and Evolution of Interplanetary Dust**
  Editors: A.C. Levasseur-Regourd, H. Hasegawa
  Hardbound, ISBN 0-7923-1365-8, March 1992
- **Volume 172: Physical Processes in Solar Flares**
  Author: Boris V. Somov
  Hardbound, ISBN 0-7923-1261-9, December 1991
- **Volume 171: Databases and On-line Data in Astronomy**
  Editors: Miguel A. Albrecht, Daniel Egret
  Hardbound, ISBN 0-7923-1247-3, May 1991
- **Volume 170: Astronomical Masers**
  Author: Moshe Elitzur
  Hardbound, ISBN 0-7923-1216-3, February 1992
  Paperback, ISBN 0-7923-1217-1, February 1992

Missing volume numbers have not yet been published.
For further information about this book series we refer you to the following web site:
http://www.wkap.nl/prod/s/ASSL

To contact the Publishing Editor for new book proposals:
Dr. Harry (J.J.) Blom: harry.blom@wkap.nl